MICROSCOPE TECHNIQUE

A Comprehensive Handbook for
General and Applied Microscopy

MICROSCOPE TECHNIQUE

A Comprehensive Handbook for
General and Applied Microscopy

by

W. BURRELLS M.R.I.

Rutherford Laboratory
Harwell Science Research Council

A HALSTED PRESS BOOK
John Wiley & Sons
New York

Library of Congress Cataloging in Publication Data

Burrells, W.
Microscope technique

Revision of edition of 1961 published under title:
INDUSTRIAL MICROSCOPY IN PRACTICE
"A Halsted Press Book"

Includes index
1. Microscope and microscopy.   I. Title

QH205.2.B87 1977   578   77-26687

ISBN 0-470-99376-6

First published by Fountain Press 1961
under the title
INDUSTRIAL MICROSCOPY
New revised edition 1977
W. Burrells 1961, 1977

Published in the U.S.A.
by Halsted Press
A Division of John Wiley & Sons Inc.,
New York

Printed and bound in England

CONTENTS

INTRODUCTION

THE ancient peoples of the world had great skill in the arts and crafts and it is clear from the writings in the Old Testament and literature of that period that their art extended beyond metal craft to that of jewellery and glass working. Cutting and grinding of stones are mentioned, while glass appears to have been better known than was generally thought up to now. Egyptian tombs opened at the end of the last century contained large numbers of glass beads and vessels beautifully wrought, now on exhibition in many museums in England and elsewhere. The Egyptians possessed the art of mixing glasses of various colours which retained their identities when the mass was blown or moulded.

The Egyptians were also expert at producing imitation jewellery, and if we may judge from the large quantity of necklaces, common head ornaments and beads found in their ruins, we may say that more artificial jewellery was worn by them than by Europeans today.

Many excellent copies of gem stones have been found. Fused glass mosaic work was a special craft and the blowpipe is often figured in Egyptian bass reliefs. Clear glass, which could have been used for lens making, was discovered in the ruins of Nineveh and its date fixed at 700 B.C., but it seems that the glass workers were more interested in producing jewellery than windows.

There is no material demanded by modern hand glass working methods which could not have been had for the picking-up during the time of the ancient Egyptians, but the climate of the Middle East does not demand glass windows of the kind we know. Buildings are even now more often designed to keep the sun out than to let light in, and when limited amounts of light were admitted a coloured window was more effective with a bright source than a plain one. It is not so difficult to see why the opportunity for a careful study of the properties of clear glass did not occur.

There still remains the problem of why the lens was not accidentally discovered while rock crystal or quartz was being worked. References to crystal often occur in ancient writings (e.g. Job, Old Testament) and the method of grinding stones of all kinds was known and practised particularly by the Assyrians. A certain piece of quartz about 80 mm. diameter and about 10 mm. thick was found in the remains of Nimrod and may be seen in the British Museum. Some qualified opticians thought for a time that this was a lens but it is unlikely that it was made as such because it has cloudy marks within. Such marks would be attractive in a piece of jewellery. They do not prevent the object producing a magnified image, though this image is poor. The curvature is shallow, giving a focal length of about 250 mm., therefore the magnification is small and the lens effect might have passed unnoticed or its importance not been understood.

It has been suggested by persons unacquainted with practical matters that much of the ancient fine work could not have been done without lens aid. The view of practical workers is that there is nothing now on view in the British Museum which could not have been done without a lens. The skill in seeing and handling small things which is developed by persons like watchmakers, microscopists and others is rarely appreciated by those in other professions. It should be remembered also that a myopic worker who is able to use his eye at a distance of 100 mm. from his work, sees about three times the size of image that his normal-sighted comrade does. The ancient engraver was doubtless as keen of sight as we are and probably knew many more tricks which helped him achieve his ends.

In classical times, Seneca (*Quest. Nat.* i.6) was familiar with the fact that a blown-glass globe full of water concentrates the light from a lamp and 'makes small things large' but he did not draw any conclusions from this. It is not surprising that little useful knowledge came from Ancient Greece because their aversion to practical matters of all kinds prevented experiments being made. While it is true that in certain rare instances philosophers have prophesied certain discoveries as a result of abstract reasoning and with the aid of mathematics, usually a fact or irregularity noticed by a person possessing great perceptive powers sets the train of thought, experiment and development in motion.

When we consider the complete coverage of industrial and domestic activities given by the Egyptian picture-writings, and the mass of ancient manuscripts mentioning the arts and crafts of various peoples, we are compelled to conclude that no artificial, recognized lenses existed until the time of Salvino d'Armato degli Armati in Florence in 1280. The first industrial lenses appear to have been made during this year for use as spectacles.

In those times an idea existed that only quartz (rock crystal) should be used for spectacle making. Glass was considered to damage the eyes and no doubt this was true if all ancient glass was as imperfect as that in the clear parts of church windows of the period.

We do not know how the medieval workers came by their knowledge of optics for spectacle making. Spectacle making remains a crude branch of the optical trade and no fine knowledge of lens corrections was or is necessary in order to make useful spectacles. Sooner or later a craftsman who had completed and polished a good clear crystal lens of great curvature for an aged philosopher would see that the lens magnified objects on the workbench, and the idea of making glasses specially for the purpose of magnifying would emerge in the fullness of time. Raphael's painting of Pope Leo X (1518) shows the Pope holding a magnifying glass.

In fact it appears from the lack of positive evidence to the contrary, that it took about 200 years for the idea to hatch, but perhaps the fact that the Florentine lens makers kept their work secret accounts for this long dark period in the history of microscopes.

It is by now generally agreed that the Englishman, Roger Bacon, in Part IV, *Opus Majus*, Jebbs publication, who lived about the middle of the thirteenth century, understood the laws of reflection and refraction. He gave a diagram of a simple lens used as a microscope. It appears that he did not make an instrument based upon his reasoning, but that kind of enquiry was definitely not encouraged in those days. Roger Bacon spent something like twenty years of his life in prison or in some kind of confinement for writing about matters of philosophical enquiry. The consequences of making a microscope are not hard to imagine and were nearly experienced by Galileo two hundred years later.

It is still not possible to say for certain who invented the microscope as a specially designed instrument. The credit is usually given to Zacharias Janssen who lived in Middelburg, Holland, about the year 1608, and in 1876 an old instrument thought by the opticians of the time to be Janssen's work was exhibited in London. This was a microscope of correct form, consisting of a short focus objective and long focus eye lens, with a diaphragm to assist the eye to keep in the right position with respect to this eye lens. Galileo mentioned through his pupil, Wodderborn, in a letter dated 1610, that he could so arrange his telescope that he could see minute objects large. Other correspondence of the time mentions that Galileo's *occhiale* magnified a fly to look as big as a hen but that the whole of the fly could not be seen at once. The instrument used by Galileo for these demonstrations was said to be as high as a dining-table. An instrument magnifying about thirty times (the relative size of a fly and a hen) and standing about three feet high was undoubtedly a telescope with a short focal length object glass used at long extension. Such an instrument with a convex object glass and concave eyepiece gives a very limited field of view.

The Author has made experiments with Galilean telescopes in an attempt to set up Galileo's microscope. An arrangement with a telescope objective of 300 mm. focus and concave eyelens of negative focus about 25 mm., both well stopped down, has a working distance of about 300 mm. and tube length of 600 mm. This is an unsatisfactory microscope and will not magnify thirty times linear. A much better one may be made from a 75 mm. focus object glass and a positive eyelens of about 150 mm. focal length working with a tube length of about 600 mm. This is a satisfactory microscope but cannot be worked as a telescope. The Author set up such an instrument made from postal tubes and simple lenses. The lenses used were large and the field of view great. When persons unused to optical apparatus used this instrument to observe a fly mounted on white paper the effect upon them was quite as marked as that upon Galileo's associates. Although Galileo may not have invented the arrangement of two positive lenses separated by a distance much greater than that used in a telescope, experiments based upon published figures show that he must have used this arrangement. He therefore made and used a compound microscope.

Galileo appears not to have hit upon the principle of microscope construction. Only the device with short focus object glass of small dimensions, working with a positive eyelens, is capable of being developed into the high aperture, high magnifying instrument that we know as a compound microscope. He seems always to have considered telescopes and microscopes as the same instrument with different adjustments.

Whether either the Dutchman, Janssen, or the Italian, Galileo, was the first to make a compound microscope is not clear, but Galileo was the first to make useful microscopical observations and to describe something unseen before his instrument was used. (Letter to Federico Cesi, 1624, describing the way flies' feet stick to glass to enable them to walk upside down.)

From that time onwards microscopes were developed but were held back by the inability of the makers to correct the lenses for chromatic aberration. By the time of Hooke, 1665, the dimensions of the microscope were very much what they are now. They used a small objective of focal length about 25 mm., a tube length of 380 mm. and an eyepiece with a field lens, the whole eyepiece being about equivalent to 'A' power of the present time, magnifying about six times. A microscope of these dimensions will magnify about ninety times. An instrument with a shorter focal length object glass or higher powered eyepiece gives unsatisfactory results when built with uncorrected lenses, also it needs more powerful illuminating apparatus than ordinary top light. Unfortunately, this power is insufficient to show the small things which cause big effects, like bacteria and yeast, and microscopy remained a study reserved for the curious until roughly the turn of the eighteenth century. The nineteenth century may be considered the golden age of discovery in microscopy as in many other sciences.

This book is issued with all dimensions given in metric units according to scientific usage but it is understood that Imperial Measure will be used for a long time to come in many places, therefore the following conversion table of much-used measurements is included.

Inches	Millimetres (mm.)	Microns (μ)
1 thou	0·025	25
$\frac{1}{100}$	0·254	254
$\frac{1}{16}$	1·59	
$\frac{1}{12}$	2·09	
$\frac{1}{8}$	3·175	
$\frac{1}{2}$	12·7	
1·0	25·4	
6·0	160·0*	
10·0	250·0*	

* commonly used equivalents for microscope tube length

CHAPTER 1

The Development of the Science of Microscopy

IT WAS clear to the great thinkers of the seventeenth and eighteenth centuries that the microscope would have to be achromatized before it could be a really useful scientific instrument. Newton came to the conclusion that achromatization was not possible and spent much time making reflecting telescopes and microscopes. No example of his microscopes exists, but one of his telescopes about 12 in. long is in the library of the Royal Society. We will see in a later chapter that a reflecting microscope is a very difficult instrument to construct and quite a different problem from a reflecting telescope.

The fact that Newton had declared achromatization impossible delayed work upon this matter for many years. It was in 1733 that Mr Chester Moor Hall of Essex designed an achromatic telescope objective. It was not until 1829, after many years of work by French, Italian and English mathematicians upon the subject of achromatization that J. J. Lister, in a paper to the Royal Society, laid down the principles of construction and use of achromatic doublets with minimum spherical aberration. Persons even today working in the laboratory do not understand that magnification only is not of much use for making small details visible. There is only small advantage to be had by printing an image upon a rubber sheet and then stretching the sheet. It is true that the picture may be made larger, and for some purposes this is useful, but no one would expect to see more detail merely by continuing the stretching, as there would come a time when all detail put into the picture would be visible at a given distance.

The magnification given by a lens is usually stated without reference to its aberrations. Magnification usually refers to the increase in apparent size of an object when the lens in question is used. The distance of most distinct vision in the normal human eye is taken at 250 mm.; therefore the focal length of the lens in millimetres divided into 250, gives the magnifying power when used as a simple microscope or magnifying glass. This focal length is only an average of the focal lengths of all the zones and of all the colours passing through these zones. The smaller the lens (assuming that the curvature is roughly inversely proportional to the diameter) the shorter the focal length. It is not difficult to fuse tiny globules of glass on the end of fine threads drawn from rods of heavy gauge glass and to get them sufficiently spherical by the aid of their surface tension to make lenses, these may have a focal length of 0·5 mm. and give a magnification of 500 diameters. Owing to the fact that these lenses had great spherical and chromatic aberrations, they had to be used with pinhole apertures and no really important work was done with them. It is remarkable that Leewenhoek, about 1680, saw bacteria but it is hardly surprising that he was not able to make out their shapes.

If the pin-hole acting as the diaphragm to any simple lens is increased in diameter, the image becomes indistinct and hazy due to the admission of light from the outer zones of the lens. The outer zone pencils in a spherical lens focus in a different plane from those of the central zone and are also much more subject to coma, astigmatism and chromatic aberration. With all these faults, it was nevertheless noticed by persons working at particular problems that the wider the pin-hole, the greater the resolution, up to the point of loss of image in haze. It is now known from the work of Abbe in 1870 that this effect of greater resolution associated with greater aperture is due to diffraction patterns spreading from the object at angles with the normal, increasing with the minuteness of the detail present. In 1800 it was only known as an experimental fact.

Lister's achromatic objective of 1829 admitted a cone of light of 50° and this was a very great improvement upon any simple systems with a pin-hole, the angles of which are usually about 5°.

The advent of the achromatic aplanatic object glass caused the scrapping of the old wood and vellum stands—which had been popular until then—and the development of the brass stand which we still see today. The higher the class of objective, the steadier and more accurate must be the stand. This is still a most important and sometimes over-looked point today.

Discoveries with the new microscopes followed fast in the nineteenth century. Brown of 'Brownian Movement' fame (1830) discovered the plant cell nucleus; Schleiden (1838) applied the cell theory to all living things and Pasteur (1860) discovered the relationship of bacteria to disease. Many other workers made important advances in these and other branches of science and industry. The advance in design of the microscope was of course the most important factor in these successes but many techniques were also invented and will be noticed in the various sections following.

General Matters Concerning Instruments and Use of the Eyes

At some time it becomes necessary to choose a microscope for a particular investigation and it is not possible to buy one which will do all things in microscopy. Detailed considerations of practical microscopical problems will be found in the chapters which follow but a start must be made upon general lines.

The choice between monocular and binocular types of microscopes must be made. This choice depends a lot upon the skill and experience of the workers, also there is a variety of types of binocular to choose from.

Most workers prefer the stereoscopic image to the flat one, but it must be remembered that when any binocular is used the head and eyes must be held still. With the monocular form the head and eyes are free to move about relative to the eyepiece and this is why experienced microscopists often prefer monocular instruments for all powers. In general, the monocular form properly used in a shaded room with both eyes open and both focussed, produces least fatigue and harm. There are, of course, many operations where the stereoscopic binocular is useful as most binocular low power instruments erect the image.

Of binocular forms, there is a choice of parallel or converging tube types and the choice is again largely one of opinion. When the tubes are parallel the eyes work at the rest position, that is, focussed at infinity. When the tubes are convergent, the eyes are focussed at a distance of 250 mm. In the human optical system the convergence of the eyes and the focal length of their lenses are connected. When the focus is at infinity, the convergence is nil; it follows that when the eyes are at zero convergence, as when using a parallel tube binocular, the eyes cannot accommodate parts of the object slightly above and below focus without attempting to converge. When they do converge, however slightly, the fusion of the left and right images is lost. The observer is not usually aware of these effects separately unless they have been pointed out to him, but he does experience fatigue due to discomfort in his eyes. If a large number of flat objects has to be observed, the rest condition of the eyes with the parallel tube instrument is an advantage. If general studies must be undertaken, the eyes will be more comfortable with the convergent form of instrument. In this case, the eyes being convergent, some accommodation in depth is possible without necessarily losing fusion, as the eyes are found to be able to wander slightly between parts of the object above and below exact focus and still maintain true binocular vision. If the instrument happens to be of the stereoscopic type, the advantage of converging tubes is greater.

The eyes of microscopical workers are not always normal. Most are able to accommodate any instrument but some cannot use parallel tubes. Inability to fuse the images is often due to errors in the focussing mechanism of their eyes, this error causes the eyes to attempt to converge when two images at infinity are presented to them. If any doubt exists as to which type of binocular should be purchased, the converging type with either twin or single objective should be chosen.

Most inexperienced persons fail to obtain the best results from optical instruments because they have not learned to look 'through' a lens system but always look 'into' it. If the observer looks through a microscope imagining it to be an empty tube 1 metre long, his eye will be nearly in the rest condition, the microscope is then focussed to suit the eye and not the eye to suit the microscope. If one looks 'into' the instrument, one is trying to see an image about 150 mm. below the eyepiece and this means that the eye crystalline lens is drawn up to great curvature, convergence of the eyes is great and strained, and the image upon the retina is smaller than when the eye is at rest. The eye, when used in the rest condition looking 'through' the microscope, presents to the brain an image about twice as large as when it is used in the looking 'into' state. These are facts which can be checked at will and should be attended to by all who engage upon microscopical work. Damage to the eyes and failure to see all that the microscope can show are the result of neglect of these principles. An experienced worker can usually tell whether another person is looking 'through' or 'into' his instrument by the difference of focal adjustments used. If it is assumed that two persons have normal eyes, the difference in focal adjustment used by them when using a microscope correctly is practically zero.

The human eye is moved about in its orbit by sets of muscles acting in opposition to each other in order to obtain an even graduated movement. When objects at a distance are observed, say farther away than 3 metres, the optic axes of the eyes move together and remain parallel, but when the eyes are focussed on a nearer point, convergence must be allowed in order that the image may fuse. This convergence is brought about by the relaxation of some of the muscles under the control of the nervous system attached to the eye. Each eye has an area on its retina known as the fovea, and when an image is received upon this spot, vision is most distinct.

As the eye moves in its socket, the pupil moves with it and remains in the optic axis. If the eye is moved up and down from the straight ahead position, the range is limited by the bony structure of the face. If the movement is in excess of (say) 30°, then the head must be moved to allow light to enter the eye. A microscope eyepiece presents an apparent angle of vision of about 60°, and often more, so when the eye is turned to look at the outer regions of such a field, the pupil aperture moves also and becomes out of line with the Ramsden disk above the eyepiece. The observer then naturally alters the position of his head in order to admit more light to the eye, but in so doing may cut off part of the Ramsden disk with his eye pupil boundary or may admit the Ramsden disk obliquely. Any interference with the Ramsden disk will interfere with resolution sufficiently to impair critical work.

It follows from the above facts that critical work should not be done with eyepieces giving large apparent field diameters unless attention is concentrated at the centre of the field only. If such eyepieces are used with binocular instruments the effects are worse, because inability to move the head and still retain fusion of the images prevents the eyes scanning the field of view. Should they naturally try to do so, fatigue and poor seeing will result. (See also 'Binocular Microscopes', page 376.)

The maximum aperture of the eye pupil in the dark is about 9 mm. and in daylight it is reduced to about 2 mm. depending upon the brightness of the day. The eye functions best with a brightness of light which demands about half the maximum pupil aperture, say about 4 mm., and the light in the microscope must be adjusted to this amount by means of neutral screens, etc. (see 'Illumination', page 16). This amount of light is about that which is reflected into the eye of a reader from a page of book-print under a 100-watt reading lamp, and it should be noted that this is about half the brightness of the light usually used by microscopists (see 'The Workroom', page 20).

The size of the Ramsden disk or 'exit pupil' of an eyepiece is related to the numerical aperture of the objective but varies much with the power and design of the eyepiece. The size of this disk may be calculated from the following formula:

$$\frac{500 \times \text{Numerical aperture of objective}}{\text{Total magnification}}$$

If figures for a typical medium power microscope be taken, we have a disk of diameter about 2 mm. It will be seen from the above formula that the size of the disk is affected by the figure in the denominator for the eyepiece magnification. Thus if the light is adjusted for an eye pupil aperture of 4 mm., the Ramsden disk must not be greater than 4 mm. and, bearing in mind the effect of moving the eye in order to scan the field of view, can best be about half the eye pupil diameter. A microscopist should, for the best results in critical work, measure the diameter of the Ramsden disk above the eyepiece by receiving its image upon a piece of ground glass and then select an eyepiece which provides a disk of not more than 2 mm. diameter. He should then incline his microscope so that he can look straight into the eyepiece. With these conditions satisfied, he is sure that the whole of the Ramsden disk enters the eye without interruption, at least when the central regions of the field of view are being studied.

The Aberrations of the Lens System of the Eye[1]

The eye has both spherical and chromatic aberrations present to a considerable degree and these aberrations are both noticeable when light enters the eye obliquely. It was mentioned above that there is a very sensitive area upon the retina known as the fovea, and this area is situated near to the optical axis of the lens system. It is clear then that parts of the retina image not on the axis of the lens do not contribute much to the brain in the way of fine detail. They are able to transmit the idea of a large field of view and when necessary the eye moves and examines with the fovea the object which attracted notice. The eye gives excellent central definition, and if an animal eye is dissected this will be seen to be the case.

It is not easy to determine the axial spherical aberration of the human eye owing to its flexible nature but the axial spherical aberration of a bi-convex lens of 30 mm. focal length is 3 mm. An attempt may be made to measure the aberration of the eye in situ by means of Young's optometer. This instrument consists of a piece of foil with four slots in it 3 mm. long, with centres 1 mm. apart, the slots must be very narrow, say a few tens of microns wide. When these are held before the pupil of the eye and a point source of light is examined, several images of the source will be seen, each formed by the part of the lens defined by a slot. When an experiment is made to determine which slot produces which image (by covering the slots in turn) it is found that spherical aberration is present but the eye is under-corrected for long distance objects and over-corrected for near ones.

The refraction produced by the eye is in two stages; the most being produced by the cornea and the rest by the crystaline lens. The surfaces of the cornea are nearly spherical, therefore they have uncorrected spherical aberration. The crystaline lens is not homogeneous, it is built up from concentric layers of stiff transparent material, the refractive index increasing in layers towards the centre. The complete crystaline lens is therefore over-corrected for spherical

---

[1] For much material in this section I am indebted to Professor H. Hartridge, FRS, 'President's Address', QMC Journal, August 1954, Series IV, Vol. 4, No. 2.

aberration. The crystalline lens corrects the spherical aberration of the cornea but is only able to do so for curvatures associated with medium to long distance vision, it is insufficiently powerful to correct completely for near vision. When an eye is used with a microscope it is under-corrected for spherical aberration.

To obtain the distance of zero spherical aberration, proceed as follows:

(i) Find the distance in millimetres by means of Young's optometer at which there is no axial spherical aberration present, i.e. a point source at this distance appears single and clear.

(ii) Focus a long-draw telescope upon this point while at the same distance as it was from the eye.

(iii) Look at an eyepiece with the adjusted telescope and move the eyepiece diaphragm up and down until it is in focus in the telescope.

If the eye now looks through the slots of Young's optometer at the eyepiece diaphragm, no sign of spherical aberration should be present.

The exit pupil or Ramsden disk of a microscope should not be larger than, say 2 mm., for the reasons given above, therefore this adjustment will not usually be necessary. So long as the pencil of light entering the eye is narrow (1 mm.) aberrations will have little effect upon its image-forming properties. In the limit, when the Ramsden disk is infinitely small, the light entering the eye is parallel and aberrations do not matter. (See 'Defects in the Eye', page 18.)

It is safe to say that few observers need to make the above adjustment unless an instrument has an exit pupil of more than 3 mm. Some persons have great spherical aberration in their eyes and because this does not show as an error in focus it remains not understood. Small matters of this kind often account for the fact well known amongst laboratory workers that some people can see things with microscopes while others cannot.

The human eye structure is not achromatic; examination of its component parts show that the error due to chromatic aberration must be large and this may be checked by experiment. If an object be viewed through a piece of glass which transmits red and blue light only, two images in red and blue will be seen. Either may be focussed sharply, with the other present as a hazy background image, but not both together. The experiment can be made with a plate of cobalt glass as the filter. The amount of axial chromatic aberration has been measured and is given by H. Hartridge in the following way:

Colour of Light	Wavelength μ	Lens required to Focus Light	
		Dioptres	Metres
Red	0·7	+0·5	2
Yellow	0·6	none	—
Green	0·55	−0·25	4
Blue	0·45	−1·1	0·9
Violet	0·4	−2·0	0·5

TABLE 1

The measurements were taken around yellow light focussed sharply on the retina. The signs + − in the dioptres column refer to conventional optical notation. Colours normally coming to a focus behind the retina require a + lens to correct them.

When an image of a point of white light is received upon the retina it is really a yellow spot surrounded by coloured haze, the haze being due to chromatic aberration. The mechanism by which the brain rejects the unwanted colours is not understood but experiments have been made to see whether or not the eye can reject colours due to the chromatic aberration of an external lens, and it is found that it can so long as the focal length of the external lens is greater than about 50 mm. A hand magnifier may have a focal length of (say) 50 mm. and not show an appreciable amount of colour while one of 25 mm. focal length would need to be achromatic, it is for this reason that high power spectacles need not be achromatic.

If a microscope eyepiece is delivering white light to the eye, chromatic aberration is present in the image on the retina and until more information is available concerning the mechanism of colour vision and suppression of unwanted colours in images, we must assume that the image handed on to the brain will be clearer if a single colour is used for illumination.

The rods and cones in the retina are packed close together and the cones in the fovea are only about 3 μ separated from each other. This separation is only about six times the wavelength of green light, therefore the haze of chromatic aberration must affect several retina cones around the one under examination. All microscopists are aware that a considerable improvement is obtainable in definition when monochromatic light in the yellow or green region is used. If red is used, the resolution of the microscope falls off at high powers and blue is a poor choice because it is difficult to see with. The improvement due to using monochromatic light is usually credited to the objective and this is often correct, but if experiments be made with the naked eye upon fine detail using a monochromatic source, the improvement will still be found to be great. When a line source is used, an improvement in the ability of the naked eye to see detail is most marked in yellow sodium light. Microscope objectives are well corrected for spherical aberration with the sodium line, and this form of illumination yields superior visual and photographic results.

It has been found by experiment that the human eye can just resolve points separated by 58 sec. of arc (Helmholtz) which corresponds to a grating ruled 14 lines to the millimetre at a distance from the eye of 250 mm. If calculations are made based upon Abbe's theory that the maximum useful magnification can be obtained at 1,000 times the numerical aperture, it is found that for quite small pupil apertures, and consequent low numerical apertures, the eye lens system is capable of resolving about twice this number of lines to the millimetre. Very accurate work is not possible owing to the changing size of the eye pupil during an experiment, but it can be seen that the limiting structure in the eye is the retina and not the lens system.

In order to obtain best microscopical seeing, the retina must receive special consideration.

The cones in the fovea have a threshold of sensitivity below which they do not respond to light, the threshold varies with dark adaptation and fatigue. If the level of light entering the eye is high, some scatter from the component parts of the eye and from the illuminated parts of the retina occurs as in any other photographic apparatus, and this amount of scatter may be above the threshold of activity of the surrounding cones. If it is, then the effects known as glare result. Glare destroys contrast and generally reduces all sensitivities; fortunately, the eye tells us when too much light is entering it; in ordinary daylight the pupil closes and reduces the light. With the microscope it closes also but the exit pencil of light is so small in diameter (about 1 mm.) that the pupil cannot contract far enough to reduce the light. When a microscope is too bright, a distinct pulling sensation in the front of the eye can be felt. The rule given above that the microscope image should be no brighter than an ordinary room in daylight should be observed, the daylight should then be shaded by means of a blind.

Level of Illumination
A figure of 100 m./candles has been given as the correct amount of light for ordinary purposes, but this amount is insufficient to bring out the full resolving power of the eye. For microscope preparation work, fine instrument adjustment and similar tasks, full daylight is desirable and most laboratories are inadequately lighted for fine

work. Five modern fluorescent strip lights over a floor area of 10 sq. metres are necessary in biological laboratories, and in other places where small objects are handled.

Many microscopists are in the habit of working with a low light in the microscope and a dark room as it is felt that this arrangement allows the eye to work at maximum resolution. In the Author's opinion it does, but there are indications that the best results can be obtained in a fully lighted room with a bright microscope lamp. Measurements using black and white objects show that the acuity of the eye increases with brightness until the brightness of reflected full sunlight is reached. It appears that the main advantage of working in a brightly lit room with a bright microscope lamp is that the eye does not experience light and dark areas alternately. It was shown by Lythgoe that resolution is affected by the brightness of the surrounding medium, and for best resolution of a test object, the surround should equal the average brightness of the specimen. This experiment indicated that the dark diaphragm in the typical microscope eyepiece is not a desirable feature, the diaphragm should be white and of controllable brightness (see Professor Hartridge's eyepiece in section 'Eyepieces', page 92).

The Author has done much work upon semi-transparent material and, without knowledge at the time of Lythgoe's experiment, adopted eyepieces without diaphragms, and limited the area of the field illuminated by means of the lamp condenser diaphragm. The optics were such that the diaphragm did not cut off the field sharply but allowed the light to trail off to the edges of the field, the edges being well out of the area of vision because of the absence of the diaphragm. The Author discovered by experiment that this arrangement allowed the eye to develop to its maximum its ability to see nearly transparent structures and the results probably apply equally to resolution tests. The difficulty of making a proper test in a fully lit room lies in obtaining a sufficiently powerful microscope illuminant. No one has suggested that the room should be brighter than the microscope, so it is permissible to darken the room sufficiently to allow both eyes to concentrate easily upon the brighter microscope image.

Resolution of the Eye

If one assumes that white light contains 1,900 waves to the millimetre, and the diameter of the disk of confusion at 250 mm. distance is 0·25 mm., a figure for the maximum resolution may be obtained from Abbe's theory. This is given as a magnification of 1,000 times the numerical aperture of the system in use and this is an estimate of the ability of the eye to separate in brightness two points in an image, but there is no doubt amongst practical microscopists that this figure may be exceeded with advantage even by workers with 6/4 vision.

In ophthalmic work the figure of 6/6 is taken as representing normal vision and is based upon an arbitrary measurement of distance from a particular card of letters. No natural system for measuring acuity is attempted. The letters on the card and the distance of observation are so arranged that in the case of a person with 6/3 vision, the letters which he can just read subtend an angle at the eye of 2·5 min. of arc. The thickness of the black lines making the letters is 1/5 the height of the letters. It is concluded that if letters can be recognized and separated from their neighbours, the circle of confusion must be about 1 min. of arc for this observer. If vision is 6/6, a figure of 2 min. of arc is obtained from the same reasoning.

If a microscopist has his visual acuity measured by an optician (Snellen's Test) he can calculate the confusion circle diameter of his own eye. (See table below.[1])

Acuity (Snellen's Test)	Size of Confusion Circle at 250 mm.	Resolution in Lines per millimetre at 250 mm.
6/4	1/10 mm.	10
6/5	1/8	8
6/6	1/7	7
6/9	1/4	4

TABLE 2

In microscopy the magnification of the object must be such that the natural blurring of the eye does not affect the sharpness of the image nor the resolving power of the instrument, in other words the magnification must be great enough to make all detail resolvable by the microscope large compared with the coarseness of the retina.

Consider the number of lines which can just be resolved for different acuities (column 3 above) and relate these figures to the resolution of the microscope. It was stated that the maximum number of lines which can be resolved is given by 0·04 × numerical aperture × number of light waves per millimetre in the light in use. Again taking white light as containing 1,900 waves per millimetre, the maximum number of lines is 3,800 × numerical aperture. If we consider an observer with 6/6 vision who can see 7 lines per millimetre at 250 mm. distance, we have

$$\frac{3800}{7} \times \text{Numerical aperture} = 543 \times \text{NA}$$

[1] Table by H. Ha

This figure gives the minimum magnification at which this observer can see all detail.

It must be remembered that the tests carried out in order to obtain the above table of results were conducted with an amplitude grating of the normal construction, i.e. black and white lines, but the microscope image is built up from diffraction patterns which have a density distribution of a sinusoidal kind. The distinction between black and white is here not sudden and black is impossible to obtain in an image without the use of perfect lenses and objects many times larger than a wavelength of light. Even if the lenses were perfect, the ratio of dark to light bands is only about 1 : 5 in intensity and this ratio varies with the distance of the bands from the diffracting objects and the conditions of recombination of these patterns. It appears reasonable to take half of the figure given in the table as the resolution in lines per millimetre of non-opaque gratings. This modification gives us a figure of 1,000 times NA for the observer with 6/6 vision.

This figure of $1,000 \times NA$ for necessary magnification only applies for persons of 6/6 or greater visual acuity and many persons can be expected to require about twice this magnification. Eye focus correction by means of spectacles must not be confused with acuity. Many persons who do not need spectacles have low acuity and vice versa.

Defects in the Eye

More observers are partially colour blind than is generally realized and many domestic disagreements concerning the colour of furnishings are due to this cause. Obviously, if a person is about to do work upon colours he must be tested. It must be remembered that all people experience defective colour vision if the coloured object is very small. One cannot define 'very small', but experiments can be made with coloured papers cut into small fragments removed to various distances from the eye. It is not likely that this effect will affect a microscopist because no high power lens system can be trusted to give a colour-free image of a small object. If the power in use is low, then a large image may be obtained by increased magnification without destroying definition.

If the healthy eye is supplied with light from an even source of moderate intensity, specks and bacteria-like bodies may be seen floating in the eye fluids. These are most visible when the pencil of light is narrow as from a high power eyepiece; they are then shadowed on to the retina. The particles are detached pieces of cellular material from the inside of the eye and are of no importance.

The eye is covered and lubricated with a sterile fluid which is periodically swept smooth and clear by the eyelids. If a microscope is used in the vertical position, tear drops of this lubricating fluid collect by gravity in front of the pupil aperture, also the internal bodies collect or tend to collect behind the pupil aperture. The fluid tends to distort the refracting front surface of the eye and the specks obscure the vision, then distortion of the lens causes fatigue of the muscles responsible for the focus of the lens. It follows that a microscope should not be used in a vertical position with the eye lens horizontal and observers who must have the microscope stage horizontal should be provided with angle eyepieces.

In old age other changes take place in the eye, such as increasing opacity of the fluids and lens, distortion of component parts, doubling of vision due to lens defect, and other unfortunate disturbances. In general such persons will not indulge in industrial microscopy for long periods of time, but any who have defective vision should use the microscope with plenty of light, small exit pencils (that is, generally use high eyepieces), and use the eye in a natural position, i.e. looking straight ahead.

The microscope will correct all focus errors in the observer's eye, but cannot correct astigmatism. It is rare to find an eyepiece of higher power than $\times 12$ with a long enough eyepoint to accommodate spectacles, the Huyghenian type cannot be made to do so but the eyepieces of Beck, Leitz and Zeiss used for their apochromatic objectives have much longer eyepoints. The old 250 mm. tube microscope needed about half the eyepiece power that a 160 mm. one needs to develop the same magnifying power, and as a $\times 10$ eyepiece can generally be made to accommodate spectacles, while the $\times 18$ necessary on the short microscope for the same magnification cannot be so arranged, an advantage is obtained with the long tube. The alternative to taking off the spectacles is to use a low power, long eyepoint ocular and a higher magnification objective. (See 'The Aberrations of the Lens System of the Eye', page 12.)

Observers who are to be employed much upon microscopical work must learn to keep both eyes open, they will then focus together, and except for the fact that the pupil of one eye may dilate more than the other, the eyes are being used together.

CHAPTER 2

Conditions of Work
and Care of Instruments

INDUSTRIAL MICROSCOPICAL work has to be done somewhere near the place of manufacture or place of handling of the objects studied. It is not practicable, for instance, to walk along corridors from a work-room to the microscope room and it is not possible to carry most kinds of microscopical materials across outdoor areas. Successful microscopical work cannot be carried on in the same room as heavy engineering or dusty operations, as vibration and dust are the chief enemies of the microscopist. The best arrangement is a microscope room partitioned off from a main laboratory or workshop. In this room no work other than microscopy and preparation for microscopy should be done.

The size of the room obviously depends upon the amount of work to be done. When there is one person responsible for microscopy, he usually needs room for five important pieces of apparatus, these are:

(1) A low power microscope and top light of the Greenough type. This equipment is used for general examinations of specimens in the rough and for preparation. The instruments usually occupy about 1 metre of bench space.

(2) A high power microscope and lamp for the study of prepared specimens. This instrument occupies about 1 metre of bench but care must be taken to allow sufficient depth of bench for the illuminating apparatus. This usually takes up a total depth of 1 metre.

(3) An optical bench of any degree of elaboration thought suitable (see 'Optical Benches', page 170) upon which photomicrography is done. It is nearly essential in industrial work to have an apparatus set up at all times for photography as little can be published without the photographic record, also much observing time is saved by skilful use of photomicrographs.

(4) A sink with hot and cold water and draining racks.

(5) A table for the microscopist which can be kept clear of other apparatus. Provision of a proper work-table might inspire the

operator to keep complete notes and records of his work. This is difficult at the best of times but is more so for a microscopist because so much of his work is done with his eyes and hands. He rarely has to prepare (say) a wiring diagram which is some form of record of work done even if nothing more is written. The microscopist often goes on observing and applying processes until perhaps the specimen is destroyed and, of course, no one believes what he says happened; the work may therefore be lost when proper notes and photographs would have made it worth while.

It is assumed that gas and electricity are available on the benches.

An effective drying rack for plates is formed by the wire grid usually to be found above the ordinary room radiator. Special air-conditioned drying cupboards may be obtained from the photographic supplies manufacturers by those who have plenty of money to spend. (See 'Photomicrography', page 177.)

It is a great help to ease and speed of work if the room can be blacked out; if this is possible, a large number of operations are simplified; for example, micrographic cameras need not have light-tight backs, then films or plates may be laid upon the guide rails of the camera direct, with only a weight of some kind to hold them down. All time wasted in loading various devices is then saved and exposed plates may be taken straight from the microscope to the developing tank without cassets and dark slides being necessary. A microscopist seldom takes large numbers of photographs at one time, and when he takes any he usually has to make a trial exposure, so he must process his negatives himself and he should not be asked to walk about a building in order to do so.

Arrangement and Form of Work Benches

The primary requirement in benches for microscopy is solidity. A structure (including a microscope itself) can be strong and stable but yet flexible. Benches with plywood tops are not satisfactory, but the old type of teak bench supported upon wall brackets is suitable. Modern laboratories are often fitted with units containing cupboards and drawers, but these are unsatisfactory for serious microscopical work because the opening and closing of drawers attached to the bench disturbs the apparatus and they are in the way of the workers. Benches should not have their underneath spaces fitted with drawers and cupboards, whether attached to the bench or not, because these prevent operators sitting with their knees beneath the bench and prevent persons standing comfortably at the bench.

A modern tendency is to undercut the plinth to make way for the feet of a person standing near.

In general, microscope benches or tables should stand clear of other furniture and apparatus where the floors are solid. Fixture to the wall of a substantial building is possibly satisfactory so long as machinery is not mounted upon the same wall, but a table standing upon a beam-supported floor is not satisfactory because of vibration. A basement is a good place in which to work so long as the atmosphere is air-conditioned. Heavy traffic makes delicate photography impossible in any building.

It is an advantage to arrange sockets for electrical connection to the mains at the back of the benches as flex may then be tucked away behined apparatus. If the sockets are placed along the front of the bench, the switches are easy to reach, but accidents to loose apparatus often occur due to cables crossing the bench edge being caught in the clothing of passers-by. The Author has found it advantageous to arrange important light switches to hang above the operating positions within easy reach of the worker. Some new laboratory benches are arranged to have electrical sockets, gas and water taps at the back with remote controls at the front, but this is not a good idea for general work owing to its inherent inflexibility and the uncertainty concerning which control operates which tap; it is easy to see when the worker is standing in front of the system, but when any parallax is introduced it becomes nearly impossible to tell which is which.

For the benches used for optical work, the tops should be of wood. For places where chemicals are used, Formica is satisfactory. Should Formica be used to cover all benches, allowance must be made for its slipperiness and its hardness, for a lens dropped upon Formica can break, whereas it will survive if dropped upon wood. Glass tops are cold and uncomfortable to work upon and sooner or later are fatal to some component accidentally knocked over.

All bench coverings in the parts of the laboratory where chemicals are used should be sealed down to the bench and to the wall at the back; if this is omitted, chemicals will find their way under the covering, contaminate other things, create smells and force off the covering from the bench. (See also 'Use of Instruments and the Eyes' and 'Microscopical Apparatus Exposed to Radioactivity', pages 9 and 27.)

Storage of Equipment: Microscopes

It is not practicable to pack apparatus away each evening in the cases provided for transport, economy demands that the set-up must

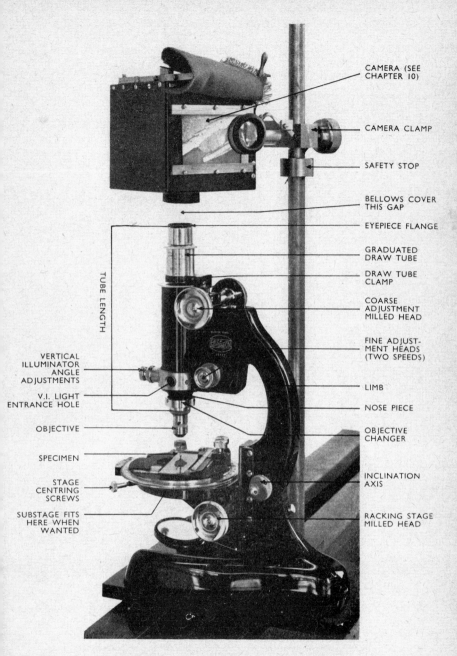

CAMERA (SEE CHAPTER 10)

CAMERA CLAMP

SAFETY STOP

BELLOWS COVER THIS GAP

EYEPIECE FLANGE

GRADUATED DRAW TUBE

DRAW TUBE CLAMP

COARSE ADJUSTMENT MILLED HEAD

FINE ADJUSTMENT HEADS (TWO SPEEDS)

LIMB

NOSE PIECE

OBJECTIVE CHANGER

INCLINATION AXIS

RACKING STAGE MILLED HEAD

TUBE LENGTH

VERTICAL ILLUMINATOR ANGLE ADJUSTMENTS

V.I. LIGHT ENTRANCE HOLE

OBJECTIVE

SPECIMEN

STAGE CENTRING SCREWS

SUBSTAGE FITS HERE WHEN WANTED

PLATE 1
Microscope and Camera Set Up to Examine and
Photograph a Metal Specimen.

be preserved for immediate work and there is no disadvantage in leaving microscopical apparatus open to the room so long as the following matters are observed:

(1) Eyepieces must always be left in microscope tubes. If tubes are left open, dust collects upon the back lenses of the objectives and this is difficult to remove and may cause damage by etching the glass. If no objective is in place, dust collects upon the inside of the microscope tubes and later falls upon the lenses.

(2) Objectives must be placed when not in use upon the bench with their rear lenses downwards. Objectives are best preserved from damp, smog, dust and mechanical accident by being replaced in their boxes, but unfortunately, the boxes will not take the objectives when they are fitted for use with objective changers. If objectives are upon changers, a small box should be made to take them. It is most important that objectives should so far as possible be protected from damp chemical-laden atmospheres. (See 'Objectives', page 54.)

(3) Cameras must have their ground glass focussing screen left in place for the same reason that eyepieces must be left in microscopes.

(4) Should it be essential to do microscopical work in a dusty place, then the instruments must be covered, as excessive amounts of dust will damage the sliding part of microscopes and microtomes. Suitable covers of plastic materials may be made or purchased, which are quite satisfactory but they will not resist the heat of a lamp so apparatus must be cool before being covered up. Plastic covers are now available with zip fasteners and are excellent for use in contaminated places.

The shutter mechanism of miniature cameras will not stand any exposure to dust and the backs must always be replaced if these cameras are temporarily laid aside.

(5) In general, the temperature of the room in which microscopes are kept should be substantially constant. If temperature changes between limits of midsummer (32° C.) and mid-winter (−2° C.) are allowed, the balsam of the lenses will sooner or later fail and condensation falling upon lenses in some conditions of temperature and humidity will cause surface breakdown of the glass. Apochromatic objects made between about 1900 and 1920 are particularly easily affected.

Many lubrication fluids experience great changes with temperature and some concoctions recommended in older books

B

on microscopy lose what powers of lubrication they possessed when made at room temperature of 15° C., when the temperature becomes, say, −1° C. This applies especially to the sticky material made from Canada balsam and beeswax (in varying proportions) used to smooth out the jumpy motion of badly fitting parts, or to increase the finger resistance of screw threads. Microscope sliding parts should be lubricated with silicone grease made by Dow, Corning & Co, and those parts requiring oil should have one of the Shell Lubricating Fluids of similar viscosity to standard clock oil. Neither of these lubricants changes in viscosity over the range of British climatic temperature changes.

Storage of Equipment: Accessories

When apparatus is stored in wooden cupboards and the doors are closed, it is partially true to say that the apparatus is lost. Staff may remember where large pieces of equipment are situated but they cannot remember the location of the hundreds of small parts associated with laboratory work.

Cupboards should be built of transparent perspex or other plastics as this material is easily sawn and filed, and may be welded really strongly with chloroform. Cupboards are more easily built up in this way than in wood.

Racks for small parts like screws may be made in the same way, or cheap screw-top bottles may be used. The saving of time in inspecting the state of the stores, in locating apparatus, and in putting things away properly, is immense. The material is more serviceable than wood and the cluttered appearance of many laboratories is remedied.

Dampness can be kept out of the atmosphere of closed spaces by means of a few grams of silica gel; about 10 grams per litre, depending upon humidity, are necessary. It is unwise to expose optical ware to sulphuric acid vapour or lime for the purpose of dessication. The attacks of mites of various kinds can be stopped by means of paraffin vapour radiated from a wick.

Storage Equipment: Tropics

Persons responsible for laboratory work in the damp tropics must expect their work to be difficult. The main troubles are as follows:

(1) Dampness, and in its extreme form, mist.
(2) The rapid growth of fungi.
(3) The presence of animal creatures of all kinds.
(4) Heat.

The above list is given assuming that the laboratory is situated in a permanent building. If the work is to be done in the dry tropics, the main troubles are dust and flies. Each part of the tropics has its own special troubles, but the Author hopes the following general remarks may help all.

Mutants of the group of microfungi *Aspergillus glaucus* will grow upon the surface of glass. They live in the thin film of water which forms upon the surface and as this film is continually dried and reformed during climatic changes each day it soon becomes a solution of glass and salts. Highly specialized mutants of *Aspergillus glaucus* are able to live in media with very high osmotic pressure, the necessary amount of water being obtained from condensation and maintained by the action of the protoplasm of the fungi. These fungi have been grown in culture by Major H. A. Dade at the Commonwealth Mycological Institute who found that they need 40 per cent of added sucrose in order to grow. When they grow upon glass they make the instruments useless in a few months by their opacity, and do permanent damage by etching the glass. Fungi grow best between uncemented components of lens systems but will grow on any glass in the damp tropics. All optical apparatus should, where possible, be made air-tight. Instruments with extending tubes like field binoculars and microscopes with draw tubes, suck in new supplies of spores and damp air each time they are adjusted. During ordinary work the nuisance must be tolerated and lenses cleaned frequently with cotton cloths, but during storage all instruments may be protected by being completely enclosed in a polythene bag together with some silica gel. So far as is known, no fungus attacks polythene. This method has the advantage over fungicides that it also excludes dust.

In general one cannot kill fungi in instruments by means of fungicides so the aim must be to keep the instruments dry and so prevent the germination of the ever-present spores.

It is necessary in the wet tropics to place all microscopical apparatus in a dessicator each night and silica gel is satisfactory as the dessicating agent and may be used in any closed container. Objectives and eyepieces may be placed in small boxes, binocular microscopes and similar larger pieces of apparatus must go into plastic cupboards with tight doors and it must be remembered that the silica gel from these boxes needs drying occasionally. Fungi may be kept off large surfaces like racks and walls by painting these with any solution containing phenol or copper. The copper need only be present in about one part in a million to be effective. Compounds containing creosote may be used internally in instruments in the field, this evaporates slowly and is not so readily removed by the air-pumping action of instruments with draw tubes.

In an emergency, instruments may be kept dry by packing them in tins with tight lids and using, as a dessicating medium, newspapers which have been exposed and baked in the sun during the same day. Newspaper is very absorbant and the paraffin in the ink is sufficient to keep away mites.

The effects of heat are fairly obvious. Damage to instruments indoors is unlikely but outside, the temperature of an instrument in the sun may rise to 60° C. and above in North Africa. At this temperature balsam cement between lenses may give way and lubrication grease of ordinary kind will lose its properties, so all instruments for tropical and arctic use should be lubricated with silicone grease. (See 'Storage of Equipment: Microscopes', page 24.)

In the dry tropics no trouble will be experienced due to fungi but dust will cause great damage to lens surfaces if these are carelessly cleaned (see the Section 'Cleaning Lenses' in Chapter 19, 'Vertical Illumination', page 361). All cleaning cloths must be kept in dust-proof polythene bags or similar containers and one cloth should be used for wiping off dust and another clean one for polishing.

When an establishment is struck by a sand-storm, a covering of dust, filth and sand of all degrees of fineness up to 12 mm. thick may be expected over everything, even in a normal closed permanent building. In field establishments with wooden buildings in desert areas, 100 mm. of dust may lodge upon the inside. When there is a warning of this kind of storm, work must be stopped and all apparatus placed in tied-up polythene bags. The amount of trouble and interruption due to dust is approximately proportional to the strength of the wind, a speed of 20 km./h. is sufficient to start the process of raising dust in an area where the sandy surface is often disturbed. Large quantities of dust should be removed from instruments which have been accidentally exposed, by blowing and the power produced by an ordinary vacuum cleaner used to blow instead of suck is sufficient. More power than this (for example, a garage air line) will certainly scour the glass surfaces with the dust raised and with the particles which passed through the apparatus intake.

Animal Pests

In both the wet and dry tropics certain groups of the animal creation are a nuisance but greater damage is experienced by the workers than by their instruments. Optical apparatus is not greatly affected by insects and mites, but care must be taken to have all flexible parts, like bellows, made of a plastic material. Leather and Rexine are both eaten by the larval stages of insects, colonies of mites are often found inside the tubes of microscopes and sometimes may be seen walking on the cross-wires in eyepieces. A trace of paraffin oil somewhere within the instrument will keep all mites away.

All the conditions mentioned above in connection with microscopes are experienced by other apparatus, in particular, electrical and wireless apparatus are much affected by growth of fungi and the attacks of insects. The high-tension circuits in radio sets and cathode ray oscillographs are readily short-circuited by damp fungus threads which maintain their own water supplies. Moths attack cloth insulation. Rats and mice eat wax insulation, so it is necessary to protect auxiliary microscopical apparatus in the same way as the instruments themselves. When such apparatus is in regular use there is less trouble because the heat generated dries the parts; it is in the store that trouble will be met.

Microscopical Apparatus Exposed to Radioactive Material

Radioactivity is a hazard which has come upon the laboratory worker in recent years and looks as if it will continue to be present for all time.

It must be understood clearly by the reader that the handling of radioactive materials of all kinds can be very dangerous and no attempt can be made here to provide an adequate picture of the problems involved in the handling of such materials. Advice must be obtained from the specialist branches dealing with health matters and the Atomic Energy Authority's atomic energy research establishment at Harwell, Healths Physics Division, will give advice to workers in this field.

Microscopes are not any more liable to become contaminated than any other piece of apparatus, but owing to the fact that the instrument is in close proximity to the face and eyes and is much handled, attention must be paid to the safety of workers.

Radioactivity is a term used to cover several types of radiation from atoms broken by various processes. The activity has a natural decay rate which cannot be affected by any chemical or physical process in the laboratory. The radiations may be of varying energies and of varying penetrations, some are stopped by the outer layer of skin, some penetrate several millimetres, and some, like X-rays, to a metre. A surface may be dangerous because the contamination upon it radiates penetrating rays or because the contamination may be rubbed off by the fingers and so placed in intimate contact with the skin or be transferred to the digestive and respiratory organs. A contaminant radiating only rays of low penetration (a few millimetres only in air) may not be dangerous as penetrating radiation but may be dangerous when rubbed into the skin or when inhaled or swallowed. Hard external skin may resist, but soft internal tissue will not. Very small amounts of the order of $\mu\mu$ cc. per litre in air of some radioactive materials are dangerous, mainly because they become localized in particular parts of the body. Solid, liquids and gases may be radioactive, all cause damage to biological systems by producing ionization directly or indirectly in vital structures in the cells, thus causing chemical and physical changes in these structures.

It is fortunate that the danger from radioactivity is now well understood and correct monitoring apparatus will always be at hand. The following remarks are therefore directed more towards avoiding loss of time due to contamination than towards protection of workers. Avoidance of contamination will usually lead to the protection of the staff.

It is very important that all workers leaving 'hot' areas have their hands checked by a monitor before touching other apparatus. It is equally important that instruments of all kinds be not swopped about between clean and contaminated or 'hot' laboratories. The main contamination to be expected is of the short range surface type which is not primarily dangerous because of radiation while outside the body, but is dangerous because it may be ingested. Microscopes must be kept covered by polythene or other plastic bags which can be removed, examined and dealt with if contaminated. A layer of easily strippable paint or a film of some medium like that used for replicas (see 'Surface Studies', page 223) may be painted upon apparatus, and should this become contaminated it may easily be stripped off and disposed of by qualified personnel.

Some radioactive materials, for example, radium, have a long life lasting hundreds of years, and almost any length of life is possible to attain given the right substance, but most isotopes have lives of only a few days. Should apparatus become contaminated in a part which cannot be stripped off, it should be placed out of the way where persons cannot come nearer than (say) 2 metres from it, and left for a couple of months. After this time it should be monitored with a Geiger counter and scintillation outfit in the hands of a competent operator, when probably it will be found to be safe. It must be remembered in this connection that a small amount (μ grams) of a substance with a short life may emit a dangerous quantity of radiation in a short time (minutes), while a large quantity of material with a long life may emit so small a total quantity of radiation over a given time (months) that it is not dangerous. Short range α radiation ionizes hundreds of times more intensely than penetrating β and γ rays.

In all cases of radioactive contamination, advice from experts in health physics must be sought. The amount of radiation (for practical purposes all kinds are bunched together) which an observer working a 40-hour week is allowed to receive indefinitely about the head and neck is 0·1 rads per week. The figure has been arrived at after experience with small mammals, painters of luminous dials and bomb victims, and does not err greatly on the side of safety.

All auxiliary apparatus for use in these laboratories should be made to fit into cases designed to present a smooth unbroken surface to contamination from outside. Enamel and plain metal finishes upon well-rounded castings are good, but designs with many switches and crannies of any sort will trap a surprising amount of radioactive matter. The surface known as 'Cryslac' or 'Crackle' must not be used because it cannot be cleaned when contaminated. The ordinary modern microscope is not bad in respect of smoothness of design, but the real safeguard is to cover everything with polythene bags. In many cases the instrument controls can be operated through the bag, and when contaminated a new bag is substituted.

Disposable Bench Coverings for Use in Laboratories
Some remarks have been made concerning benches under the heading 'Choice of Workroom' but where much chemical work, possibly in contaminated surroundings, is done, it is desirable to be able to remove the surface and replace it with a clean covering. Most materials have been tried, ranging from white paper to sheets of Perspex 6 mm. thick. Most are unsatisfactory in some way and Mr R. C. Chadwick of AERE, Harwell, investigated the matter and his findings are as follows:

As a radioactivity decontamination test, the coverings were spotted with a solution of fission products in dilute nitric acid at pH 1. They were allowed to dry for twenty-four hours and then scrubbed with Teepol and water. Radiation counts were made before and after cleaning. Other tests were made using products at pH 6, i.e. as near neutral as possible without precipitation occurring.

As a chemical resistance test, the materials were spotted with three molar solutions of hydrochloric acid, nitric acid, sulphuric acid and sodium hydroxide, and with concentrated hydrochloric, nitric and sulphuric acids. Hexone was used to represent the organic solvents. All chemicals were allowed to act for 24 hours and the results studied.

The materials tried were as follows:

White polyvinyl chloride acetate copolymer on a white backing paper and green polyvinyl chloride acetate copolymer on brown backing paper (white PVA paper and green PVA paper) made by Leonard Stace Ltd, of Cheltenham, and bitumenized brown paper of the type used for tropical packing supplied by the Stationery Office.

Results of the exposures to contamination showed that PVA papers retained less than 0·2 per cent radioactivity, while bitumenized brown paper showed 5 per cent at pH 1 and 34 per cent at pH 6 even when the surface had been rubbed away.

Results of the chemical exposures showed according to the following table:

	PVA Papers		Bituminized Paper
	White	Green	
3M Hydrochloric acid	No effect	Slight discolouring	Slight discolouring
Conc. ,, ,,	,, ,,	,, ,,	,, ,,
3M Nitric acid	,, ,,	Discoloured	Discoloured
Conc. ,, ,,	,, ,,	Paper attacked	Paper attacked
3M Sulphuric acid	,, ,,	No effect	Surface charred
Conc. ,, ,,	Blistered	Paper attacked	Surface charred
3M Sodium hydroxide	No effect	No effect	Paper attacked
Hexone	Soluble	Soluble	Discoloured

TABLE 3

The life of the coverings varies with the job, but at Harwell it has been found that a life of six months may be expected from the PVA papers. This is twice that of bituminized paper under similar conditions. It will be seen that PVA paper deserves trial in conditions where there is no great heat and where it is not likely to be torn by sharp corners. Polythene and PVC resist water better, the paper backing of PVA being the weak spot. PVA may be glued down if required, in which state it makes a smooth durable covering of a semi-permanent kind.

CHAPTER 3

General Microscopy

The Optical System

THE COMPOUND microscope consists of three lenses or their equivalent corrected combinations supported in a mechanical framework which may vary very much in design.

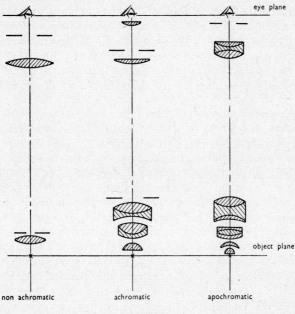

Fig. 1.

The magnifying power of a microscope can be calculated from the following formula:

$$\frac{(\text{Tube length}) \times (\text{Distance of distinct vision} (=250 \text{ mm.}))}{\text{Focal length of Objective} \times \text{Focal length of Eyepiece}}$$

The tube length is measured in practice between the top of the microscope tube (where the eyepiece shoulder rests) to the shoulder of the objective (where it contacts the bottom of the tube when screwed home). It is usually measured in millimetres but was measured in inches. The objective focal length is stated on the mount and care must be taken that objective focal length and tube length are in the same units. Either may be in millimetres or inches. The magnification of the eyepiece is usually stated upon it. The focal length of an eyepiece is not easy to obtain when it is of the Huyghenian form as shown at B Fig. 1. This eyepiece is called 'negative', i.e. it cannot be used as a magnifier and its focal length must be obtained by comparison in magnifying power with a simple lens used as an eyepiece upon a microscope. It is also difficult to measure tube length between the correct points in eyepiece and objective assemblies.

It will be seen from the above very brief outline of the structure of the parts of a microscope that although certain points are chosen for calibration purposes, calculation of magnifying power should not be attempted when doing practical microscopy. The methods for obtaining the cardinal points and positions of principal planes in combinations of lenses will be found in the text books upon Optics.

Measurement of Magnifying Power
This is always done by direct methods. An instrument known as a Stage Micrometer may be purchased from the manufacturing opticians and consists of rulings upon glass at intervals $0 \cdot 1$ and $0 \cdot 01$ mm. This slide is placed upon the microscope stage and used as an ordinary object. A second scale known as an Eyepiece Micrometer, engraved or photographed upon a disk of glass about 20 mm. diameter, is placed in the eyepiece. The eyepiece scale is arranged to be in the focus of the eye lens of the eyepiece at the same time as the object is in focus. The spacing of the stage micrometer rulings being known, the value of the eyepiece graduations also becomes known and will remain so as long as the same eyepiece objective and tube length are in use. (See 'General Microscopy: Micrometry', page 48.) The apparatus is used as follows:

(1) If a known division of the stage micrometer occupies a certain length of eyepiece scale, the magnification clearly is: Eyepiece scale length ÷ Stage scale length.

   The eyepiece scale length may be calibrated by placing a piece of celluloid ruler on top of the micrometer scale and observing it through the eye lens. It is best to work in millimetres because these divisions are easier to accommodate in the eyepiece.

(2) An image of the stage micrometer may be projected on to a
screen 250 mm. away from the eyepiece and the image
measured with a ruler.

Magnification and Resolution

There is no limit to the amount a microscope can be made to magnify.
It is necessary only to decrease the focal length of objective and
eyepiece and increase the tube length. Attention to all or any one of
these factors will increase magnification. During the nineteenth
century before the mechanism of image formation was understood,
it was usual to make objectives with focal lengths 1/50 and even
1/100 in. and to use them on tubes 10 in. long with eyepieces magnifying about six times. The resulting total magnification was about
5,000 times and more, but it was soon discovered that magnification
is not the only factor governing ability to see clearly with the aid of
a microscope.

When a pencil of light passes through an object a certain amount
of the light appears scattered all round the illuminating pencil due
to the diffracting action of the object. If the object is a diffraction
grating the spread of light is easily seen because the grating is sufficiently regular to form the light into a spectrum, several orders of
which may be clearly separated if the grating is about 400 lines to the
millimetre spacing. The grating acts upon light in a directional way,
but if several transparent gratings (or replicas) are placed on top of
each other at random, scattered, spread spectra result. A typical

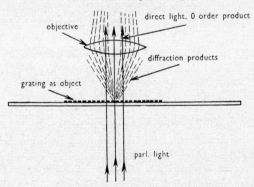

Fig. 2. *Note:* Diffraction shown from one applied light pencil
only. Grating not to scale.

object consists of many small details which may or may not be
regularly distributed but may be compared with a random placing
of gratings. The result of passing light through such a structure is to

produce incalculable masses of mixed up spectra surrounding the central beam.

If we now return and make experiments with a known grating made up of light and dark lines (known as an amplitude grating because it forms diffraction patterns due to its effect upon the amplitude of light) we will find that a clear picture of the grating cannot be obtained in a microscope unless an objective is used which has sufficient aperture to accept all the diffraction products.

Abbe made an objective mount which allowed him to place diaphragms of various shapes behind the objective by which means he could cut off orders of spectra before they reached the eyepiece. He was able to show that resolution of detail was affected by the number and distribution of the orders of spectra employed in forming the final image, and that the results followed theoretical predictions.

When a diffraction grating with lines of wide spacing is used (say twenty or so lines to the millimetre) the spectra do not spread far and a lens of low aperture can receive them all and produce an accurate image of the coarse grating. This corresponds to the use of a low power objective used to look for coarse detail, its focal length can be long, perhaps 25 mm., and its aperture low, perhaps subtending an angle of 30°. Such conditions automatically give plenty of space between the object and objective lens and such an objective is used to obtain large fields of view and general impressions of a specimen's structure. There is no point in making its lenses large and corrections correspondingly difficult when a shorter focal length for higher magnifications makes aperture and correction easier to obtain.

If a grating of a great number of lines to the millimetre is used as a specimen, 600 lines for a manufactured grating, and 5,000 for certain test specimens, the spread of diffraction patterns is very great and an objective which can accept light at nearly 180° angle is necessary in order to obtain an image containing all details in the specimen. This corresponds to the use of a high power high aperture combination of lenses intended to reveal all possible detail in the finest structures.

In objectives of 4 mm. focal length or thereabouts, and shorter, all efforts are concentrated upon increasing the aperture, the aim being to grasp as many diffraction components from objects of mixed structure as possible and so reveal as much detail as possible. It will be shown that the maximum useful magnification which can be used with visible light is 1,500. There is, therefore, no point in making objectives of shorter focal length than 2 mm., as such an objective can be made to possess the maximum aperture and hence the maximum resolving power.

Most microscopical specimens have insufficient regular structure to diffract light at angles to the main beam pencil sufficient to fill the available aperture of a high power lens, so in these circumstances something must be done to assist the object. It is done by condensing the light upon the underside of a transparent specimen by means of a lens system of aperture equal to that of the objective; light then falls upon the specimen at a considerable angle of obliquity, and

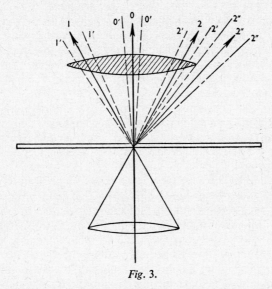

Fig. 3.

whereas the spread of diffraction components for the central parallel pencil would only have been 0' without a condenser, the rays 1 and 2 now are present also and have their diffraction components 1' and 2' which also enter the objective. In this way diffraction components associated with beams at all angles are produced and made to yield their information when the final image is formed.

It will be observed from the sketch that the condenser angle need not be quite as high as that of the objective because the diffraction bands from beams 1 and 2 spread an amount dependent upon the nature of the specimen, therefore beams 1 and 2 need not enter the objective at its periphery, but at a little distance in to leave room for the diffraction components. It is true that if a beam 2″ were present, some of its inner diffraction components would fall within the lens aperture but the beam itself and the outer diffraction components may be beyond the angle with which the objective can cope and be reflected within the objective mount and cause mischief such as fogging the image. These are the main reasons why the Nelsonian teaching, empirically derived, states that an objective should not be illuminated with light at a greater angle than will fill its aperture up to 2/3 of the diameter of its back lens as seen by looking down the microscope tube with the eyepiece removed.

When a dense, semi-transparent object is examined, so much light is scattered by the lower layers of it that the uppers parts are illuminated by light arriving upon them at angles possibly of 70° to the optic axis and such objects do not need the help of great condenser aperture. If a typical specimen of this type, say a leaf cuticle, is set up and the objective back lens examined without the eyepiece being in place, the back lens will be seen to be full of light whatever the condenser aperture and the microscope will, under these conditions, be giving a good performance.

When an opaque object is examined by top light, the object scatters light in all directions and so fills the aperture of any objective completely so a microscope working upon opaque specimens is always working at full aperture and will therefore develop its full resolving power.

Not all objectives are well corrected in their outer zones and many do not work well on opaque specimens for this reason, but when the specimens are transparent, the substage condenser diaphragm is closed until the outer zones of the objective are cut off. In this way a very poor objective may be made to yield a reasonably clear image by stopping down its aperture either directly or by reducing the substage condenser diaphragm. Following the reasoning above, such cutting out of some diffraction products prevents the lens resolving the full amount of detail that its aperture suggests it should and its image will not stand the same amount of eyepiece magnification as will the lens working at full aperture.

In the above discussion it has been assumed that the workmanship of the lens grinder is good, i.e. the lenses are all true parts of spheres, all are centred on the axis of the combination, and all the surfaces are of good smooth optical polish. Optical polish is difficult to define but is understood by those performing the work. A lens may appear sparkling clean but its surface may be wavy, or it may have minute scratches upon it. Experience tells the worker when the surface is as smooth as the particular glass can be made, and study with a magnifying glass of the lens surface in the light from a window which has bars across it, should show no structure where the light and dark reflections join. It is unusual to find a lens of poor workmanship even amongst combinations fifty years old, but workmanship still makes the difference between a top-class objective and an exceptional one. It is common to find surface distintegration in lenses of the apochromatic kind more than forty years old. (See 'Objectives', page 54.)

The Homogeneous Immersion System

The history of the immersion system does not show clearly the reason for its invention. It may have been invented for the purpose of doing away with the need for adjustment for cover glass thickness or for the purpose of allowing more oblique rays of light to enter an object glass. In either case, the matter of homogeneous immersion is intimately connected with Numerical Aperture and correction of spherical aberration. Both objectives and condensers may be of the immersion type and, as the theory and purpose of both is identical, attention will be given mainly to the objective.

In the immersion system the front lens of the objective is connected to the specimen directly or through the cover glass by means of cedar-wood oil. The oil has its refractive index and dispersion carefully corrected to be the same as crown glass from which the objective front lens and slide cover glasses are made. If we further assumed that the object is mounted in Canada balsam, again of the same refractive index and dispersion, a pencil of light will pass in a straight line from the object to the posterior surface of the objective front lens without reflection or refraction.

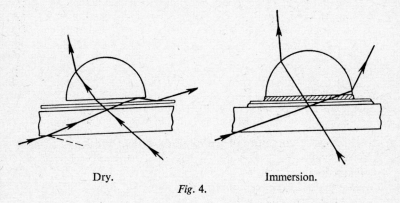

Dry.   Immersion.

Fig. 4.

In addition to the fact that rays of obliquity up to nearly 90° with the axis in oil, can now enter the objective, it is possible to design the front lens to have no spherical aberration, as shown below.

It must be noted that angles measured in oil (RI 1·5) and in air (RI 1·0) cannot be compared directly and this led to the use of the term 'Numerical Aperture' (NA), which is explained under that heading. Numerical Aperture is a number and any objective may be so measured. NA represents resolving power and any objectives may in this way be compared regardless of their focal lengths. The highest NA practicable is 1·4 and this is usually found only in the best-corrected 1/12-in. (2 mm.) apochromatic objectives.

Structure of the Immersion Objective

In addition to the extra numerical aperture possible with an immersion objective, an advantage over other systems is possible in the direction of correction for spherical aberration.

It is possible to make a hemispherical immersion lens quite free from axial spherical aberration by arranging matters according to Fig. 5. The lenses following in the combination have much less aberration to correct (ASA is proportional to zone radius squared) and may be of conventional form with the greater part of design effort applied to colour correction.

Consider Fig. 5.

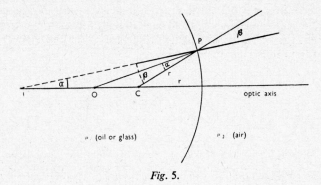

Fig. 5.

r=radius of curvature of rear surface of the front lens
o=object
β=angle of refraction
α=angle of incidence
μ_1=RI glass or cedar oil
μ_2=RI air   Light travels from μ_1 to μ_2
$OC\mu_1=r\mu_2$ this determines the position of the object O with respect to the posterior glass-to-air surface of the lens.

In $\triangle$ POC $\dfrac{OC}{\sin a} = \dfrac{r}{\sin \lfloor POC}$

$\therefore \dfrac{OC}{r} = \dfrac{\sin a}{\sin \lfloor POC}$ but $\dfrac{OC}{r} = \dfrac{\mu_2}{\mu_1}$

$\therefore \mu_1 \sin \lfloor PIC = \mu_2 \sin \lfloor POC = \mu_1 \sin a$

PIC and POC are similar

By Snell's Law applied at P

$\mu_2 \sin = \mu_1 \sin a \therefore$ using equations above

POC $= \beta$ and PIC $= a$

$\triangle$ PIC and $\triangle$ POC are similar and

$\dfrac{OC\,(a)}{r\,(\beta)} = \dfrac{r}{IC} \quad \therefore IC = r\dfrac{r}{OC}$

and $\dfrac{r}{OC} = \dfrac{\mu_1}{\mu_2}$ i.e. $IC = r\left(\dfrac{\mu_1}{\mu_2}\right)$ and $OC = r\left(\dfrac{\mu_2}{\mu_1}\right)$

That is, the position of I is determined; IC must be made equal to the right-hand side. This expression is independent of angles and so is true for the largest possible angle at O. The distance OC, object to centre of curvature, fixes the distance away from the lens at which the object must be placed and this position and all others to the left of the glass-to-air surface upon which P rests, are in the homogeneous medium cedar oil or glass, these two substances being interchangeable optically.

In order to accept the maximum possible angle of light from the object at O, the lens is often made greater than a hemisphere. The front lenses are normally cemented in and even where a hemisphere is used (giving slightly less aperture) great care must be taken that they do not strike the slide or they will be pushed out. For the same reason solvents like spirit must not be used to remove hardened oil; this is best removed by soaking in fresh fluid oil.

The light emerging from the front lens is still diverging considerably and must be followed by another lens designed to give minimum spherical aberration. It is not possible to make this second lens work in oil because the posterior surface of the front lens would then not exist optically. The problem may be tackled as follows with an aplanatic surface in air, Fig. 6.

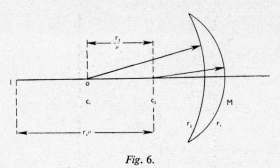

Fig. 6.

r of Fig. 5 now becomes r_2, i.e. the first surface,
O object must be the centre of curvature of r_2 and

$$OC_2 = r_2\left(\frac{r_2}{\mu}\right); \mu_1 \text{ and } \mu_2 \text{ are now 1 (air).}$$

It must be arranged that the first surface presents no refraction to
rays starting from O, hence O is used as the centre of curvature for
lens surface C_1. If the lens system is set up according to these dimensions, the first spherical aberration occurs at the fourth lens surface
when the rays are nearly parallel. After this point ordinary colour
correcting components follow.

The diagrams, Figs. 5 and 6, show the layout to obtain a system
with no spherical aberration. If, however, the medium μ_1 is something other than oil or glass, the light will appear to come from a
different direction than from 1 and will not be presented to surface r_2
on the second lens according to the diagram along a radius. One way
of correcting this is to move the first lens to and from the second
until a position is found where the light does pass along a radius into
the second lens first surface and this is the basis of correction of
tube length by correction collar. In practice, the rear components are
moved bodily to and from the front lens as this is safer than moving
the front lens when it is so near the specimen. The tube length of the
objective may be set on test by adjusting the distance the front lens
screws on to its mount and this is very critical both as to distance
and centration.

The type of front assembly described is typical of all high and
medium power objectives even though some be dry and in the dry
lens most of the advantages shown to be obtainable in the immersion
type can be partly obtained with the smaller apertures of dry lenses,
but in these the front lens is not free from spherical aberration.

The complete structure of the high power objective is as shown in Fig. 7.

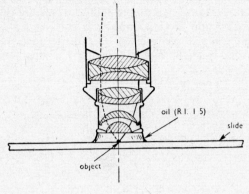

Fig. 7.

A few other interesting points arise in connection with immersion objectives. Because the first surface of the front lens does not exist optically, it need not be carefully worked and scratches upon it are of little importance. As explained above, the first refracting surface in the objective is the front lens posterior curve and this may be considered as a device present to make the top of the immersion and mounting media spherical. So long as this posterior surface is spherical, the flattened front may be of any finish. It has to be flattened only to obtain working distance and successful front lenses have been made in recent years by the Author and others by fusing a glass bead on a fine tube of soda glass and roughly grinding and polishing a flat on it, surface tension having made the globule substantially spherical. In the dry and water immersion systems, the first surface of the front lens does refract and so must be properly worked.

Numerical Aperture
In order to be able to compare the resolving powers of various objectives regardless of their working medium and magnification, the term 'Numerical Aperture' (NA) was invented. NA is a number which bears a linear relationship with resolving power and in microscopy, resolving power is usually measured directly in lines per millimetre.

In Fig. 8 the paths of rays of light at the objective front lens are shown for the dry and immersion conditions. When air is present between object and objective, ray AB passes through slide and object into the objective, making an angle p with the optic axis. If oil is present the ray will not experience refraction when leaving the slide and will make an angle q with the axis. The illuminating angle r is, in the case of oil immersion, compressed into angle q at the objective and so the diffraction products resulting from its passage through the object are likewise compressed. In this way more rays and diffraction products can be included in a given maximum aperture

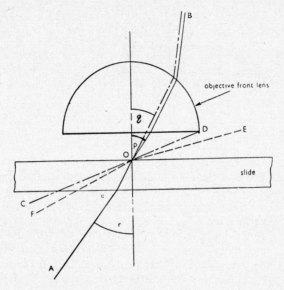

Fig. 8.

CD when oil is the medium. Following the same reasoning, if the oil were removed, ray COD could not enter the objective but would be refracted outside to E, and again the maximum angle of illumination possible in the dry condition which permits a ray to enter the objective is indicated by FO. If the front lens of the objective is brought nearer the specimen in an attempt to capture ray OE, the process will soon be limited by the total reflection and refraction occurring at the glass to air surfaces.

In order to provide a ray of sufficient obliquity to use to the full the oil immersion objective aperture, the condenser must also be immersed, but owing to the fact that most objects scatter a quantity of light themselves, considerable advantage is to be had from the increased light grasp of an oil immersion system also when the illumination is dry.

Image Formation and Diffraction Effects

There is little difference between the image-forming mechanism of a telescope and that of a microscope. In the telescope the input rays are parallel and an image is formed by the lens in its focal plane, but matters are reversed in the microscope, the object is where the telescope image is and the image is formed a long way from the objective so that light may be considered to emerge parallel. The same considerations regarding theoretical resolving power can be applied to both instruments but it is easier to show diagrammatically how the result is arrived at if the telescope be considered.

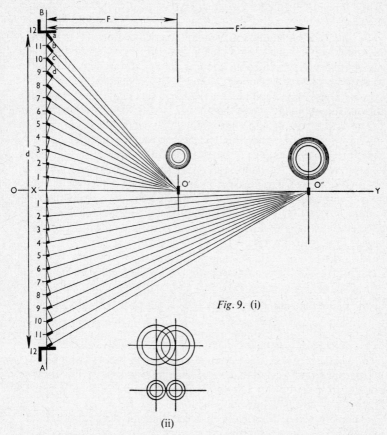

Fig. 9. (i)

(ii)

In Fig. 9 let AB represent a wave front passing through an aperture AB and consider this aperture divided into zones 1 to 12. Consider the conditions obtaining at a point 0′ at a distance F along the direction of travel of the front. Light arriving at 0′ from zones approaching the periphery of AB has farther to travel than light from the zones nearer the axis *xy* but the light in all zones is derived from a single wavefront, therefore it is in a condition to interfere. As the distance of the zone from the axis increases a point is reached when the path length difference between the aperture plane and focal point is ½ wavelength and cancellation of light takes place in this region. Going further towards the periphery the path length difference becomes one wavelength and addition takes place. As we proceed towards the periphery alternate dark and light rings are produced in the image plane at each multiple of a wavelength of path length difference.

In Fig. 9 the path length differences for two distances F and F′ are shown in heavy line at *a b c d* etc., and it will be noticed that towards the periphery the path length changes are rapid, this causes the rings to become close together and ultimately to run together and disappear. These effects can be clearly seen in an astronomical telescope focussed upon a star where the aperture AB is that of the objective.

The upper half of Fig. 9 (i) shows that when the aperture of a system is increased, the diffraction pattern becomes smaller owing to the greater rate of change of path length differences as the periphery is approached (F′ 0′). Two systems of rings (representing points in the image, or two stars) are considered separable if the bright centre disk of one falls no closer to the centre of the other than upon its first dark ring [Fig. 9 (ii) upper] and if increase of aperture causes the diffraction pattern to be smaller, a smaller separation of points is visible [Fig. 9 (ii) lower].

Much elaborate mathematical work was done upon the observations by Airy but it is omitted here, the results were that the radius

of the first dark ring is given by $\dfrac{1 \cdot 22 F \lambda}{d}$ where F is the focal length

of the object glass, λ the wavelength of light and *d* the linear measurement of aperture AB (the objective diameter in practice).

It must be understood that Fig. 9 is in no way a ray track drawing nor is it to scale, but refers only to a point in an image and serves to illustrate the mechanism of diffraction.

There are some differences between the microscope image and that of the telescope, as in the microscope the light coming from two points close to each other can be in a condition to interfere because the light comes through the condenser from points near by each other in the illuminating source. However, the amount of interference possible appears to be very small and the figure of $\dfrac{1 \cdot 22\lambda}{d}$ for distance between points which can just be resolved, is accurate enough for practical purposes. Also with reference to this figure for resolution, the accepted limit of separation of two points in an image stated on the previous page is not a definite measurement and in practice features may be seen separately which are closer than this,

If $2a$ Fig. 10 is the angle subtended by the objective front lens diameter when focussed upon the image, $\tan a = \frac{1}{2}\left(\dfrac{d}{F}\right)$

transposing we have $d = 2F \tan a$

substituting into $\dfrac{1 \cdot 22 \ F\lambda}{d}$ we have

$$\text{resolution} = \dfrac{1 \cdot 22 \ F\lambda}{2F \tan a}$$

$$= \dfrac{0 \cdot 6 \ \lambda}{\tan a} \qquad\qquad \text{(i)}$$

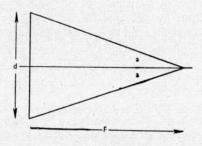

Fig. 10.

Mainly as a result of experiments with a microscope Abbe[1] concluded that the resolving power is nearer

$$\dfrac{\lambda}{2\mu \sin a} \qquad\qquad \text{(ii)}$$

In equation (i) the refractive index of the medium in which the system is working is 1 (air) but in equation (ii) the value μ had to be included owing to the possible presence of a medium other than air, in fact, oil or water. In equation (ii) all factors concerned in the resolving

[1] His mathematical work is inaccessible.

power of objectives of all kinds are included in the denominator and equations (i) and (ii) are remarkably similar.

The working medium air, water or oil, has its refractive index taken care of by the term μ and the physical size of the lens is represented by sin a. No other factors in the lens structure itself (except workmanship) can affect its resolving power, so we have then an expression μ sin a (Numerical Aperture) which will satisfactorily represent the resolving power of any objective.

If we assumed that the expression $\dfrac{\lambda}{2\mu \sin a}$ represents the smallest distance apart that two points can be seen as separate objects by a normal eye, it is easy to calculate what magnification is necessary in the microscope as a whole to develop this resolution. The practical limit of NA is 1·4 and a typical wavelength of visible light is 5×10^{-5} cms. With these figures inserted into the formula we have

$$\frac{\lambda}{2\mu \sin a} = \frac{5 \times 10^{-5}}{2 \times 1 \cdot 4} = 1 \cdot 41 \times 10^{-5} \text{ cms.}$$

A normal eye can separate points 2 min. of arc apart and at the distance of most distinct vision (10 in. or 25 cms.) this angle is subtended on the retina by a distance of 10^{-2} cms. nearly, hence it is necessary to separate the points in the microscope image by means of magnification until they fall at least 10^{-2} cms. apart at the retina.

$$\frac{10^{-2}}{1 \cdot 41 \times 10^{-5}} = 700 \text{ nearly}$$

In this calculation the figure for maximum NA was taken but the figures which refer to the human eye are constant for one observer, therefore the answer of 700 is the amount by which any numerical aperture must be multiplied to give the minimum required magnification necessary to show all contained detail to the normal eye. (See 'Instruments and Use of the Eyes', page 9.)

In these computations almost all figures vary with circumstances. Light has a sinusoidal distribution in an image, therefore there is no definite point at which all eyes fail to separate points. The separation in practice is performed by recognizing tiny differences of brightness of two points, Fig. 11, and not by resolving an optician's black and white card. For this reason, eye resolution in a microscope does not increase linearly with brightness of image except up to a definite level of brightness, after which resolution falls off with increasing brightness. The preference often shown by experienced microscopists for the steady low intensity paraffin lamp as a source of light is explicable in this way. Acuity of vision varies greatly between individuals and in the same individual at different times of the day (it is usually greater in the evening). For these reasons the figure of 1,000 times the NA is taken as a guide to the necessary amount of magnification. A person's visual acuity is not related to short or long sight.

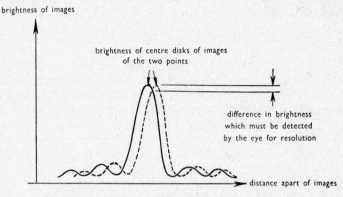

Fig. 11. Distribution of light in the images of two points just resolved.

Choice of Magnification

It has been shown in the last section that magnification must be great enough to show all resolved detail in an image but it cannot, however, be increased indefinitely even if the intention of the observer is merely to make existing detail appear larger. Some extra magnification known as 'Empty Magnification' in excess of the amount demanded by the NA of the objective, may be helpful to those observers whose eyes are not good and to those who require only bold images of well-known structures. Low power objectives up to about 12 mm. focal length will stand about four times the proper eyepiece power before the image becomes dark and breaks up, but high powers such as the dry 4 mm. and immersion 2 mm. should not have more than two to three times the proper power applied. The high power system is producing an image of a series of points near to the limit of resolution in visible light, therefore extra magnification only spreads the already fine gradations of light and shade which make up the image. Loss of light also takes place as magnification is increased and if the brightness of the illuminant is increased accordingly, artifacts are apt to be produced. A skilled observer who knows that his instrument cannot show him detail less than $\frac{1}{2}$ micron for an NA of 1·4 is not likely to be misled by artifacts, but the photographic plate is very likely to show as solid detail all kinds of artifacts and great care should be taken as described under 'Photo-micrography', page 190.

Micrometry

The measurement of size of microscopical particles is always done by direct means. Cross-wires in an eyepiece are calibrated from a stage micrometer as indicated above under 'How to Measure Accurately', page 52.

Eyepiece Micrometer

This is a composite disk of glass composed of two disks, one of which has a scale photographed upon it with the other as protection. The scale is such that it covers about two-thirds of the eyepiece field diameter. The micrometer rests upon the diaphragm of an ordinary Huyghenian eyepiece and the diaphragm is normally in the focus of the eye lens; it therefore supports the scale in focus also. Should the scale not be in focus, the diaphragm can be pushed up or down the tube with a cylindrical piece of wood, in size just less than the inside diameter of the tube. The scale must be so placed that when the eye looks through the microscope with the object clearly and comfort-ably in focus, the scale is also in clear sharp focus.

The gradations on the scale have no particular value and the scale may be placed in any eyepiece. All micrometers in eyepieces are calibrated directly from accurately ruled slides known as Stage Micrometers, commercially obtainable with rulings separated 0·01 and 0·1 mm. Examination of a Stage Micrometer with a 25 mm. objective will readily show the distribution in groups of the lines.

To calibrate an eyepiece micrometer proceed as follows:

(1) Set up the microscope with the stage micrometer and eyepiece micrometer in their places.

(2) If practicable, adjust the tube length of the microscope so that the magnification is such that an even number of large divisions in the eyepiece micrometer span an exact number of divisions on the stage scale as seen in the microscope (a or b below).

   If it is not possible to move the draw tube sufficiently far to arrange this (bearing in mind the objective tube length), arrange that the stage scale spans an exact number of small divisions in the eyepiece scale.

(3) Note the value of the stage scale division used, in millimeters and count the number of eyepiece divisions necessary to span this stage scale division. Dividing the value of the stage scale by the number of divisions in the eyepiece scale gives the value of each eyepiece division for that tube length and objective.

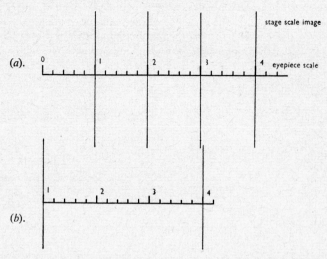

Fig. 12.

(4) This procedure must be carried out for every objective.
(5) It is usually best to sacrifice exact tube length adjustment of an objective in order to have an easy figure to work with in the eyepiece scale calibration. A tube length adjustment error of 25 mm. or so will interfere with the best definition but will not make an important difference to a micrometric measurement. It is the work of a moment to set the tube length ready for measurement from a previously drawn-up chart.
(6) Stage micrometers are now so accurately ruled that it is unnecessary to take the average spacing of a number of rulings when calibrating an instrument.
(7) It is convenient to set up one eyepiece with a micrometer and keep this for use on all objectives. An eyepiece magnifying ×8 is convenient.

Modification of Eyepiece Micrometer for Photomicrography

It is essential in technical work to attach a scale to photographs and this is most easily done by using a micrometer eyepiece scale in the microscope eyepiece. When the image is projected on to the screen by the microscope, the eyepiece diaphragm and scale are not in focus but this can easily be rectified by changing the position in the eyepiece of the field diaphragm and with it the scale. The diaphragm should be adjusted in position on trial until a sharp image of the scale crosses the projected image.

When high powers are in use the centre of the field only gives the best definition and the presence of the scale may be a nuisance there. The Author made the following small modification to avoid this. The field diaphragm was cut out, as shown at (A) Fig. 13, with a fine file.

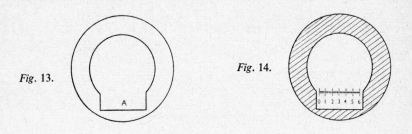

Fig. 13.

Fig. 14.

An eyepiece micrometer was made by Messrs Beck with the standard scale offset so that it rested in the gap (A); the appearance of the field is then as shown in Fig. 14.

Stages used for such purposes as track analysis commonly have engineers' micrometers to drive the stage in X and Y directions so allowing direct measurements to be taken regardless of the optical conditions. An attachable stage should be mounted on top of the object stage for use as a finder. See page 145.

Once the diaphragm of an eyepiece has been set up for the camera it is desirable to use this eyepiece always, not only because the scale is always present but because a clear field limited by the eyepiece diaphragm is projected upon the plate. It has been found on trial that the change in magnification over the field of view is negligible, therefore no inaccuracies arise as a result of offsetting the scale.

The Screw Micrometer Eyepiece

This apparatus fits into the upper end of the draw tube and consists of two parallel spider lines in the focus of a Ramsden or Huyghenian eyepiece. One line is moveable across the field by means of a micrometer screw and graduated cylinder. Marks known as 'teeth' are provided each 1·0 mm. along the travel of the spider line and are visible in the eyepiece field. The cylinder is graduated in 1/100 of a millimetre and is read from outside the eyepiece.

Although the calibration of these micrometers is very accurate, they do not read the dimensions of an object directly but must be related to the magnification of the microscope in the same way as must a simple eyepiece micrometer.

The method of calibration is as follows:

(1) Set up the microscope focussed on a stage micrometer.
(2) Set the micrometer eyepiece preset adjusting screws so that at 0 on the graduated cylinder of the micrometer screw, the fixed and moveable spider lines lie exactly coincident.
(3) Set a line of the stage micrometer to lie under the eyepiece micrometer lines.
(4) Rotate the micrometer screw until the moveable line encounters the next ruling of the stage micrometer.
(5) Note the number of teeth and graduations in the eyepiece micrometer passed through during the operation, and read them as 1 mm. per tooth and 0·01 mm. per cylinder graduation, say 10·41 as a typical figure.
(6) This is the calibration of the micrometer for this objective and tube length and any change in either demands a fresh calibration. In the example above, 10·41 read from the micrometer scales equals (say) 0·01 mm. (stage micrometer ruling) therefore each division on the graduated cylinder equals 0·01 mm. (100 microns) divided by 1,041, equals 0·096 microns. Each tooth is 100 times that=9·6 microns. The note under 'Eyepiece Micrometer' (page 48) regarding a convenient choice of draw tube setting is here illustrated.

Stability of Screw Micrometers

Any mechanism mounted high upon the tube of a microscope is liable to cause instability when vibration is present and further instability is caused by the hands of the operator when they are used to adjust milled heads on apparatus so placed. (See 'Devices to aid Stability of Microscopical Apparatus', page 174.)

The first cause of instability can be minimized by building such apparatus, in this example, the micrometer, of light materials like aluminium. The second cause, namely, the weight of the hands, usually cannot be eliminated even in the best and most solid microscopes. Screw micrometers are best mounted upon separate heavy stands so that movement is not transmitted to the top of the microscope body tube, but this arrangement makes quick and accurate setting of tube length difficult, particularly as the micrometer must be carried clear of the draw tube to allow the coarse adjustment to operate.

The Author's method of overcoming the difficulty was to use a separate micrometer stand and measure the calibration of the micrometer at several tube lengths. A graph of tube length in milli- metres against Micrometer Division Values was plotted and the tube length measured with a ruler each time a micrometer measurement was taken, so by referring the tube length in use to the curve, the micrometer reading in μ per graduation obtaining at that tube length could be read. On microscopes having focussing stages the micro- meter can be operated remotely by miniature Hooks joints, but this form of stand is not always more stable than the conventional one and much depends upon the weight of the stand being adequate. In the Author's experience the simple eyepiece scale which can be used without being touched by hand is quite adequate for all micrometry.

How to Measure Accurately

When the micrometer wires are applied to the image of an object seen under high power, difficulty may be experienced in deciding just where the edge of the object is situated owing to the surrounding diffraction bands. The rules are as follows:

(1) Use as large an objective aperture as possible, even though the image may become foggy, for this will define the edge of the object even though internal detail is dimmed.
(2) If the specimen is of higher refractive index than the mounting medium, measure across the innermost dark diffraction bands.
(3) If the specimen has lower RI than the medium or is a hole in the object, measure across the outermost dark band. (Note that the diffraction effects are inside the boundary of the specimen when it has lower RI than the surrounding medium. The student should examine an air bubble in a balsam mount.)
(4) Measure the specimen several times, putting the micrometer out of adjustment between each measurement, and average the results.

In engineering and technical work the best results are undoubtedly obtained by the use of the micrometer-controlled, direct reading, mechanical stages described on pages 50 and 144, but these are very costly and a special instrument is required.

Projection Methods

Any scale or cross wire arrangement may be placed near the lamp condenser or diffuser and projected by the substage condenser into the plane of the object. A coarse scale drawn upon glass with Indian ink is projected as a fine and workable pattern when reduced in size by the substage condenser and two advantages are obtained by using the method:

(1) The micrometer is in a much more stable part of the optical system, and it and its movements are scaled down not up as in the conventional position.
(2) The micrometer image is actually on the specimen, therefore there is no parallax error.

The disadvantage is that the projected lines are not so sharp nor so well defined as eyepiece lines, and well-corrected condensers must be used or definition of the rulings is very poor.

How to Measure Areas

Eyepiece micrometers are available commercially with squares engraved or photographed upon them and the dimensions of the squares are evaluated in the usual way with a stage micrometer. When the squares are superimposed on the object the squares occupied are counted taking each square more than half covered as 1 and each less than half covered as 0. The objective magnification should be set so that a number of squares, say about fifty, are covered by the object to be measured. Tiny objects are easily estimated when they cover a portion of one square. There is very little point in making fussy measurements of area of microscopic objects because they are usually irregular and the microscope sees only an optical section, and it is rare for an object to present a flat surface to be measured. (See also section 'Methods of Counting Particles', page 323.)

All microscopes used for micrometry should have their fine adjustments calibrated in microns for depth measurements. Scales are easily attached to the milled heads and can be calibrated from pieces of cover-glass, etc., previously measured with an engineer's micrometer, also standard engineers' feeler gauges when new and clean are satisfactory. A feeler gauge should be chosen of such a thickness that focussing from top to bottom of its thickness employs about one turn of the fine adjustment milled head. To be sure of seeing accurately the bottom edge of the gauge, a small part of its edge should be ground to an angle. The gauge should then be clamped between two pieces of slide with cedar oil between. Having obtained the travel of the periphery of the milled head scale for (say) a focal change of 0·03 of a millimetre, the rest of the scale can be divided on a lathe or by geometry. A direct reading scale is here useful and practicable because it is accurate for all magnifying powers. If the specimen is dry the calibration is different.

CHAPTER 4

Objectives

THE OBJECTIVE is so called because it is the lens or combination of lenses which is close to the object. It is the main image-forming and light-gathering component of the microscope and is therefore a well-corrected assembly of lenses.

Objectives are of three types regardless of their focal lengths and the media in which they work; namely, (1) Achromatic, (2) Semi-Apochromatic or 'Fluorite', (3) Apochromatic. These names refer to the colour correction of the objective only.

An Achromatic objective is one in which two colours in the spectrum of visible light are brought to the same focus in the image. If the lens were a simple one, the image in red light would be formed farther away from the lens than that in blue light; owing to the fact that glass refracts rays a greater amount as the wavelength of the light decreases, the images in the various colours are spread out as the total amount of refraction increases with lens curvature. If an object is focussed in one colour, say blue, then it has a red diffusion area surrounding it and vice versa, but in the achromatic combination the red is caused to come to a focus in the same position as that of the blue. Other colours in the spectrum are more or less affected by the required combination of glasses and the result is that although a chosen pair of colours come to the same focus, the others are not far away and a good working image results.

The colours chosen for correction are not always the same pair. Usually for visual work the brightest are chosen with due regard to their positions in the spectrum, thus apple green and red, or orange and blue, may be selected. Much depends upon the characteristics of the optical glass available, and manufacturers can and do choose different pairs of colours. An objective meant for photography is corrected so that the high energy, high frequency colours are best corrected and in this example green and violet may be chosen. Most photographic objectives are used with a filter to cut out the low energy colours below green.

Only the pair of colours chosen for correction form images in exactly the same position and, although the images due to the remaining colours are near, a fringe of colour from images not in exactly the right place is apparent and is known as Secondary spectrum. The colours in this secondary spectrum are different at points above and below focus and vary with different objectives depending upon the pair of colours chosen for achromatization.

An achromatic objective always works best with a filter to cut out more or less completely all colours except one band, say apple green. Without filters they work well when used upon large stained specimens like histological sections but do not do well upon small slightly coloured objects as the imperfect colour correction of the combination is apt to cast spurious colour into fine details. However, an achromatic system is capable of resolving test objects up to its theoretical limit of resolution in white light so for ordinary industrial use where known structures are looked for, achromatic objectives are sufficient and, if used with a filter, only expert microscopists can tell the difference in performance between these and aprochromats.

In the high power form, an achromatic combination usually has less aperture given to it than an apochromatic example, but they have all the aperture they can usefully be given and the slight reduction makes for a more secure mounting of the front lens, there being more room for the bezel at the lens edge.

A Semi-Apochromatic or Fluorite objective is really an achromat but the mineral fluorite is used in the construction owing to its unusual optical property of high refractive index with low dispersion. When this mineral is used together with optical glasses which match the fluorite, a superior achromatic combination results. Technically, only two spectrum colours are brought to the same focus but all the others are so near that they may be considered corrected for all ordinary purposes and when used with a colour filter, modern fluorite objectives are indistinguishable from apochromatic ones.

It is usual to design fluorite objectives to work with compensating eyepieces and these are explained under *Apochromatic Objectives* (below). Fluorite objectives made earlier than about 1914 often did not require compensating eyepieces.

Apochromatic Objectives
In these combinations several types of optical glass and the mineral fluorite are used to bring the images formed in three colours of light to the same focal plane. As in achromatic systems, the colours chosen for correction vary with manufacturers but are always chosen so as to be evenly distributed along the spectrum. When three colours are thus corrected, the remaining ones are also very close to correction, and in practice the secondary spectrum can be considered absent. Technically it is still present and can be demonstrated. It is one problem to arrange to form the images in three colours in one plane,

C

but another to arrange all the images to be of the same size. It is usual to make objectives with a solid front lens which does the magnifying and collects the light, the other components in the combination correct the aberrations of this front lens. It is difficult with this construction to make the images in all colours the same size in all parts of the field, but the error is easily put right by designing the eyepiece to provide greater magnification for the colours requiring it. Apochromatic systems are usually designed to work with these compensating eyepieces but a system can be made apochromatic without their aid, but their use allows the easier solid front objective construction to be adopted. All reflecting systems are truly apochromatic and refracting condensers may be made apochromatic but in general this is unnecessary because the condenser is not an image-forming device. The degree of compensation required by objectives varies and it is therefore desirable to use objectives and eyepieces by the same designer. The effect of the compensating eyepiece can be seen only in white light and away from the centre of the field of the microscope. (See 'To Set up a Microscope to the Correct Tube Length', page 62.)

Because the apochromatic objective is practically free from spurious colour, its spherical corrections are also taken to a high degree of accuracy. Older apochromatic objectives, say earlier than the year 1900, were not specially well corrected for spherical aberration and would not work well at full aperture, but the modern apochromat works perfectly at full aperture and its performance is measured by the manufacturers, with interferometers. All specimens sold now will yield perfect images up to their theoretical maximum resolving power.

Aprochromatic systems made between about 1900 and 1930, particularly those by the Continental firms, contained glasses which were unstable and very few of these otherwise excellent objectives are of any use today because surface disintegration has taken place within the objective and they have become opaque or nearly so. An apochromatic objective made before 1936 should be purchased only after expert examination and it is to be hoped that the modern glasses will not decompose within twenty years.

The aprochromatic objective is the finest production of the optician. For research purposes when photography is often necessary, the apochromatic system should be used as the visual and photographic foci are in the same plane and whatever the colour of the specimen, the results, visual and photographic, will be equally good. Owing to the accurate correction for spherical aberration they will stand higher power eyepieces than achromatic objectives and though high power eyepieces are rarely useful as an aid to resolution, on some occasions when the specimen is in some way inaccessible they help considerably, particularly in photography.

Objective Tube Length

All properly computed objectives are made to give a fully corrected image at a particular distance from their rear focal planes. Any point of reference may be taken for practical work and the objective shoulder which screws against the nosepiece is that chosen. The upper point of reference is the flange on the eyepiece which rests upon the microscope body or draw tube. Tube length is a matter of spherical aberration correction.

The objective tube length is nothing to do with any eyepiece; it is a property of the objective alone associated with a specified quantity of known material between it and the object.

Old objectives were made with a tube length of 250 mm., the modern ones are 160 mm. but a large number of laboratory workers, particularly petrologists and metallurgists, prefer a tube length of 200 mm.; more objectives are being made now to this specification. Any transparent material, other than a gas, between the objective and specimen affects the distance at which the best image is produced, i.e. it is said to affect the tube length of the microscope. If the objective was computed to work in air on uncovered specimens like metals, the given tube length engraved upon the mount need only be set up on the microscope and the instrument can be trusted to work at its best. Most specimens have to be studied through a cover glass and this exerts a disturbing effect proportional to the aperture and focal length of the objective in use. In general, low power objectives of less than 16 mm. focal length will work with and without cover glasses and through 2 mm. of glass like the side of a trough. The effect of a cover glass is clearly apparent with most 12 mm. objectives which have an aperture of about NA 0·5. It is recommended that laboratory workers try the effect of using a 4 mm. objective meant for uncovered specimens, through a cover glass, in order to see the effect of tube length error.

When the tube length is stated of an objective meant for covered specimens, the thickness of cover is usually not given. Invariably a No. 1 cover is meant which is 0·15 mm. thick but the absence of this figure is not an important omission because the depth of the object in the mountant is of equal importance in tube length correction and can rarely be measured. An object in water (RI 1·33) under a glass cover (RI 1·5) will give a different result from a similar object in styrax (RI 1·6) under the same cover glass.

Several courses of action are open and all are used to overcome the cover glass and depth in mountant difficulty. They are as follows:

(i) To use a homogeneous immersion system where all the media between the objective and object have the same refractive index and dispersion.

(ii) To build into the objective a device which moves the correcting components of the combination to and from the front lens. This is known as a 'Correction Collar' and is useful on all objectives of shorter focal length than 6 mm. This action affects the correction for spherical aberration and can be made to off-set that due to the cover glass and mountant. (See 'General Microscopy', page 31.)

(iii) To adjust the length of the microscope until the point at which the image is best is found by trial.

(iv) To insert an optical combination or simple lens into the path between objective and eyepiece and use this to correct the image for a particular tube length.

The immersion system has been touched upon under 'Magnification and Resolution' and 'General Microscopy' and only a few practical details need to be added. Not all studies can be carried on with immersion objectives as only some objects can be immersed in oil or mounted in materials of higher RI than $1 \cdot 5$, and living creatures cannot be treated in this way. When objective focal lengths become greater than 4 mm. it becomes impossible to maintain oil contact between the front lens and specimen owing to the distance between them. Packing pieces of glass may be used but other advantages are small. Of course, for the highest resolution and magnification a homogeneous system must be employed.

The Correction Collar

The correction collar is the proper device for adjusting tube length, but in recent years it has often been left off objectives of 4 mm. and higher power, the excuse being that it interferes with the accurate centring of the components. The most suitable reply is that many objectives made eighty years ago still have correction collars working perfectly and not introducing centring errors. All objectives of 6 mm. focus and above intended for microscopy, as opposed to recognizing objects and comparing them with text-book figures, should have correction collars. An apochromatic objective of 4 or 3 mm. focal length and NA greater than $0 \cdot 7$ is useless when used at more than half its aperture on most specimens unless it is fitted with a correction collar, the exception is of course when the specimen is hard against a cover of exactly the correct thickness. Some help can be obtained by the use of the draw tube (below) but the range of movement on a modern microscope is usually insufficient to effect correction.

Most correction collars have a range, assuming a fixed tube length, extending from an uncovered object to one with a thickness of cover glass equal to the working distance of the objective. For a naturalist or anyone who works with specimens in water, such a range is valuable, but for those who work mainly with laboratory prepared specimens less range is adequate and it is hardly necessary to have adjustment extending to the uncovered position. A test slide should be at hand covered with glass twice the No. 1 thickness and any correction collar should be able to cover this increase; the test specimen must be in contact with this thick cover or misleading results will occur. Should the cover be thin, it can always be accommodated by increasing the length of the microscope but one cannot always shorten a microscope sufficiently to accommodate a thick cover.

Correction Collar on Oil Immersion Objectives
Old, first-class objectives were so fitted in order to allow the objective to be adjusted to work upon a microscope of any length between about 5 in. and 12 in. and this enabled apparatus like binocular attachments to be used without loss in objective performance, also if a specimen is mounted under a cover glass in a medium which has a refractive index other than 1·5, the tube length will need altering even though the objective be of the oil immersion kind. Media of RI less than 1·5, e.g. water, require a tube length increase (some resolution is lost due to using such media), while dense mountants like styrax and realgar require the tube to be shortened. The lengthening and shortening in the examples mentioned are only up to about 20 mm. each way on an objective corrected for a 160 mm. tube length but accurate figures cannot be given because there is no easy way of telling what depth of mountant is above the specimen.

Correction Collar of Water Immersion Objectives
These objectives are not homogeneous, therefore they need correction for the effect of the cover glass. The immersion medium, water (RI 1·33) does a lot to offset the effects of the glass (RI 1·5) but proper correction is necessary, and is more than can be obtained by means of the draw-tube alone, therefore all high power water immersion objectives must be fitted with a correction collar.

Tube Length Accommodation by Adjustment of the Draw Tube
If the specimen is beneath a cover glass or its equivalent thickness of

mounting media nearly correct for the objective to work at its given tube length, it may be accommodated by adjusting the length of the microscope body by means of the draw tube. This tube is placed there for this purpose and is not intended to be used to vary magnification. Good class instruments have a rack and pinion controlled draw tube. This refinement is not necessary in connection with accuracy but is most useful for preventing the draw tube slipping when apparatus is at the eyepiece. Draw tubes may have to be adjusted to an accuracy of a few millimetres and this can be most easily done by means of a rack and pinion, though a sliding fit is better mechanically than many rack and pinion assembles one finds in use. Tube length is always set up by trial but the rule is 'shorten for thickness, draw out for thinness', thick and thin referring to cover glass dimensions. When shortening a microscope with a sliding draw tube for use on a thickly covered preparation be careful to steady the instrument by the coarse adjustment milled heads or there is danger of pushing the whole body on to the slide with disastrous results.

It is important when high resolution work is attempted to select a microscope which can be shortened to at least 140 mm. if 160 mm. objectives without correction collars are to be used. The sensible alternative is to choose objectives corrected to work at a tube length of 180 mm. or more. (See 'Microscope Stands', page 140.)

Adjustment of Tube Length by Means of Lens Systems
A device known as the Jackson Tube Length Corrector was marketed in recent years and may still be found in use. It fitted between the objective and nose-piece and consisted of two achromatic combinations, Fig. 1.

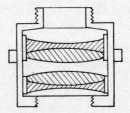

Fig. 1. Sectioned view, not to scale.

which could be moved to and from each other. The system provided controllable positive and negative aberration, which was used to offset that due to cover glasses of incorrect thickness. The lenses were insufficiently large in diameter to cover the back lenses of many high and medium power objectives and the apparatus has now passed into disuse.

A fixed amount of tube length adjustment may be had optically by placing in the draw tube a simple lens of spectacle quality of power varying between $\frac{1}{2}$ and 2 dioptres (see 'Instruments and Use of the Eyes', page 9). To make a 160 mm. objective work at 250 mm., a negative lens of about 1 dioptre is satisfactory. When it is necessary to reduce the tube length a positive lens is used. Fortunately it is rarely necessary to reduce the tube length of an objective as a positive lens is sensitive to centring and therefore would have to be mounted accurately. The concave lens used to increase the tube length, is often required and is not sensitive to centring; it may be dropped into the system in any position, say upon the nosepiece of the draw tube. Lenses of a suitable kind may be purchased at an optician's shop. This method of correction is not entirely satisfactory.

It should be noted in passing that owing to the use of 160 mm. tube length objectives, a lens of the kind mentioned above is to be found in most binocular body assembles to increase the working tube length of the objective sufficiently to allow for the extra mechanical length of these bodies.

To Set Up a Microscope to the Correct Tube Length
In practice this has to be performed upon the specimen under examination, but in order to gain experience of the appearance of an object at correct and incorrect tube length and to understand the theoretical side, the following experiments should be made.

When an objective is working at the correct tube length for the cover glass in use, that objective is giving the best performance it can with respect to correction for spherical aberration. If the lens is reduced in effective aperture by diaphragms in the substage or elsewhere, its spherical aberration is also reduced as it is in a simple lens. It follows that an objective of low aperture, whether constructed that way or stopped down, is not sensitive to tube length and it is quite possible to construct or stop down a 4 mm. objective so that no change in resolution and contrast are apparent over the whole range of the draw tube of the microscope and many old naturalists' lenses prior to 1900 were of this kind. The conclusion is that unless an objective has a numerical aperture greater than (say) $0 \cdot 5$ it is not sensitive to tube length, that is, assuming it is working through a standard cover glass, the microscope may be 100 or 250 mm. long and the image will not deteriorate. The same lens will also work satis-factorily through a trough of water.

The larger the aperture of a lens, the more important does spherical aberration become, whatever be the cause. Modern objectives are very good in this respect and at correct tube length may be used at full aperture without loss of contrast. Should the objective be defec-tive in either workmanship or design, spherical aberration (or other aberrations) will be present and clearly such a lens cannot have a tube length; at high aperture the image is fogged by aberrations of various kinds, while at low aperture spherical aberration is reduced to invisibility. It will be seen from the above that it is of no use making tube length experiments upon any lenses except those of good class, of aperture greater than 0·5 and preferably NA 0·8.

Aberration in an image is light out of place. If a lens is set to produce an image of a flat black object, the image will also be flat and black if the lens is perfect, but if spherical aberration is present the black will be fogged and the fogging will be different in shape and intensity above and below the best focal point. Fig. 2. Black represents the wanted image and shading the fogging.

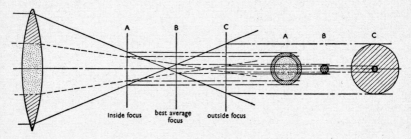

Fig. 2.

If the lens is perfectly corrected the rays of light from a point object will form a point image (ignoring diffraction effects which are small compared with aberrations). The blurred image then has exactly the same appearance when out of focus by the same amount above or below true focus, also the image of the black object is black when in focus, and not shaded over with haze.

A suitable test object may be made as follows. Procure a small quantity of aluminium paint which must be the kind used as under-coat, or the material sold for gold-painting ornaments known as 'Avenue Gold'. Both these materials are made of a suspension of ground metal sheet in laquer; the particles are of all sizes between a micron and several hundred microns in diameter, but are much less than a micron thick, have jagged edges and are quite opaque to light and form perfect flat black objects for testing objectives. A dab of this material should be placed upon a number of cover glasses of different thicknesses and mounted in balsam, metal downwards, on ordinary slides. The metal particles are certain to be in contact with the cover and true cover thickness is being observed.

A specimen of the above should be examined and a piece of metal lying flat, about 1/20 of the microscope field in diameter, should be brought to the centre of the field and focussed. Probably at full objective aperture its image will be covered with a layer of fog unless by chance the objective is very good and the tube length exactly right for the cover-glass chosen. If it is fogged, reduce the effective aperture of the objective by means of the substage diaphragm until the fog clears. Inspection of the objective back lens by removing the eyepiece will show by how much the aperture had to be reduced to clean up the image. Open the diaphragm again and adjust either the draw tube, correction collar or cover-glass thickness until the fog clears. At this point the tube length is nearly correct. Now alter the microscope focus above and below the best position approximately the same amount and observe the out-of-focus effects. After practice it will become easy to make the final tube length correction such that these out-of-focus images are identical.

Should it be found impossible to obtain the above results, the quality of the objective must be suspect and the same experiments should be tried with a small reduction in aperture.

An objective of NA above 0·5 which is not sensitive to tube length is of poor quality. An objective of the apochromatic kind NA 0·85, focal length 3 mm., working on a nominal tube length of 180 mm. at full aperture should be sensitive to 2 mm. tube length change when checked upon the above test object. An achromat will not be so good but should be sensitive to about 5 mm. change. When used for practical work such sensitivity will not be apparent owing to the nature of typical objects, but the test is a good one for objective quality.

Tube Length Adjustment for Ordinary Specimens
Correct adjustment of tube length on a good objective upon the specimen described above is, when once seen, perfectly clear, but it is not so easy to perform upon an ordinary object. It can be suggested that if wrong tube length is difficult to see, it cannot matter much and this is partially true, mainly because on typical objects the objective aperture is reduced in the interests of contrast in the image and depth of focus. But it is also true that if the tube length be correct

the aperture can be considerably increased without loss of contrast. In many thick specimens an average tube length must be used but, in general, the tube length should be set correctly for any object or part of an object under examination and if this adjustment is not attended to, resolution and contrast will be lost.

If the object has in it a black particle or small dark area at the required level, this may be used as a test point in the same manner as the test slide described above and several modern diatom mounters leave such a particle in their mounts. If no such particles are present, a coloured part of the object may be used in a similar way, the idea being as always, to obtain maximum clarity of the feature at the highest aperture. If the specimen is made up of semi-transparent line structure, the image above and below best focus will show clearly soft fuzzy lines and hard white lines or vice versa, and the adjustment must be continued until the out-of-focus images are nearly the same. Should the specimen be composed of transparent material of different refractive index to the mountant, say a diatom, the difficulties are greatest because all features in the object behave as lenses and give out-of-focus effects dependent upon their optical qualities. Black and white 'dots' in diatoms are simply out-of-focus images of holes. In these cases everything should be done to try to set the tube length upon another object at the same depth in the mountant, but should there be no other object, the edge of the diatom should be used for the above-and-below focus test. It is unwise to try to set up tube length by looking for best visibility of fine details as it is easy to set up artifacts by this means.

Graphical Determination of Tube Length

The best way of tackling this problem is to measure by means of the calibrated fine adjustment the actual thickness of mountant and cover-glass above the feature in the object upon which interest is centred, for once this distance is known, a graph showing the tube length demanded can be consulted and the draw tube or correction collar set accordingly.

The graph is constructed as follows:

Obtain a live box or compressorium with parallel action, place a dab of aluminium paint on the lower glass and when dry, fill the box with cedar oil. Set it up on the microscope with the objective concerned in place and adjust the box so that the objective can just work through the thickness of glass and oil. Set the tube length to be correct, read the tube length in millimeters or the correction collar graduations, and the figure on the scale on the fine adjustment head. Reduce the thickness of oil by operating the compressorium and repeat the readings. A graph may then be plotted of tube length against cover thickness change. About six sets of readings should be taken to enable a smooth curve to be drawn.

The last reading taken is when the upper glass of the compressorium (of known thickness) is in contact with the specimen. If the reading of the fine adjustment scale at this point is taken as the starting point, the readings of cover glass thickness change may be added to the starting reading and the graph plotted cover glass thickness against tube length.

In practice it is often necessary to count divisions passed through on the fine adjustment head when the differences are being measured because the head may pass through several revolutions during the experiment. It does not matter whether the fine adjustment is right- or left-handed, it will only alter the direction of the line and once the readings are taken the graph is drawn to fit them.

The Table of Results should look like that following. Column No. 4 is computed from the other two. All figures are only representative.

Reading No.	Divisions Passed on Fine Adjustment	Tube Length for Best Image (mm.)	Cover Glass Thickness (μ)
1	0	130	150+15
2	6	140	150+ 9
3	10	150	150+ 5
4	15	160	150μ

TABLE 4

Reading No. 4 is when the cover, 150μ thick, is in contact with the object, and so its thickness may be entered in column 4. The rest of column 4 is then built up and the results in columns 3 and 4 plotted against each other.

A different graph must be plotted for each objective.

The tube length reading is correct only for objects in a mountant of RI 1·5 and a similar graph should be plotted with water in the box, but the graph will then be correct only for a cover of particular thickness used over water.

When opaque specimens are under examination on metallurgical microscopes, tube length adjustment becomes very critical owing to the fact that the objective is working at full aperture and in this case a bright part of the specimen may be used for the above-and-below focus test. In metal specimens and objects of the opaque kind generally, the contrast in the image is the best guide to tube length correction, but the objectives for this work must be good and well set up. (See 'Vertical Illumination', page 208.)

Special attention must be paid to tube length setting when high power dark ground illumination is in use. Lack of success when attempting to examine bacteria by this method may often be traced to too thick preparations and insufficient range of adjustment on the microscope draw tube, for semi-transparent filmy particles must be studied without interference from any stray light due to incorrect tube length. Fortunately, being bright against a dark ground the above-and-below focus test may easily be applied to the specimens and unless the object can be made to show a change during adjustment from a sharp disk above focus and fuzzy disk below, to the opposite of this, the range of adjustment is insufficient to cover the optimum position of same appearance above and below. Should it not be possible to obtain the mid point, the microscope tube length must be altered or an auxiliary lens fitted in the tube.

Tube length is a matter which affects objectives only. When the eyepiece is applied, whatever type it may be, the microscope image is completed and it does not matter how far away the screen or photographic plate be placed from the eyepiece, the image will not be affected by any extra aberrations. If photography or projection is attempted without an eyepiece, the objective must be arranged to have infinite tube length or have low aperture. It is generally not possible to alter an existing medium or high power objective to make it work at tube length greater than say 300 mm., because its front lens comes into contact with the next component, but large alterations of several inches can be made optically by the use of additional lenses mentioned above.

Objective Focal Length

This is merely the focal length of a simple lens equivalent in magnifying power to the objective combination. The expression is used in microscopy only to indicate magnifying power and although focal length and aperture are connected in manufacturing technique, they should not be confused. It is possible to make an objective of 0.2 mm. focal length, giving a magnification of 10,000 when used with a 10-times eyepiece, yet the whole may have only the resolving power of 16 mm. modern objective. It is also possible to give to a 16 mm. objective a NA of 0.5 but this is pointless and is not done, because impossibly high eyepiecing would be necessary to employ its powers fully.

A quick estimation of the magnifying power of a microscope can be made by dividing the distance of most distinct vision (250 mm.) by the focal length of the objective and multiplying the result by the stated power of the eyepiece, but these given figures must not be relied upon for accuracy (see 'Magnification and Resolution', page 33). Objectives have been made in almost all focal lengths from 6 in. to 1/100 in. The metric system is now usually used and objectives are sensibly limited to focal lengths 16 mm., 8 mm., 4 mm., 3 mm. and 2 mm.

Objectives of longer focal length than 4 mm. are known as Low Powers and those shorter than 4 mm. as High Powers. There is no definite ruling about this and it is somewhat unfortunate that short focus, high magnification lenses are still known as 'high powers'. The power of course is dependent upon aperture and not focal length, so objectives should always be referred to by their focal lengths and apertures. Occasionally one sees much longer focus objectives still about and these were used mainly by naturalists for viewing flowers and insects and still have considerable industrial use. Modern examples are usually incorporated in binocular microscopes and may not have standard objective threads upon them.

All objectives of shorter focal length than 3 mm. are of the immersion kind. Some 3 mm. objectives are made to work dry with a numerical aperture of up to $0·95$ and these represent the highest resolution obtainable with any dry lens. There is, therefore, no point in making them magnify more. The total magnification may be altered at will by eyepiece changes. The 2 mm. oil immersion objective with NA up to $1·4$ but usually $1·3$, gives maximum resolution with a $\times 10$ eyepiece on a 160 mm. tube length microscope, and if high aperture were given to, say a 4 mm. immersion objective, a $\times 20$ eyepiece would be necessary for most observers to bring out all resolved detail. Such an eyepiece has a large light loss and an inconveniently close eyepoint, so such an arrangement would be unsatisfactory. It was not unusual to manufacture (1·5 mm.) oil immersion objectives of NA $1·4$ for use on long microscopes with low power eyepieces as this combination has the advantage that the low power eyepieces allow more room to use spectacles. Most of these objectives were constructed between 1900 and 1920 and are of very fine quality when the glass remains clear.

Working Distance
This is closely coupled with focal length but again must not be confused with it. Invariably the 'working distance' between the objective front lens mount and the top of the cover glass is much less

than the focal length of the objective. The working distance of all objectives intended to work through a cover glass varies with the thickness of the cover and mounting medium. Homogeneous immersion systems can have their working distances measured directly between the object and objective front lens as the presence of a glass cover does not alter the figure though it does of course occupy space. A dry objective meant to work without a cover also can have its working distance measured directly.

Apart from the variations mentioned above, objectives can be designed to have long or short working distances for the same focal length; in general, the longer the focal length, the greater the working distance, but an immersion system of any kind has nearly double the working distance of a dry lens of similar characteristics and it is often possible to work through a thick cover glass with 2 mm. oil immersion objective when a 4 mm. dry one will not reach.

For practical work it is still necessary to refer to the quality known in the past as 'penetration'. If an objective is made with aperture greater than that needed for normal use it is uncomfortable and misleading to use, mainly because it focusses only a thin layer of the object. Such a lens may show very well a test object (page 62) but does not assess a thicker structure so well as does a higher power of the same aperture. Over-apertured objectives also have short working distances.

Objectives vary over wide limits depending upon many features of design. High apertures have less working distance than low ones; apochromats less than achromats. Objectives with very short working distances are a great nuisance because they will not reach through even a No. 1. cover glass if the object is even slightly deep in the mountant and shortness of working distance in high aperture objectives generally, seriously limits many techniques (see 'Reflecting Objectives', page 70). Manufacturers rarely state working distance in their catalogues.

To Measure Working Distance
This measurement is a plain mechanical one. In most objective constructions the front lens mount projects a little in front of the front lens both for mechanical reasons and for protection and the measurement of working distance must take into account this projection.

The most direct method is to set up the microscope upon a test specimen covered with the normal cover glass and to insert strips of card between the objective and cover, or, owing to the fact that there is a metal projection on the mount, there is no harm in inserting feeler gauges into the space. An oil immersion objective may be set up, the oil drawn away with filter paper and the feeler gauge inserted as before without altering the focal adjustments. If card is preferred, it must be measured with a micrometer after withdrawal. Card and paper are both manufactured under great pressure, therefore both are relatively incompressible and may be used for packing.

A calibrated fine adjustment may be used for this measurement but care must be taken that the slide or nosepiece can give when contact between objective front and slide takes place. A safety stage such as that described under 'Stages', an inverted microscope or an improvisation as shown below may be used to prevent damage. (Fig. 3.)

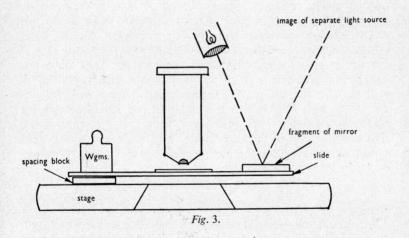

Fig. 3.

If great delicacy is demanded, a piece of mirror may be rested upon the upper surface of the slide, then when contact is made, the deflection is instantly apparent by the shift of any image in the mirror. The reading should be taken of the fine adjustment graduations when the objective is focussed and tube length corrected. The objective is then advanced until it just contacts the slide, then the difference between the readings gives the working distance for the test conditions.

The Water Immersion Objective

These are similar in design to homogeneous immersion objectives but the water immersion medium allows a numerical aperture of only 1·25 maximum to be obtained. The advantages of the system are that water is easily removed from apparatus without leaving stains, it is non-poisonous to microscopic animals and some chemical experiments such as etching may be carried on in the immersion fluid. Such objectives can be made in the achromatic and apochromatic series and their performance, allowing for their reduced aperture, is identical with the oil immersion type. They must have correction collars for use when a cover glass is present.

The main use of these objectives in ordinary work as opposed to special experiments is when specimens on slides under cover glass are stored for long periods in humidity chambers. (See 'The Study of Objects in Liquids', page 256.) The covers often cannot be wiped clean owing to risk of disturbing the specimens, and if oil is upon the surface it rapidly becomes opaque in the humidity chamber, it is also poisonous. When dry objectives are used, time is lost while condensation dries off, so it appears that the water immersion objective is the answer to both of these problems.

Another type of water immersion objective is the 'Tank' objective. This has a focal length of 50 to 100 mm. and is intended to be used on a water-tight tube immersed in the aquarium or other tank, the eyepiece end protruding. It is used to study objects alive *in situ* on the bottom or the sides of the vessels and being of low power it is not sensitive to salinity of the water and so may be used in marine tanks or in rock pools. Some examples have built-in electric lights beamed on to the focus of the objective.

Reflecting Objectives

The reflecting microscope as a whole is treated in a chapter to itself. Under the heading 'Objectives', a type of reflecting objective marketed by Messrs Beck which will fit upon an ordinary microscope is here described, the function of it being to increase working distance as the short working distance of all high power objectives causes much difficulty in use. Becks' objective is similar in layout to a Cassegrain telescope, the main reflecting part is an aluminized spherical concave mirror (C) giving a magnification of unity. A semi-aluminized plane mirror is placed at A between the concave mirror and specimen. A convex mirror reflects the image from the concave one into the microscope at B, and this image is magnified by a standard eyepiece system if required. Fig. 4.

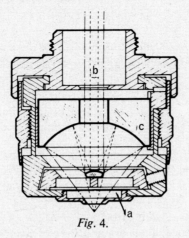

Fig. 4.

An aperture of NA 0·65 and working distance of 25 mm. is provided
by this system. Slight falling-off in definition and contrast are present
but in the example used by the Author no loss in resolution could be
demonstrated when compared with an ordinary refracting objective
of the same focal length and aperture. Such an objective is of great
use to laboratory workers who must conduct experiments and
manipulate materials under microscopes.

Cleaning and Repairing Objectives
An objective may have several faults which impair its performance,
even though properly computed and made in the first place, and a
list of these faults follows:
  (1) The back lens may be very dusty causing a hazy image to be
      produced.
  (2) Immersion oil may have dried on the front lens and become
      dirty so that an optically uneven medium is present when new
      oil is used.
  (3) Immersion oil may have been used on a dry objective and not
      very carefully cleaned off.
  (4) The surfaces of any lens in the combination may have become
      decomposed giving a frosted look to the glass but this is not
      always apparent until the component is isolated from the
      combination.
  (5) The balsam cement between lenses may have failed.
  (6) Odd pieces of glass, lenses or Polaroid may have been dropped
      on to the back lens in connection with previous experiments.
  (7) A lens may be cracked due to mechanical violence. This

damage is not so obvious as may be expected unless the lens is shattered.

(8) It may be wrongly assembled, or a component may have slipped in its mount when hardening the cement after a repair.

(9) A liquid like immersion oil might have fallen on to the back lens and permeated the whole combination.

There are in addition numerous optical faults from which an objective may suffer but as these are points of design they will not be considered in this section.

When an objective, obviously faulty is taken from the microscope, steps must be taken to find out what is wrong. The tests are tabulated below and are strictly practical ones which can be carried out on the spot.

(1) Naked Eye Examination:

Arrange a small bright source of light like a clear electric light bulb to be visible about 30 metres away in a poorly illuminated place. This bulb can then be considered to give parallel light at the position of the observer.

Clean the objective with a cotton cloth carefully removing any immersion oil, place the objective front lens to the eye and look through it backwards at the light. A large dim field of light will be seen, and if this field is quite clear of specks of any colour when rotated, the lens is free of dirt, cracks, corrosion and balsam failure.

(a) If a frosted appearance is present, the frosting being made up of dark spots, the lens is either very dirty or more likely is corroded on a surface.

(b) If there are large dark specks the back lens is dusty.

(c) If the specks or larger areas are lighter than the background the balsam has failed, leaving areas between cemented lenses unconnected optically. The lower refractive index of these areas causes the lighter spots. When the lens is rotated the light spots give that known as 'flock of sheep' appearance.

(d) If a crack is present it will be seen as an ill-defined line across the field.

The naked eye examination will not show any other faults.

(2) Microscopical Examination:

Set up the whole objective, front lens upwards on the microscope stage with transmitted light and look at and through its components with a 50 mm. objective on the microscope. Corrosion of the glass, cracks and balsam failures can often be located in position by this test without dismantling the objective.

The rear lens (a regular offender) can be examined by turning the objective under test upside down.

(3) Clarity Test:

An objective should, when looked at through its back lens against a clear light, be sparkling bright. A system, e.g. an apochromatic one with many components, often shows rings of light within its structure due to reflections from the surfaces of lenses but in no case should the objective show cloudiness. Should it do so, the above tests must be applied.

To Clean an Objective

Objectives like many things are often discarded when they only need cleaning to restore them to working order. At the beginning of this section it should be pointed out that there is considerable risk of damage when an objective is dismantled. They have some components which require accurate centring and others which have critical spacing. If due care is taken by an operator used to fine work, ordinary repairs and cleaning can be carried out in the laboratory, but if there is any doubt about the skill available, laboratories are recommended to return the objective to the maker, whatever the cost may be. It is often cheaper than buying a new one.

It is unwise to try to clean the back lens of an objective by pushing materials, including camel hair brushes down the mount as it is rare to find only dust there. Cleaning usually involves polishing off greasy substances in which dust is embedded. If a cloth is pushed down and rotated, circular marks are made on the glass which do more harm to the image than the even film of grease and there is a considerable chance of scratching circles on the lens with particles of dust on the cloth. (See 'Cleaning Glass', page 361.)

The optical part of an objective can usually be unscrewed from the mount and in this optical part will be seen the back lens exposed near the top. When in this position it can be wiped clean and polished in the proper way and the proper way is worth repeating. Use an old, perfectly clean, often-washed, cotton handkerchief, wipe the surface of the lens firmly from one side to the other to sweep off hard particles. (Wash the fingers first or grease will come through the cotton to the lens.) Change the part of the cloth used and polish the lens with it lightly in a common sense way, change the position of the cloth again and finally polish. Check for cleanliness by holding the lens in the focussed beam from the high power microscope lamp. The light passing normally through the lens will be scattered by any dirt and irregularities present in or on the lens and become clearly visible to the naked eye at the side of the beam.

Some objectives are assembled in the manner shown in Fig. 5.

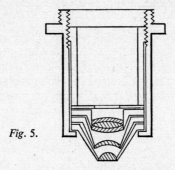

Fig. 5.

Each component is mounted in its own collar and all the collars slide into the barrel, being held in place by a funnel stop at the back. This type is dismantled by removing the screwed-in stop and pushing out the components from front to back. Do not press on the front lens, but press with a cork drilled to clear the front lens and press only on the brass. Such an assembly may be dismantled without fear of disturbing the centring because there is no way of adjusting the centring after the lenses are secured in the separate cells.

When each component is mounted in a separate screwed-on cell, there is some room for adjustment of centring, but generally the screws go home very accurately and errors are likely to be in the mounting of the lenses in the cells. The centring and squaring-on is attended to on a lathe or turntable described below. If the intention is only to clean the lenses, centring troubles after reassembly are unlikely to arise in an objective constructed after the year 1900.

Tube Length

The spacing between the solid front lens and the next component is very critical as any change in this spacing affects the tube length of the objective. Should the screw thread carrying the front lens be left slack, the objective will operate on a reduced tube length and if it is arranged to screw further on to the mount than before, the objective tube length is increased up to a point when spherical aberration becomes apparent. Great care must be taken when making these changes that the front hemispherical lens posterior surface does not contact the next component when the front is screwed home. It is usually possible to adjust an objective tube length between 160 mm. and 250 mm. by this means, and if the objective has a correction collar it is often possible to re-arrange its tube length by setting up the screw collar to engage on a different position on its multi-start thread. This must be tested by trial and error.

Correction Collars

This fitting fell into disrepute a few years ago but fortunately the need for it has been recognized and it is now fitted again by the manufacturers of the best equipment. If a correction collar is to be useful it must operate easily between the fingers when the objective is on the microscope. If it is stiff it is usually because the grease lubrication has dried.

A collar is easy to disconnect. Somewhere on the mount will be found two small screws and when these are removed the collar will slide off. The mechanical parts should be separated and wiped quite clean, and if corroded, polished with common metal polish, the system should then be reassembled with a film of silicone or ordinary grease as lubricant. Oil must not be used because capillary attraction causes binding of the parts and the oil may spread to the lenses.

It is easy when reassembling to engage the screw mechanism of the collar on the wrong thread so that either the objective works on a different mean tube length or the collar will only rotate through a part of its proper travel. A trial may be necessary after experimental assembly and it is never difficult to put it back in the right place.

Balsam Failures

When the components of an objective are separated it is easy to see whether or not the balsam cement between the lenses has failed, if it has, there is no great difficulty in doing something to repair it. Most doublets or triplets from which an objective is built up are held in their cells as a group by an edge of metal turned over upon them called a 'bezel'. Fig. 6.

Fig. 6. (A) Bezel Open.   (B) Bezel Closed.
(Exaggerated).

The thin rim of brass A is only a few tens of microns thick and is turned in on a lathe by pressing a hard smooth object against it. Before the lens can be removed this bezel must be opened and this is by far the most difficult task associated with the repair. The bezel may usually be opened by pressing the end of a matchstick against the edge in such a way that the end grain of the stick works its way under the bezel as in Fig. 7.

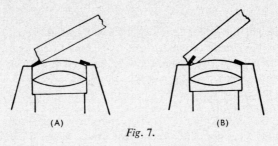

(A)   (B)

Fig. 7.

Some perseverance may be called for because a number of objectives are made with rather heavy metal parts, but do not insert pins under the bezel or the lens edges will be cracked. If a bezel has to be taken right off to remove the lens do not worry about it as the lens will remain in place stuck with the new balsam, the only extra care necessary is with the squaring-on, the bezel helps considerably to square the lens group in the mount.

When satisfied that the bezel is open, the lens group must be either warmed on a hot plate until the whole can be pushed out from the back, or it can be soaked for a week in benzol until the old balsam is dissolved. It is usually necessary to heat the part even when soaked in benzol. Note the arrangement of the lenses and place the lot in a tube of benzol to dissolve away all traces of old balsam. Dry them and inspect them carefully. If they are clear and show no surface corrosion or damage, clean them with a cloth as described above and make ready to reassemble. For this purpose use raw balsam so that when heated there is the minimum of contraction due to driving off volatile components.

Place all the components of the group on a sheet of paper on a warm plate at such a temperature that they may comfortably be handled without burning the fingers. Assemble the lenses in a stack with the minimum quantity of balsam between them which will easily cover the interfaces when pressed together. Insert the pile into the mount and press down to exude all spare balsam. Wipe off the gross surplus, place the group on a ringing turntable, centre, and rotate. If the top lens is properly squared-on, a reflection of a light in the room will remain quite motionless as the turntable is revolved. If it does not, press the top lens with a matchstick until it does, but do not try to clean up the last traces of balsam at this stage. Bake the component on the hot plate at about 212° F. for some hours until waste balsam on the mount is tough but not brittle to the thumb nail. Remove from the plate, cool and clean up with benzol or xylol on a cloth dampened with it only and polish with a clean cloth.

If a lathe is available the bezel is closed upon the warm assembly by gently turning it back with a steel roller such as that used for spinning brass. Usually a bezel will go back properly and square-on the lens automatically, but sometimes it is broken and will not. If the lens is well squared-on without the bezel it may either be left open or the bezel gently pressed back when the balsam is hard, but do not try to turn-in a bezel when the balsam is soft unless a lathe is available upon which to do the work as the lens cannot be centred by the bezel unless it is rotating on its optic axis. All lens systems may be re-cemented in this way, only practice is required when dealing with very small microscope lenses.

Do not try to join lenses with the absolute minimum of balsam. When this is done failure usually results due to the surfaces springing apart later.

Caution with Apochromatic Objectives

Some apochromatic objectives have odd materials in their assemblies used to help correct chromatic aberration to a near perfect extent. A particular example is the use of potash alum as the centre double concave component in the rear triplet of some Zeiss oil immersion lenses. If the assembly is heated to remove the lenses from their mounts, the alum is destroyed, therefore these should be soaked out even though it may take weeks; it is wise to soak out all apochromatic combinations. Alum may be cleaned in the usual balsam solvents and polished very lightly with a silk or cotton cloth. It will stand heating for re-cementing with balsam so long as the temperature does not rise above 212° F.

Also in an apochromatic system is a double convex lens made of fluorite. This mineral is never entirely free of bubbles and faults but suitable fluorite is selected by the manufacturers and unless the fluorite has surface damage it is acceptable as found in an objective. The fluorite component may be found in one of several places in the system depending upon the computation; it does not require any very special treatment and it rarely decomposes, but it is softer than glass. Owing to the fact that it has a different refractive index from glass and balsam, any scratches on its surface remain visible when it is cemented in, therefore care in handling and polishing must be specially observed when cleaning any double convex lens because it may be a soft one.

Lenses with Corroded or Decomposed Surfaces

Decomposition of glass is different from tarnishing of the surface. Old optical glass, say about 100 years old, always shows an oxidized layer on its surface which may approach $\frac{1}{4}$ wavelength of light in thickness. If this layer appears on a telescope or microscope objective, it may even improve slightly its light transmission qualities because of the blooming effect of the layer. The layer is well seen on really old Roman glass and can be put on cheap moulded glass artificially, giving a coloured lustrous effect. When such a glass is studied with a microscope, no surface structure can be seen. Most lenses are now coated to prevent reflections.

Decomposition is quite different and the effect is confined to glasses made from unusual materials with odd optical characteristics such as those used in apochromats. The Jena glasses of 1900 to 1920 were great offenders, the surface of the lenses breaks down into a crazed, rough structure as if fungi are growing over it. Sometimes layers decompose and break off and it is not only the exposed surfaces that fail. There is no treatment for this except to replace the component or rework the surface.

Replacement is not easy for although the curvature can be measured it may not be easy to obtain glass of exactly the right characteristics to fit the combination. If glass can be found, the lens may be made up according to ordinary optical techniques.

If it is decided to try to save a valuable objective, an attempt to work the surface may be made and the Author has developed the following laboratory method which has yielded satisfactory results as a one-off process.

Polishing a lens which has a corroded surface

   (1) Soak out the components of the group which can be seen to be cloudy.

   (2) Wash carefully in xylol each lens and study it when dry and locate the faulty surface.

   (3) Obtain some opticians' pitch and fuse a ball of it on to a wooden stick about 25 mm. long like a piece of pen holder.

   (4) While the pitch is plastic, press the lens on to it, thus obtaining a mould of the correct curvature, and allow to harden.

   (5) Stick another ball of pitch on a peg in a lathe chuck, preferably a vertical one, and stick the lens on this. Centre the lens with the lathe rotating and the pitch plastic by using the lamp reflection test described under 'Balsam Failures' above. Allow to harden.

(6) Take the first pitch block, that is the 'tool', and scratch across the face a few irregular grooves with a needle point so that some clearance is available for particles.

(7) Wet the tool with saliva, charge it with rouge (obtainable from a jeweller) and work the tool over the lens. The method is to press the tool firmly on the lens and move from side to side with the centre of swing at the centre of curvature of the surface being treated.

(8) Re-charge the tool with rouge and saliva occasionally and continue operations until the lens surface is seen to be clean and polished. If it will not clean up after a few minutes of this treatment, it is unlikely to clean up at all, but if further work is contemplated grinding must be employed.

To Grind a Small Lens

It matters little whether an old lens is being re-ground or a new one made, the operation for one-off lens grinding is the same. First a tool must be made of the same radius of curvature as the surface required but when the tool has to carry grinding paste it must be hard, not yielding like a polisher. The materials available are tinman's solder, brass and iron, the glass, being a brittle substance, can be ground by almost any other material strong enough to hold the abrasive (see 'Preparation of Metal Specimens', page 472). If the tool is for regular use, brass or iron must be used, but solder, being easily cast, is satisfactory for one-off jobs. Two tools, male and female, are always required and are usually made one from the other.

Parts of spheres are the only shapes which can be moved over each other in all directions and fit exactly in all positions. If then they are ground one upon the other, ultimately two spherical surfaces of the same radius result. This is the principle behind all optical grinding and is the reason why spherical surfaces only can be made accurately on a commercial scale. Though the tools be only roughly spherical when they start working upon each other with abrasive, they are truly so after a short while and the lens, when it replaces a tool, becomes accurately spherical for the same reasons.

The first job is to obtain the radius of curvature of the surface required and when this is known, a standard steel ball from a ball bearing may be obtained and used as the standard because these balls are accurately spherical. If the radius must be measured, a template technique should be used; obtain a pair of heavy, short-legged compasses, fit a drill-shank into the lead holder and grind a knife edge on the drill-shank so that the compasses cut rings in cardboard or thin soft metals. Obtain the curvature of the lens by trial and error with templates and when a template is found to fit, measure its radius with a ruler and preserve it for checking purposes.

For grinding the small lenses in question a steel ball can always be found which fits the curve, this ball becomes the male tool. The female tool is made from the male by pressing the ball into a disk of molten tinman's solder. When the solder is hard and the ball is mounted on a stick, male and female are ground together with 500 carborundum and water until both show even surfaces. The glass lens then replaces one of the tools but the tools are ground together occasionally during the operations to preserve their sphericity.

To grind the surface of an existing lens, proceed as follows:

(1) Obtain the radius of curvature by means of template trial or the method described in (2) below.

(2) Select from a stock a steel ball which has the same curvature. With care this can be done without a template by slightly wetting the lens and observing the area of contact when the ball is applied to the surface.

(3) Melt a pool of solder in a small tin lid, say about 25 mm. diameter and 5 mm. deep, and press the ball into it to about two-thirds of its radius and allow to harden.

(4) Mount the ball on a stick with pitch, charge it with 500 carborundum and water and work ball and socket together by hand until they both have an even surface touching each other at all points.

(5) Mount the lens (or blank) on a piece of wood as for polishing, mount and centre the required tool in the lathe, and work lens and tool together with the carborundum, swinging the tool or lens as described under 'Polishing' above.

(6) Every few minutes grind the male and female tools together to preserve the sphericity of the surfaces.

(7) When the lens can be seen to be evenly ground, stop the process, thoroughly wash the tools and job, and proceed to polish.

(8) Prepare the polisher from pitch moulded to shape by the appropriate opposing tool, and proceed as described above under 'Polishing'. When the lenses are more than 8 mm. in diameter, the pitch is best used as a covering to the grinding tool but it is still moulded to the correct curvature by the opposing tool.

It is clear that very little glass can be removed from a decomposed surface or the optical qualities and physical size of the lens will show important changes. If a new lens has to be made, the disk must be trepanned out to size first. (See 'Laboratory Methods of Working Glass', page 498.)

With larger lenses such as those of cameras and eyepieces it is permissible to polish a surface which has become scratched by forming a tight wad of cotton muslin about 10 mm. diameter, gripping in the fingers, moistening with saliva charging with rouge and working over the surface with a circular motion. About half an hour's work will restore the polish without sensibly altering the curvature of the lens.

Making Components of Other Materials than Glass

Where the material is not brittle it may be turned by means of a cutting tool and for small lenses a steel ball may be used as follows: Grind a groove across a ball of the correct radius to give a section as shown in Fig. 8 (A).

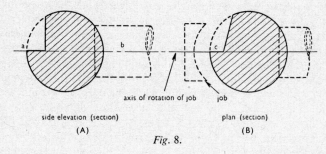

side elevation (section)　　　　　　　　　　plan (section)

(A)　　　　　　　　　　　　　　　　(B)

Fig. 8.

Fig. 8 (B) shows the plan view. At (C) the tool is backed off to provide clearance for the part of the work moving upwards when in the lathe. The shaped ball is mounted in a tube (*b*) with pitch. The job is mounted in a lathe and centred in the usual way, the cutting ball is mounted in the tail stop chuck in the position shown in Fig. 8 (A) and the cutting edge (*a*) is advanced into the job. Brittle materials like alum can be shaped accurately concave in this way but convex shapes are best ground in the conventional way. Polishing is achieved by wrapping a cotton cloth over a similar (uncut) ball and charging this with rouge and oil or dry rouge. Convex polishers are made by pressing with a ball, a cloth into plastic pitch.

For methods of making lenses over about 25 mm. focal length, see 'Laboratory Methods of Working Glass', page 496.

A convex spherical surface may be turned with a hardened steel tube which has a squared off end. This when swung from side to side will cut a convex surface equal in diameter to its own inside measurement.

CHAPTER 5

Microscope Eyepieces

M OST EYEPIECES are now made in standard sizes, the lenses are assembled in a tube of 23 mm. diameter which drops into the top of the microscope draw-tube. They rest upon a flange and the distance between this flange and the objective shoulder is the tube length of the microscope. Some eyepieces are still made to a larger size of 32 mm. diameter and many old ones of all diameters up to 40 mm. are still in use on microscopes which will accommodate them.

Eyepieces should be an easy fit in the draw-tube except in portable microscopes, where they should be gripped lightly by a sprung tube. It must be possible to lift eyepieces in and out of a set-up microscope without the use of any force which may alter the adjustments and the test for correct fit is as follows. Place the eyepiece in position with the draw-tube extended and an objective on the nose-piece. Place the microscope in the vertical position, hold the coarse adjustment, and push in the draw-tube when the eyepiece should pop out of the microscope. If it is too tight it will not slide out, while if it is too loose the imprisoned air will pass round it and again it will not come out. If it is too loose there is trouble due to it slopping about when the microscope is in use. If the microscope has a number of holes in the body to take fittings like quartz wedges, the test cannot be made without the trouble of closing holes.

The component lenses are always fitted into mounts and secured to the tube by screw threads and they may be unscrewed for cleaning without danger of derangement. Within the tube of the eyepiece is a diaphragm made from brass and blackened, this limits the field of view of the eye lens and is at its focus. The diaphragm rarely needs adjustment except under certain circumstances when an eyepiece micrometer is in use or it is used as a projection eyepiece. Should it be necessary to move it, a piece of wood should be turned to fit inside the eyepiece tube and rest upon the diaphragm squarely. A sharp hammer blow on the wood will start the diaphragm sliding but after this has been done it may be necessary to reblacken the interior of the tube.

When choosing paired eyepieces for binocular microscopes, the instrument should be set up and each eyepiece revolved in turn to check that their lenses and diaphragms are concentric.

Optical Construction: Huyghenian Type

The usual eyepiece in use on all microscopes is the Huyghenian pattern, constructed as shown in Fig. 1. The upper lens is called the 'Eye lens' and the lower the 'Field lens' and both are plano-convex. The ratio of focal lengths is not critical but should be about 1 : 3. Both are constructed of ordinary crown glass and are separated by a distance equal to twice the shorter focal length. Typical focal lengths are 20 mm. and 55 mm. for a $\times 8$ eyepiece. High powers can be made but the short focal length of the eye lens causes the eye to be unpleasantly close to its surface and the steep curvature field lens becomes small and transmits little light. The highest generally useful power of a Huyghenian eyepiece is 15 and the normal 10. Very low powers are not satisfactory because the apparent field of view is small owing to the low power of the eye lens, so it is usual to use special wide field eyepieces for low power searching because these have large real and apparent fields of view. They have not, however, the good defining power of the Huyghenian form.

This eyepiece gives a white image and is usually considered achromatic. There is, however, nothing in its structure which is achromatic, but it does give images in the various colours of the spectrum which are of the same size, Fig. 1, and for practical purposes it can be called achromatic inasmuch that it does not contribute colour to the objective image.

If the focal length of the eye lens is f_1 and the field lens f_2, the focal length of the combination is $\frac{3}{2}f_1$, i.e. it behaves like a simple lens of this focal length.

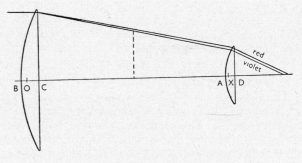

Fig. 1.

Huyghens found that when a pair of lenses of the same glass were placed experimentally at such a distance apart as to give the best achromatic image, the value of A O, so obtained, agreed with the value of d calculated from Airy's more complete formula for achromatism, viz.:

$$d=(f_1+f_2)\left(2-\frac{f_1}{u}\right)^{-1}:$$

where d=distance between the lenses: f_1f_2, the focal distances of field and eye lenses respectively; u, the distance between the field lens and the centre of the objective.

The Huyghenian eyepiece may be used with success upon any microscope and may be had in low and medium powers in the compensating form. It may be used to hold a micrometer (see 'Micrometry', page 48), but owing to the fact that it is a 'negative'[1] combination, the micrometer must be placed within the structure so as to be in the focus of the eye lens. Because the eye lens is a simple one, the physical length of the eyepiece scale must not be more than about one-third the apparent field diameter long, as if this figure is exceeded, distortion of the scale occurs and makes calibration difficult.

If a collection of plano-convex lenses is available, eyepieces of the Huyghenian form may be built up in the laboratory. If a very wide apparent field of view is required and best definition is not of primary importance, a double convex 'crossed' lens, having a ratio of curvature of its surfaces 6 : 1, may be used as the eye lens, the separation of eye lens and field lens, and the ratio of focal lengths of these lenses remaining the same. The crossed lens should have its more convex side towards the field lens, as in this position it has least spherical aberration, while it magnifies the field lens more than does the simple plano-convex type. The apparent field diameter may be doubled by this means without introducing undue distortion and loss of definition. Such an arrangement is often found on wide field binoculars and on target telescopes.

The Ramsden Eyepiece
This is a positive eyepiece arranged as shown in Fig. 24. Its main use on a microscope is to magnify a micrometer scale, it is 'positive', i.e. it can be used as a magnifier and its field is flat, therefore a micrometer scale can be placed at the focus of the combination instead of at the focus of one lens only. It does not give such good definition as the Huyghenian form but can be made truly achromatic. The ratio of focal lengths and spacing of the components are

---

[1] 'Negative' in this sense merely means that the system cannot be used as a magnifier. The focus is within the system.

$f=f_1=$distance apart.[2] The eyepiece has certain useful qualities when used in association with the Huygehnian form. For a given position of focus on a microscope the Ramsden eyepiece stands higher in the tube than the Huyghenian because the image is outside the field lens of the Ramsden and inside in the Huyghenian. For eyepieces of power $\times 8$ the difference is about 40 mm. and more for lower powers and it is possible to use this difference to accommodate apparatus such as prisms in one leg of a binocular microscope where these pieces of gear occupy a length of optical path. The arrangement was used in the Abbe binocular eyepiece, Fig. 2.

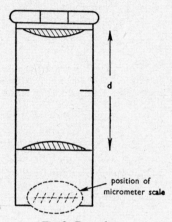

Fig. 2. Ramsden.

Ramsden Eyepiece

$d=\frac{2}{3}$ focal length of either lens
lenses *a* and *b* are equal, plano convex.
Focal length of combination is

$$\frac{f_1 \, f_2}{f_1+f_2+d}=\frac{3}{4}f_1$$

The position of the micrometer if used is set by trial at the focus of the combination.

The Ramsden eyepiece is now seldom used; the Huyghenian form having become almost universal except in the compensating type.

---

[1] It is usual to make d less than this ideal amount so that particles of dust on the field-lens do not show sharply in focus in the eyepiece, FIG. 2.

Abbe Binocular Eyepiece

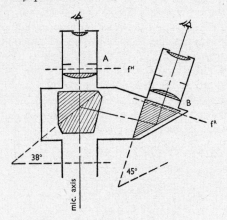

Fig. 3. Abbe Binocular Eyepiece.
(A) Huyghenian   f^H focal plane Huyghenian.
(B) Ramsden   f^R focal plane Ramsden.

The Kelner Eyepiece
This is a wide field eyepiece rarely used on microscopes of ordinary construction but is often found on binocular telescopes and low power binocular microscopes. Its construction is shown in Fig. 4.

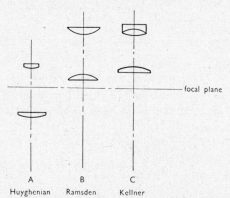

Fig. 4. Position of Focal Planes of Eyepieces.
(All eyepieces shown are working at the same tube length)

The field lens is in the focus of the achromatic eye lens and so no limiting diaphragm is necessary. Little colour and distortion are introduced by the achromatic eye lens and compensation may be arranged for the small amount of chromatic aberration in the field lens. The ratio of focal lengths of the lenses is the same as that for the Huyghenian form and is not critical so it is usual to reduce the separation to place the field lens within the focus of the eye lens so that particles of dust on the field lens are not in focus when the microscope is used. The focal length may be obtained from the formula

$$\frac{f_1 f_2}{f_1 + f_2 + d}$$

The eyepiece does not give the best definition and cannot be used to carry an eyepiece micrometer.

Compensating Eyepieces

All apochromatic objectives remaining in service, and most fluorite objectives, need compensating eyepieces to complete their colour corrections. As mentioned under 'Apochromatic Objectives' it is the construction employing the single front lens which requires further correction by the eyepiece. It should be noted that it is only the regions of the field away from the optic axis which need the correction given by the compensating eyepiece, an ordinary Huyghenian eyepiece will give as good results in the centre of the field. The amount of compensation required varies with objectives from different manufacturers and while the differences may not be apparent when objects are being viewed by ordinary transmitted light, differences will show when oblique and annular illumination are employed. Few people properly understand the functions of a compensating eyepiece and it is recommended that the test object described on page 61, in the section 'To Set Up a Microscope to the Correct Tube Length', be set up with an apochromatic objective and white light. If now various eyepieces are tried, starting with a low power Huyghenian, colours are seen surrounding the black objects. The margin of colour increases in width as the periphery of the field is approached and the symmetry of the colours around the axis is a sensitive indication of the centration and squaring-on of the objective to the microscope tube. The student is recommended to loosen the objective in the nosepiece and move it in such a way that its accurate squaring-on is interfered with, then the effect upon the image can be seen immediately when the region of the good images of the black objects does not coincide with the centre of the eyepiece field.

When the right degree of compensation is applied, the colour fringes disappear right up to the edge of the field of view and when the correct eyepiece for the objective in use is found, it should be marked or remembered. The compensation is a matter between objective and eyepiece only and is not affected by ordinary changes in tube length nor by the nature of the specimen.

Any kind of eyepiece may be designed to magnify in the red less than in the blue, thus an objective image which has red larger than blue can be made to give a correct final microscope image. A compensating eyepiece can always be recognized by holding it to the light and observing the colour of the fringe round the edge of the diaphragm, if the fringe is blue, the eyepiece is ordinary, but if red, it is compensating.

Types of Compensating Eyepieces
The Huyghenian form is made with the field lens of dense flint glass and eye lens of ordinary crown, it works as shown in Fig. 5A. No other changes are necessary.

Solid Eyepieces
These are usually of the compensating type but are sometimes used as high power eyepieces for ordinary objectives. They work like simple lenses, but are usually composed of three cemented lenses to give achromatic characteristics. If they are over-corrected, they are compensating. They require to be mounted in such a way that the eye is kept in the position required with respect to the lens and this position is usually about 20 mm. above the upper surface of the lens in the optic axis where a perforated diaphragm is placed. For powers below about ×8 there is no advantage to be had from the more expensive structure of the solid eyepiece, but for higher powers there is considerable advantage to be had from the long eyepoint, it is also easier to design them for the correct amount of compensation. Micrometers may be used as in the Ramsden form and there are only two lens surfaces to keep clean. The focal length of the cemented combination may be measured directly but it must be treated as a thick lens. Long eye-point eyepieces are of the greatest use to those who wear spectacles and the high power solid eyepiece allows plenty of room for the spectacles to be used with the microscope.

The Adjustable or 'Holoscopic' Type
These are designed to provide adjustable compensation and consist of an eye lens mounted in a separate graduated tube so that it can slide to and from the field lens, the amount of compensation is in this way varied. The device is adjusted on test preferably with the test slide described above. Fig. 5C.

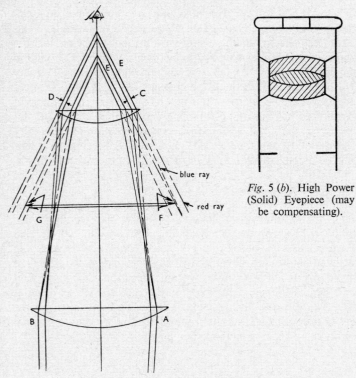

blue ray

red ray

Fig. 5 (*b*). High Power (Solid) Eyepiece (may be compensating).

Fig. 5 (*a*). (B) Ordinary Huyghenian. (A) Comp. Huyghenian.

A Dense glass in compensating type (flint).

C Greater spread of colour due to dense glass A compared with that at D.

E Greater refraction experienced by red at regions of lens further away from axis than blue.

F Greater size of red image due to E above, compared with D and G.

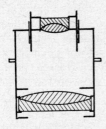

Fig. 5 (*c*).

In general it is best to provide a set of eyepieces and objectives by the same manufacturer. The best results can be obtained only in this way as all dimensions of the eyepieces are then optimum and the chance of wrong compensation or adjustment is removed. Some old achromatic objectives work better with a compensating eyepiece though they seldom need the full compensation of the apochromat. The adjustable type is useful in this case and where a varied range of objectives is in use but an adjustable compensating eyepiece is not easy to set up, one reason being that its magnification changes with the compensating adjustment. The manufacturers recommend that it should be preset when the characteristics of the objective are known.

Miscellaneous Eyepieces

The 'Projection Eyepiece' is simply a Huyghenian or Ramsden type of very low power, say ×2 or less. From the section 'Tube Length' it will be seen that an eyepiece is necessary in a projection microscope, but if an ordinary eyepiece is used, a very large screen field of view is produced. It is usually necessary in demonstrations to throw the picture a long distance without greatly increasing the field size and decreasing the brightness of the picture. The projection eyepiece does this and also provides a focussing adjustment on the eye lens in order that a clear sharp image of the eyepiece diaphragm surrounds the projected picture. (See pages 94 and 396.)

Erectors

These may be constructed from prisms or lenses but the lens type only is described here, the prismatic variety being included in 'Binoculars'.

The lens erector is the same as that used in the telescope. Two similar lenses are mounted in a tube, each about 70 mm. in focal length, uncorrected, and about half the sum of their focal lengths apart. A perforated diaphragm is placed in the centre of the combination to cut out extremely oblique rays. The lenses merely invert the image and pass it on to the eyepiece. Some distortion and loss of definition always results from the use of this kind of erector but it is never used in high definition work. Its function is to help in manipulative operations by causing movement in the image to be in the same direction as the movement of the operator's hands and needles, but an erector is not an advantage to every worker, because those used to compound microscopes soon develop the art of reversing their hand movements to suit the inverted image seen by the eye, so an erector is of no help to such workers and will probably be a nuisance.

Another useful property associated with erectors is that if the draw-tube be moved in and out, a great change of magnification results.

The function once performed by the lens type erector is now almost wholly performed by the low power binocular microscope properly constructed for dissection. (See 'The Greenough Binocular', page 383.)

Inclined Eyepieces

These are really only inclined mounts for standard eyepieces but they are always called by the name 'eyepieces'.

It is usually convenient to use a microscope with its stage horizontal but this is a difficult position for the worker and bad for the eyes (see 'Instruments and Use of the Eyes', page 9). The inclination eyepiece goes some way toward solving the problem by turning the light through about 60° so that an observer seated upon an ordinary stool looks through the microscope with his head inclined at a comfortable angle, there is then no tendency for tear drops to collect over the pupil nor internal particles to gravitate there.

On first sight it might be thought that an ordinary first-surface mirror mounted as shown in Fig. 6 would perform the operation. It will incline the image but it will also reverse the picture top to bottom but not from side to side, so if reflecting arrangements of this kind are designed, it must be remembered that the distribution of features in a specimen is not according to the image, and a correction must be made.

All commercial angle eyepieces are corrected and use for this purpose a prism in which two reflections take place in the vertical plane, thus forming an inverted image as in the ordinary microscope, Fig. 7.

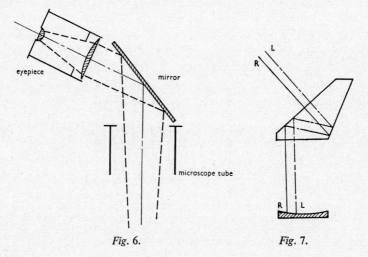

Fig. 6.   *Fig. 7.*

A lens is sometimes built into the angle assembly to compensate for the extra tube length introduced.

There are many variations of the angle or inclined eyepiece and as a description of them is to be found in the makers' catalogues, a list only is given here.

The Demonstration Eyepiece is an angle eyepiece with a tube about 200 mm. long to carry the eyepiece proper, and is revolvable upon the microscope in the optic axis, thus it may be turned from one observer to another without disturbing the microscope or the observers.

The Comparator Eyepiece is a combination of two angle eyepieces (usually right angled) so that the images from two identical microscopes can be combined into one eyepiece.

Binocular Eyepieces may be composed of one of several arrangements of prisms designed to divide the light from the objective into two parts, thus forming a non-stereoscopic binocular system. The optical arrangements are quite sound but mechanically these devices are not recommended because they add too much weight and height to the microscope and make it unstable. A tube length corrector is necessary owing to the much increased length of the microscope. (See 'Binocular Microscopes', page 380.)

Other eyepieces and eyepiece apparatus, e.g. spectroscopic, micrometric, analysing and petrological, are explained in the appropriate sections dealing with these special subjects.

Hartridge's Eyepiece
Reference to this eyepiece is made under 'The Aberrations of the Lens System of the Eye' and a diagram, Fig. 8, shows the construction. The Author has constructed such an eyepiece and finds it remarkably comfortable to use for long periods of time.

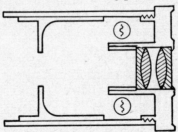

Fig. 8. The use of a bright surround has been found to improve the visual acuity of the eye. The eyepiece shown in the above figure has a diaphragm which is painted white, and this is illuminated by means of small electric lamps which are supplied with current from a battery, or a transformer, through a variable resistance.

MISCELLANEOUS EYEPIECE APPARATUS

Pointer Eyepiece

It is possible to buy a Huyghenian eyepiece with an externally controlled pointer in the plane of the diaphragm, which is used to indicate to other persons points of interest on a specimen and on a projector with a long throw it can be most useful at lectures if the instrument is sufficiently stable at the eyepiece end. It is usual to make the pointer of simple form so that it swings on one pivot, its point travelling in an arc across the field.

Camera Lucida

This is an old apparatus usually taking the form of a diagonal plate of glass above the eyepiece, so arranged that part of the light of the image is cast upon a sheet of paper. With electric lamps in use, a mirror or prism will project an image of sufficient brightness for drawing when the paper is looked at directly with the eye, but when the light is less intense, the eye is placed above the partial reflector, thus seeing some direct light from the microscope but with the paper superimposed upon the image. The observer can learn with some trouble to see the pencil point on the paper as well as the image and so trace it on the paper. One advantage of the apparatus is that it can be used by a class in daylight.

Exit Pupil of an Eyepiece

Mention has been made of this under 'Instruments and Use of the Eyes'. In microscopy the matter is of little importance except where the observer has eyes with faulty lenses, when a large exit pupil will embrace more faulty areas than a small exit pupil.

With telescopes the matter is of greater importance. The maximum size of the normal eye pupil when the eye is fully dark adapted is about 9 mm., therefore if an eyepiece transmits a pencil of light of larger diameter than this, the light is wasted because it does not enter the eye. If, however, the instrument is such that the exit pupil is smaller than that of the eye, the telescope image is darker than the naked eye image. The diameter of the exit pupil is determined by the size of the object glass and the magnification of the eyepiece,[1] and the best telescope is that which will give the largest exit pupil with the largest magnification. No optical instrument can create light, therefore the best that a night glass can do is collect as much light by means of as large diameter objective as possible and magnify the image as much as possible so that the exit pupil is 9 mm. in diameter.

---

[1] Ignore the presence of the erector which may have a diaphragm.

The object will then appear as bright in the night glass as to the naked eye but closer by the extent of the magnification. In daylight, when the pupil diameter is about 3 mm., the magnification may be increased by 9 mm. ÷ 3 mm. = 3 times, before the image is darkened.

The size of the exit pupil of any instrument can be observed when it is illuminated by standing about three feet away and looking at the eye lens. The exit pupil is then seen as a bright disk in the eye lens which can be measured with a ruler.

The Periplan Eyepiece by Leitz

This is a compound eyepiece designed to give a large well illuminated apparent field of view without the disadvantage of a close eyepoint. Fig. 8. Its apparent field of view at 250 mm. distance is about 250 mm. diameter. Definition is excellent but the magnification of 25 is too much for most microscope systems. With low power objectives the results are good and the eyepiece is useful for obtaining depth of focus in photographic work. It works well with apochromatic systems.

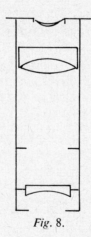

Fig. 8.

CHAPTER 6

Polarized Light

THE INTRODUCTION of films known as 'Polaroid' made possible polarizing devices of great size compared with the old Nicol prism arrangements. The largest diameter of polarizing material required in a microscope is 38 mm., that is, the substage diameter. Polaroid is available in several different qualities and for microscopical work the best should be obtained. This quality is sold in 50 mm. squares and one square, when cut up, is sufficient to make a polarizer and analyser of adequate diameter for all purposes. The efficiency of the best quality material is equal to that of a Nicol prism.

The polarizing disk should be placed in the stop carrier of the substage and should be equal in size to the back lens of the largest diameter condenser in use. The analyser may be placed anywhere in the optical system behind the objective. The Polaroid film and its supporting celluloid is only about 1·5 mm. thick, therefore the easiest place to put the analyser is above the eyepiece and it may be mounted in a cardboard cap of the pill-box kind which can be placed over any eyepiece. If a more permanent arrangement is required which can be calibrated, the Polaroid may be placed upon the diaphragm of an eyepiece used specially for this purpose and secured at its edge by means of adhesive. The eyepiece itself will then rotate fairly firmly in the draw-tube and a circular scale may be attached to it if required. Polarizer and analyser are identical materials.

Polaroid material is sold secured between two thicknesses of celluloid and it is unnecessary to further protect the film unless scratching is feared. It is quite satisfactory to enclose the film between glass disks but it must not be stuck there by Canada balsam or any other cement. If this is attempted the heat disturbs the delicate structure of the film and greatly reduces its efficiency as a polarizer, also the transparent cement follows the contours of the celluloid, which is seldom flat, and so creates lens effects between it and the flat glass plates which spoils its ordinary optical properties. It can be obtained optically true on glass.

Polaroid materials consist of a mass of microscopic crystals embedded in a transparent film, usually of polyvinyl alcohol. The crystals operate after the manner of tourmaline, that is, they separate light into the two planes of polarization and absorb one component plane. During manufacture the crystals are arranged to lie in one direction only and so their effects are added and the film in this way transmits plane polarized light with comparatively small attenuation of the wanted component. When the disks are crossed in a microscope system normally illuminated, nearly complete blackness should result and if the field is only grey or otherwise coloured, the disks are not of the best quality.

The completeness of polarization does not affect the accuracy of results in a direct way. Incomplete polarization only prevents the more delicate effects being observed.

When some substances like many crystals, cotton and a host of other things are placed on the stage of a microscope and examined, they will be found, when the analyser and polarizer are crossed, to still transmit light and often colours while other substances will be found to have no effect upon the darkness of the field. Yet another group will be found to transmit light as above, but the light can be extinguished by rotating the analyser to another position.

Substances which transmit light through crossed polarizers are called 'doubly refracting' or 'birefringent'. An example is cotton fibre.

Materials which have no effect at all between crossed polarizers are called 'optically isotropic'. Unstrained glass is an example.

Substances which require a movement of the analyser to bring about extinction of the light they transmit are called 'optically active'. An example is sodium chlorate.

When the polarizer is transmitting light which is made up mainly of vibrations in one plane only, and the analyser is set to transmit vibrations in a plane at a right angle with it, the result when both are in the optical path is darkness. When the polarization planes of the polarizer and analyser are at 90° of rotation with each other, they are said to be 'crossed' and the light transmission through the pair is at a minimum. The light which passes through due to the presence of a birefringent material can do so only because the material in some way alters the plane polarized light from the polarizer and gives it components which have a different plane of polarization and so can, in part at least, get through the analyser.

Birefringent substances have a different refractive index for each of their two component rays, therefore the rays experience a difference in speed of transmission through the substance. Because the two rays are derived from one ray which enters the substance, they are, when recombined, in a state in which they can interfere with each other. If the substance is sufficiently thick for the speed difference to cause an appreciable difference in phase between the rays where they emerge, cancellation of some frequencies of white light occurs and colours result.

The interference is made visible by the action of the analyser which is as follows:

Examination shows that when polarizer and analyser are crossed, transmission of light is zero only at one position which has to be accurate to a degree of arc. If a birefringent substance receives a ray from the polarizer when thus set up and introduces the smallest change in the plane of polarization of one of its components, light will pass through the analyser. The birefringent substance divides the rays into ordinary and extraordinary rays, which are transmitted at different speeds as mentioned above, also they are in different planes of polarization with respect to each other and sometimes with respect to the polarizer. The analyser can transmit rays only in one plane, therefore from the mixed rays coming from the birefringent substance it selects only the components of those which are vibrating in the plane in which the analyser is set. These components are selected from rays of mixed polarization which started into the substance as one ray in one plane of polarization. The selected components therefore are in an optical condition to interfere with each other and produce colours.

There is a further matter in connection with this. The pairs of rays produced by the birefringent substance travel at different relative speeds in different directions in the substance. A crystal is the clearest example and may be turned in the polarizing apparatus so that a position is found in which the greatest speed difference between ordinary and extraordinary rays is exhibited. (Ordinary substances like cotton are too complex to be used for demonstrations.) At this position the extraordinary ray is retarded most and the greatest refractive index difference is experienced between it and the ordinary ray. There is then more of the ordinary ray transmitted by the analyser than at any other position of the crystal and the crystal appears at its brightest. The retardation measured between the ordinary and extraordinary ray for the crystal in this position is an identification feature of the substance, but the thickness of the specimen must be known. When typical specimens like rock sections and cotton fabrics are examined in white polarized light, many parts will be seen to be brightly illuminated in white and coloured light. As the specimen is rotated the colours will change and other parts of the specimen light up. The effects here observed are combinations of the above-mentioned effects. Many substances are made up of birefringent components of all thicknesses and complexity of construction, lying in all directions on the slide and the effect, upon first observation, is one of astonishing beauty (for example, a slide containing crystals of salicilic acid or a section of sandstone about 50 microns thick), but it is not so easy to obtain meanings for all the effects seen.

When isolated chemical crystals are being examined, some rules can be laid down. When parallel polarized light is applied to a crystal placed between crossed polarizer and analyser, the following effects will be seen in the analyser:

Effect (between Crossed Polarizer and Analyser)	Substance
No effect when rotating specimen, dark in all positions	Cubic system of crystals or amorphous substance
Transmits when held in all positions except one	Uniaxial, system hexagonal or tetragonal
Transmits when held in all positions except two	Biaxial, system monoclinic, rhombic, triclinic
Shows dark crosses on bright particles	Starch
Bright rod or tube roughly 100μ diameter	Animal hair
Flat twisted fibre, bright and coloured	Cotton
Bright sharp-sided particles	Mineral dust
Highly coloured radiating 'dartboard' effect	Crystallization of salts on glass slide

TABLE 6 (a)

The above list is intended mainly as a guide to specimens which should be made up by persons intending to learn analytical work of any kind. Details of the groups of crystals must be sought in books of chemistry.

In the above descriptions, a birefringent material between polarizer and analyser has been dealt with, but a birefringent substance is a polarizer in its own right and if its physical characteristics are suitable it can be used for this purpose. If the polarizer and specimen only are examined without an analyser, normally no effects different from ordinary light examination are seen, the ordinary and extraordinary rays are present but they are not combined in a way that allows interference. When a specimen is coloured it may appear slightly different in colour, depending upon whether it is seen in the ordinary or extraordinary rays of light passing out of it. As the substance is rotated in the polarized light incident upon it, there are points at which the extraordinary ray is maximum and minimum in intensity, thus the substance changes colour slightly at two points in its rotation. Such a substance is said to be 'Pleochroic' or to exhibit pleochroism.

Unfortunately, most coloured crystals exhibit this effect but it is still useful as an aid to recognition of groups of substances. Compounds are described as dicroic or tricroic in the books of tables prepared for analysts. Particles of calcite are a good example.

Devices for Manipulating Crystals in a Polarizing Microscope
If a polarizing apparatus is set up of such a size that a large crystal, say a piece of quartz, can be manipulated in the hands, it is relatively easy to discover by turning it about where the axes are, but many microscopical jobs are done upon minute pieces of material and some form of manipulator is required. Not very much can be learned about the crystalline structure of a rock when it is simply ground into a thin section and looked at with a polarizing microscope. Certainly its structure can be seen but it is unlikely that any particular crystal will be so placed that a measurement upon it in polarized light can be made. It is usually necessary to break up the section and study the pieces.

Some petrological microscopes were built with rotating and tilting stages coupled with the Dick system of coupled polarizers described under 'Petrological Microscope' (page 110) and in this way a specimen can be tilted into the correct planes within limits, but it is better to mount the crystal on a needle after it has been taken out of its surroundings.

The apparatus used is described as follows:

A simple goniometer is arranged in which a crystal, stuck upon a needle, can be turned so that another needle mounted permanently above can be advanced to the crystal and caused to stick to it, thus holding the minute crystal in the position considered most advantageous. The crystal is then detached from the first needle and studied upon the second which is duly mounted in the microscope. A simple microscope and polarizer is arranged on the goniometer to help in the mounting.

A simple groove is arranged on the microscope in which the needle carrying the crystal is rested, the needle may then be turned on an axis at a right angle with the axis of the microscope.

Handling Technique

The goniometer may be of the simplest form and the Author has found a simple ball joint as used on stage forceps and mounted upon a mechanical stage set in the vertical plane forms a satisfactory goniometer. The second needle is simply held in a piece of bent wire vertically above the apparatus. The sticky wax used by dentists is used to secure the crystal fragment to a matchstick, which is then held in the stage forceps. By using the mechanism of the stage and forceps the fragment is presented to the second needle at the angle required.

The second needle is steel (a darning needle in a wooden handle) coated at its tip with sticky wax. When the crystal fragment is presented to the point, a hot iron or electric element is brought near the steel, when the wax fuses and connects the crystal with the steel. The wax on the steel cools quicker than that on the matchstick and the needle and crystal break away from the wood easily. Fig. 1.

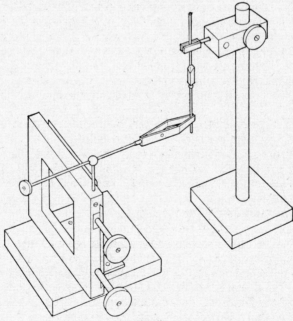

Fig. 1.

The steel needle and crystal are now transferred to the microscope. Fig. 2.

Block (*a*) may be of a plastic material with a file groove along it and

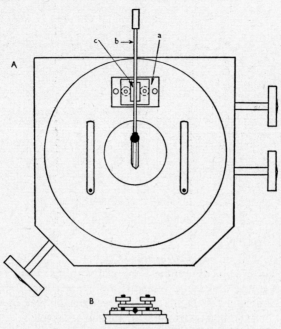

Fig. 2. (*c*) Cut-out to prevent needle rocking.
(B) Enlargement of fitting (*a*).

a top cap to clamp the needle (*b*). Rotation is given to it by means
of the wooden handle while the normal mechanism of the stage
moves it in all other directions.

It is usually necessary to immerse a fragment of crystal in a liquid
of the same refractive index in order better to see through the outer
layers and to kill reflections (see table, page 252), and this is best
done in an open top trough with the microscope stage vertical.
Fig. 3. The construction of such a trough is described in the section
'The Study of Objects in Liquids', the glass strips should be about
1·5 mm. thick and the cover glass a large No. 2, and with this arrangement the crystal can be positioned without the liquid being present.
When the liquid is added the cover will adhere by capillarity. The
trough should be wide at the mouth so that about 90° of stage
rotation is possible before the needle fouls the sides of the trough.

In addition to the polarized light application of this accessory, a
lot of useful identification work can be done as liquids in the trough
can be changed easily by means of a pipette or filter paper and
identification of minerals from their refractive indices is possible.

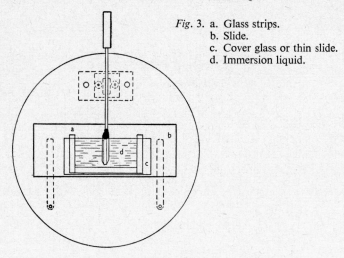

Fig. 3. a. Glass strips.
         b. Slide.
         c. Cover glass or thin slide.
         d. Immersion liquid.

Preparation of Sections of Minerals

The first move when examining a mineral is to make a translucent section of it and almost all minerals are brittle when thin, so the following notes give only the basic procedure for making sections and must be modified by trial to suit various substances.

Under the heading 'Metallurgy' some techniques of grinding were described. Most rocks are of the brittle material type where grinding is not a cutting but a breaking operation, i.e. when a hard particle rolls between hard brittle surfaces, points of high pressure develop as shown in Fig. 4, and flakes of the substance are pressed out.

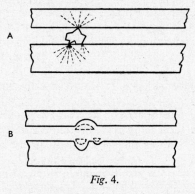

Fig. 4.

The sharper and more irregular the particle, the quicker the grinding action, because the greater are the forces involved when the particles roll between the surfaces. The roughness of the surface resulting is only approximately proportional to the size of the grinding particles.

Minerals can be ground satisfactorily on hard laps like glass and cast iron plates, while metals cannot. The mineral requires to be supported by a glass plate as soon as it becomes thin, the procedure then is as follows:

(1) Detach a piece of the mineral about 20 mm. square and 4 mm. thick.

(2) Prepare a sheet of plate glass or a rotating cast iron plate with water and carborundum, and rub down the fragment until one face is sensibly flat. Use the finest carborundum which will cut at the speed required, coarse carborundum will cut fast but may shatter the structure. The finish should be a partial polish and only trial with a particular mineral will give the correct answer.

(3) Stick the fragment on to a slide with hardened balsam and let it remain on the hot plate without boiling the balsam for several hours. This is to make sure that the balsam has penetrated the surface of the mineral.

(4) When cold, grind the mineral in the same way as described above. The glass slide is held in the hand by sticking a block of pitch on the back. Reduce thickness with the finest emery (500) plenty of water and light pressure until the section is about 20 μ thick (for average examination).

(5) Wash and mount in balsam under a cover glass.

(6) It is possible to dissolve the balsam away in zylol and gently shift the specimen to a clean side and remount it, but if this is attempted, the old balsam must not be heated but must be allowed time to dissolve. The section can then be floated from slide to slide with aid of a biologist's section lifter. Needles must not be used or breakages will certainly result.

Grinding Softer Minerals, e.g. Coal

The method used here is to soak the material for several days in either new liquid balsam or synthetic resin. Of these, the older treatment by balsam is the best because it does not harden readily and it can be dissolved out. Other materials can be experimented with and techniques worked out.

When balsam is used the mineral fragment should be soaked for a day in hot turpentine, and transferred hot and wet into balsam, where it remains for several days. It is then removed, rested on a slide, and heated on a hot plate until it is hard enough for the balsam, when at room temperature, to just take the impression of a thumb nail. The grinding operations described above may then continue with plenty of cold water. When the first surface is flat, another slide must be applied to it and the first slide warmed off. After this transfer, made while the section is still thick, no further changes of slide must be attempted.

In work of this kind with soft substances there is always a considerable risk of particles of abrasive sticking in the substance to be cut so causing confusion and damage. It is often an advantage to grind the specimen embedded in balsam or other penetrating hardened substance upon a new oilstone under a running tap as described later.

Mineral Saws

In general it is not necessary to employ diamond-armed saws to cut ordinary rock sections though such saws are desirable for sections of teeth. A mild steel disk fed with carborundum and water will cut any rock and most artificial substances easily. The sheet should be about 250 mm. in diameter and 1 mm. thick or less, mounted to rotate on its axis in the vertical or horizontal plane. There is some advantage in having the disk horizontal because sludge can then be drip-fed more easily on to the wheel and centrifugal forces throw the carborundum and water to the rim. Splash guards are also easier to arrange. The best holder for the rock is an old instrument lathe bed and cross slides. The rock itself may be stuck to a block of wood with a ball of pitch at any angle required, the wood is then mounted in the tool post. Carefully made slides, etc. are a waste of money on such machines because of the wet and grit to which they are exposed. The whole should be mounted in a large bath made of galvanized iron or of wood lined with roofing felt and after a cutting operation the whole should be hosed down. The electric driving motor should be connected to the disk shaft by means of a long belt so that all wet, grit and the bearing exposed to it, are separate from the motor. A good bearing which withstands water lubrication is made from ordinary electrical industry S.R.B.P. working with a steel shaft. The saw should revolve at about 500 r.p.m.

If a fully automatic machine is required, the lead screw of the lathe may also be used to propel the saddle along the bed, but in general this is not very effective unless the speed can be set to suit a particular pure mineral, for example, quartz. It is usually better and simpler to arrange a rope and pulley to pull the saddle of the lathe against the wheel. The force is adjusted by means of weights and the non-positive drive causes less shattering of the material being cut. With such an arrangement it is usually found after small experience with individual machines that a weight can be found which suits all cutting operations, the saw removes material in inverse proportion to its hardness and so looks after itself. A stop must be arranged, and this is conveniently placed under the weight to arrest its fall.

When thick specimens of crystals are prepared for examination it is desirable to have the surfaces flat and parallel. A suitable template for use while grinding is a hole in a piece of Perspex the required thickness. The specimen must be a sloppy fit in the hole but not slack enough to bang about and break, it may then be ground between two sheets of iron or glass.

Some machinery for grinding very thin sections of hard materials like quartz works on this principle. An apparatus due to J. Sayers consists of two cast iron disks, the lower of which is fixed, and the upper rides upon it being revolved by a drilling machine eccentrically, but resting upon the lower disk by its own weight. Between the disks is a diaphragm plate, the thickness of the thinnest sections required, made of bronze or sheet iron. In apertures in this plate rest loosely the sections to be finally reduced in thickness. The machine is pump-fed with carborundum and water and is left to itself. The diaphragm plate takes up its own motion between the grinding plates and produces a very regular result, while side guides prevent the diaphragm plate working out of the system. The best results are obtained (down to about ·05 mm. thickness) when many sections are in the plate.

There appears to be no limit to the shape of the sections cut, it is only necessary that they roughly fit the shape of the diaphragm plate apertures. For shaping water-soluble crystals cotton-seed oil or paraffin oil should be used in all operations in place of water.

Delicate sections of water-soluble crystals are best cut by passing a thin reciprocating wire over them like a cheese wire, under a pressure of a few ounces and wetted with water from a damp pad. A more elaborate apparatus consists of a continuous loop of wire 50 to 100 microns thick which passes over slightly tapering rollers vertically placed, one of which has a soft surface kept continuously damp. The taper of one of the rollers causes the wire loop to tend to run down, thus providing a little pressure on the crystal. The dampness of the driving roller communicates dampness to the cutting wire which gently dissolves its way through the crystal Fig. 5.

The only practical difficulty in the system is to obtain or make a continuous wire belt of 100 microns thickness and join it without a lump forming. The matter is mentioned under 'Metallurgy'. The Author has experimented with fine stranded wires and has found that with special care they can be spliced under a bench magnifier. After splicing they should be soft soldered by applying a hot, clean, large electric soldering iron near but not touching the joint until the flux runs. A piece of solder beaten out to a needle can then be applied while the wire ends are held in a jig during the operation.

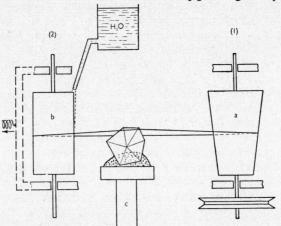

Fig. 5. a. Driving roller.
      b. Driven roller. b must be mounted on swinging arm spring loaded to tension the wire.
      c. Chuck holding crystal stuck in pitch or wax.

Several tries will probably have to be made so it saves time if the spring loaded cylinder, (2) Fig. 5, be made to have several inches of travel to accommodate a wire belt which may have a length different from that with which it started. The bump of soft solder can be gently rubbed down with glass-paper as emery paper is too likely to cut the rest of the wire as well. The strength of the joint is in the splice, the solder only keeps it together.

Compensators

In the previous paragraphs light has been considered to be vibrating in one of two planes, i.e. it is either vertically or horizontally polarized. It is, however, quite possible to have conditions of polarization which are combinations of the two planes and an electrical analogy may again help with a picture of the state of affairs (see 'The Phase Contrast System'). If we connect to the X plates of a cathode-ray tube, the sinusoidal wave output of an oscillator, the resulting trace on the screen will be simply up and down. If the connection is made to the Y plates, the trace or vibration is from side to side and the conditions may be considered to represent vertical and horizontal polarization. It must be noted that the terms vertical and horizontal are only relative directions and if the tube with all its connections is turned through a few degrees, vertical and horizontal will have different meanings, but the relative positions are the same.

Elliptical polarization does not mean that the planes of vibration are tilted. It means that the planes are combined. If the sinusoidal waves from the oscillator are connected both to the X and Y plates, the amplitude on each being equal, an inclined trace at an angle of 45° is produced on the screen. If the phase of the wave is altered in either the X or Y connection, an elliptical trace results. If the phase difference between X and Y plates is 90° ($\frac{1}{4}$ wavelength), we have a circular trace. When light passes through some birefringent materials, the ordinary and extraordinary rays are so produced and recombined that elliptical and circular polarization results.

If the analyser (which accepts only plane polarized light) can be used to produce blackness when an object is illuminated in polarized light, then the object is of the plane polarizing kind, but if darkness cannot be produced with any setting of the analyser, then the light is elliptically polarized. Referring to the cathode ray tube picture, a straight dark band (the analyser) used as a shutter, cannot cut out an ellipse or a circle as it can a straight line. It is for the purpose of sorting out these effects that the Compensator is mainly used.

The Babinet Compensator consists of two wedges of quartz, one of which is fixed while the other may slide in front of it, thus maintaining the composite plate parallel sided. Cross wires are usually fitted near the wedges. Quartz (not fused quartz but natural crystal) has the property of producing from a parallel beam of light two components, one polarized at 90° with the other. Also one beam travels faster through the quartz than the other, therefore if the quartz is of sufficient thickness, one beam will become 90° out of phase with the other, and if the axes of the crystals are chosen as in Fig. 6 they may be made to do opposite things to the beam of light. One may retard the first component while the other retards the second. If then the wedges are cut with their axes as shown and the combination is given the correct physical thickness by means of the adjusting screw, both components leave the quartz plate in the same phase. Should they have started in different phases, the setting of the wedges must be altered but a position can be found where compensation is possible.

The compensator must be set up so that its effect upon the phases

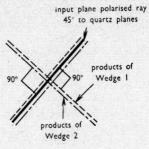

Fig. 6.

of the applied rays at the point marked by the cross wires is known and this requires that a scale be present on the milled head of the micrometer screw of the moveable wedge.

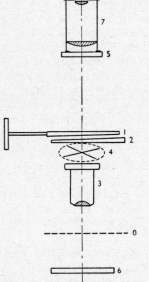

Fig. 7. 1, 2. Quartz wedges of compensator.
        3.   Objective.
        4.   Sighting cross wires.
        5.   Analyser.
        6.   Polarizer.
        7.   Eyepiece.
        0.   Object plane.

The plane of the polarizer is rotated until plane polarized light from it is applied to the wedges in a direction 45° to their axes. The wedges then each produce from this ray two components polarized at 90° from each other and at a certain position along their lengths where the wedges are the correct thickness, the rays will emerge in phase and also in the same polarization plane, Fig. 7.

It is only where the thickness of the quartz is such as to produce zero phase difference (or multiples of 2π, i.e. multiples of a wavelength difference) that this state of affairs exists. If the original incident rays (plane polarized) are extinguished by rotating the analyser, the positions of dark bands of extinction are seen crossing the field at each multiple of a wavelength. Rays in different parts of the field traverse different relative thicknesses of quartz and so experience zero, one, two, three, etc. wavelengths of retardation, thus producing the series of dark bands. One of these dark bands is moved on to the cross wires and the milled head of the micrometer marked. The difference in milled head graduation between one dark band and the next represents one wavelength retardation of the light used and the zero phase difference band is the only one which appears black in white light and so can be recognized.

In order to recognize elliptically polarized light, the procedure used is largely the reverse of that described above in connection with Fig. 7.

The compensator is set by means of its calibration to give a retardation of $\frac{1}{4}$ wavelength. The whole compensator is rotated in the optic axis of the instrument until a dark band is seen to cross the wires. Where the band exists there is zero phase difference between the output rays, and the axes of the ellipses of vibration of the light must lie along the long axes of the wedges. The analyser is adjusted to make the dark band as dark as possible.

The angle ϕ which the analyser makes with the axis of a wedge is related to the ratios of the axes of the ellipses. If, for example, the angle is 45° with the wedges, the light is circularly polarized (page 106). The actual ratio of the axes of the ellipse can be obtained by geometry:

$$\frac{\text{Length of axis parallel to wedge}}{\text{Length of axis perpendicular to wedge}} = \phi$$

The wedges are cut from natural quartz by means of a mild steel circular saw charged at its edge with diamond particles and cooled with a jet of pumped oil. Final grinding is usually performed by hand with carborundum and soda water on a cast iron lap. Polishing is by rouge and water on pitch. The thinner the tip the better.

Only a few pieces of quartz are optically good enough for the job. The quartz to be tested is immersed in a glass tank of benzyl alcohol (which has the same RI as quartz) and polarized light is applied. When the system is examined with an analyser, the optical axis can easily be found and any faults show up instantly in colour. The taper of the wedges need not be any exact figure but they must be optically flat and a taper of 1 in 70 over a length of 25 mm. is suitable. To avoid trouble due to rounding of the edges of the quartz during polishing, the wedges should be made much wider than the beam diameter they are intended to handle. Compensators are obtainable commercially. The wedge is cemented to glass for grinding and mounting.

A quarter-wave plate is a very thin lamina of mica, talc or selentite, of such a thickness that $\frac{3}{4}$ wave of relative retardation takes place between ordinary and extraordinary rays for a particular colour of light when light strikes the film normally. The material is split with a needle and pieces are tried until one is found which, when examined between crossed polarizer and analyser in the particular colour of light chosen, shows no light change when either is rotated. A quarter-wave plate is really a form of compensator but it is not adjustable, but several plates of the correct thickness for chosen colours of light are useful as compensators. A simple quartz wedge showing say 7 orders of Newton's Scale is quite satisfactory and is self calibrating.

The Petrological Microscope

The instrument is little different from most other microscopes, but it must have the following fittings:

(1) A polarizer in the substage which is rotatable with or without the substage, with graduations. It must be removable easily without disturbing the microscope adjustments.

(2) An analyser somewhere between the objective and eyepiece but preferably on top of the eyepiece so that it can easily be removed. It also must be rotatable and graduated.

(3) A fully rotating graduated stage.

(4) A slot or similar device in the body tube to carry a compensator. This should also be rotatable through 180° (see 'Dick Microscope').

(5) A swinging arm to carry a magnifying lens (say 25 mm. focal length) above the eyepiece to study the Ramsden disk.

Other fittings such as coarse and fine adjustments, mechanical stage, condenser changer slides, eyepiece and objectives, are normal.

Binocular microscopes cannot be used for serious work with polarized light because the light-dividing apparatus introduces selective effects upon polarized light.

The Dick Microscope

No mechanical stage is so good mechanically that a minute specimen can be rotated exactly on the cross wires of the eyepiece or compensator. The Dick design couples the polarizer and analyser together through sliding rods and links so that both may be rotated about the specimen; they may also be rotated independently. This design allows the smallest crystals to be studied and makes the setting-up of a compensator easier. It is more helpful if the stage also has independent rotation. (See Babinet Compensator, page 107.)

For a detailed study of the use of polarized light in mineralogy, the reader is referred to a work on petrology, but the details given above of the behaviour of polarizing apparatus should help in matters of simple analysis and should show the possibilities of the method.

Optical Strain Observations

Several materials like glass and Perspex are optically isotropic until they are strained. If a strip of Perspex is twisted while being viewed in a polariscope, colours appear and if a model of an engineering structure is built to scale and stressed, the parts under most load show up in colour before the lesser loaded parts, also if glass parts like the base of an electronic valve be examined, strain in the glass due to too rapid cooling shows as colour bands.

Evaluation of the actual force applied usually has to be done by applying measured loads and making rough comparisons of the results. In, for example, work upon beams, mathematics can give only average results and cannot indicate what is happening near small irregularities, but if a simple model is set up between large polarizers, known stresses can be set up by means of spring balances, etc., and the colours noted for reference when the complicated model is examined.

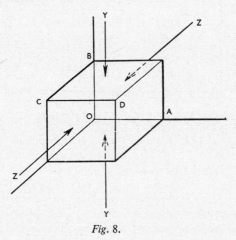

Fig. 8.

The following formula for computing stress in a regular block was used by Coker.

The block ABC is compressed by Y and Z lbs. per square inch in the directions shown. Fig. 8.

A wave applied to face BC (left-hand face in diagram) is decomposed into two waves polarized in planes 90° apart, i.e. in planes vertical and horizontal in the diagram. After passing through the block the components have a phase difference in radians given by

$$(\theta) = c(Y - Z)l$$

where l is thickness of the block and c is a constant which must be evaluated by experiment. In many cases either Y or X is zero and colour observations give the stress directly from a calibration. Glass requires very high stresses to show the effects required but Perspex and Zylonite show them more easily. Quantitative work can hardly be called satisfactory but qualitative experiments are most revealing.

It may be useful to know that a metal mirror produces elliptically polarized light from an incident plane polarized beam except when the incident plane of polarization is at right angles to or parallel with the mirror surface plane. In this case the reflected light is polarized in the same plane as the incident light.

Table showing colours resulting when certain component colours are removed from white light:

Colour Removed	Colour Produced
Blue	Yellow
Green	Magenta
Orange	Violet
Violet	Yellow
Red	Light blue
Green-yellow	Purple

TABLE 6 (b)

Rings and Brushes in Crystals

Certain crystals exhibit other interference figures than those associated with polariscopic colours. If incident light is applied to a crystal from a condenser in such a way that the light converges during its passage through the crystal, the rays occupying the outer parts of the condenser cone travel a longer distance than those occupying the centre parts and when they meet at the focus another set of interference conditions is thus set up. These conditions are superimposed upon those due to retardation differences in the crystal axes described above.

If the light is monochromatic, then the usual black and white interference rings are formed but should the light be white, then colours due to cancellation of components appear as in the case of ordinary polarized light. When a study is about to be made of convergent polarized light, it pays to select a crystal which shows the effect well and to use it until the method is mastered.

Select a crystal of either calcite, muscovite, borax, phosphate of soda, nitre, ferrocyanide of potassium, or bichromate of potash. In all these a section with or across an optical axis is required. The axis may be found as described above but if a longitudinal section of a borax crystal or a transverse one of phosphate of soda is taken, a suitable object is obtained. An ordinary rock section may be used.

A slice of crystal about 0·25 mm. thick is required, and this is prepared by honing down according to the methods described above for preparing sections. In a thick crystal like this, the surface finish is of little importance as long as it is reasonably flat and polished, and the sides are mechanically parallel. Some water-soluble crystals must be lapped in paraffin. The sections should be mounted in a deep cell of some kind such as a hole in a sheet of Perspex and water soluble crystals should be mounted in watchmakers' oil.

Place a good achromatic condenser in the substage, insert the Polaroid disk below it and illuminate the specimen in the usual way with convergent light from the substage condenser. The microscope objective is not used in the ordinary way because the interference rings sought exist in its posterior focal plane, but the objective must have a good aperture to grasp the effects and a 6 mm. objective is suitable as higher powers operate too near the specimen. The objective is not forming an image of the specimen and so should be placed nearly touching it and if its back lens is now examined colours will be seen there. The easiest way of making these more visible is to look at the back lens with a telescope made from a 75 mm. objective mounted in the bottom of the draw-tube and a normal eyepiece, the draw tube is then focussed upon the back lens of the objective. Should the microscope not have an objective thread on its draw-tube, the image of the back lens of the objective (the Ramsden disk) can be examined above the eyepiece with a magnifying glass.

Do not forget to insert the analyser in place before seeking the rings and brushes.

The magnifying power of the system must be adjusted by varying the eyepiece power as the objective must have good aperture and cannot be much reduced in focal length. An 8 mm. is about the lowest power which normally has a suitable aperture for this work.

When the worker has the hang of the method it may be used for analysis of crystal types. An optically isotropic body causes the rings to be circular or nearly so, while a uniaxial doubly refracting crystal shows concentric rings with a cross between them. Biaxial crystals (axistropic) show two sets of crosses, the rings being of a varying figure-of-eight shape, depending upon the orientation of the crystal axes with respect to the microscope axis.

The effects seen depend upon the relative positions of crystal and polarization plane of the light, the crystal axis with that of the microscope, and the position of analyser and polarizer (which should be crossed before inserting the crystal).

It will be seen from the above that description is only useful to give the layout of the apparatus. If the method is to be made use of, known crystals must be cut along different axes in known directions, and the effects studied and recorded. For most analytical work on crystals of thickness, say 1 mm. and greater, it is necessary when using ordinary polarized light to apply parallel light to the specimen and all the effects discussed in the earlier part of this chapter refer to parallel incident light. If, however, very thin microscopical sections such as those prepared for mineralogy, are being studied, it does not matter much whether the light is parallel or convergent, because the material is not thick enough to affect path length in ordinary convergent light, nor is the microscope set up to observe interference effects at the objective posterior focal plane. In such work the presence of a condenser below the specimen often usefully brightens the image without introducing complications.

If mineralogical specimens of 20 to 50 microns thickness are being studied a section of coarse sandstone should be chosen. In this, grains of quartz felspar and possible mica lie at all angles, and if the microscope is set up in ordinary polarized light, the specimen rotated on the stage, a grain found which stays grey during rotation, that grain is a base section of quartz. If now the condenser is opened to full aperture, a 6 mm. objective brought close to the specimen, a 12 mm. lens held over the eyepiece to show the Ramsden disk, the specimen rotated, the interference figure will be seen. This may be repeated for several grains until the varying effects due to a grain being tilted are seen.

CHAPTER 7

Microscope Lamps

OF SHAPES, sizes and types of microscope lamps, there is no end and many of the types were based upon misconceptions concerning the need to focus the illumination upon the specimen but since the factors governing the formation of the optical image are now better understood, the problems of illumination can be much simplified. A matter often overlooked is that it is the brightness of a lamp which is important in magnifying instruments and not the quantity of light. There is never any difficulty in producing a sufficient quantity of light but as magnification increases, brightness decreases and it is not easy to increase the brightness of a source beyond its natural standard. Examples are as follows:

A 100-watt lamp gives a lot more light than a 40-watt lamp but the difference in brightness of the tungsten filaments is not great, therefore the microscope image is little brighter when the 100-watt lamp is used. A lens-fronted bulb is very bright at 2 volts.

A mercury vapour street lamp gives out a lot of light but its brightness is little better than that of a paraffin lamp. A very small carbon arc produces as bright a light as a large one and this source of whatever size is the brightest artificial source known and should always be used for high magnification projectors. The brightest ordinary lamp is an over-run motor car headlamp bulb.

The important matters in illumination from the point of view of microscope optics are as follows:

(1) The whole of the objective back lens must be evenly filled with light when the microscope is properly set up.

(2) Only that part of the object which is in the field of the microscope should be illuminated.

(3) The brightness of the light in the microscope should not be greatly different from the light in the room experienced by the, open, unused eye and should a binocular be in use, the same amount of light should be used as if it were a monocular.

(4) The specimen should be illuminated with an even, structureless disk of light against which fine detail can be seen.

In connection with (3) and (4) see 'Instruments and Use of the Eyes', page 9.

From the point of view of resolution of detail, (1) is the most important matter and Nelsonian microscopy was intended to achieve this end by simple means.

The Nelsonian Method

If a source of light be focussed by means of a substage condenser upon the object, the objective will be filled with light to an amount equal to the aplanatic aperture of the condenser (see 'Corrected Substage Condensers', page 342). Should the objective be of smaller aperture than the condenser the substage diaphragm is used to reduce the working aperture of the condenser until it is the same as that of the objective· When a microscope is set up in this way the observer can be sure that the condenser is being fully used and that its diaphragm is controlling the light in the proper way.

If a point-source of light is used with a good condenser the above effects are retained but the illuminated area of the object is so small that ordinary observation is not possible, therefore it is usual to use a large source of light, for example the flame of a paraffin lamp or an area of diffusion like an opal electric bulb. Such a source is better than a point in all ways for if the source is, say, 25 mm. diameter and the substage condenser a normal one, an adequate amount of the microscope field is illuminated, the condenser need not be particularly well corrected and no structure will appear in the back lens of the objective. This, the Nelsonian method, is the best way of illuminating a microscope and is least likely to produce artifacts, but there is one important disadvantage in the system: it is difficult to obtain enough light from a flame or diffuser for use with colour filters and with high power objectives.

The way of overcoming this difficulty is to use a small bright source like a motor car headlamp bulb and render the beam from it as near parallel as the geometry of the filament permits, by means of a good lamp condenser. In this system the lamp condenser must be good because the filament is small, and when the lamp condenser is imaged by the microscope condenser it must show as a structureless disk of light. A controlling diaphragm is mounted close against the lamp condenser and limits the size of the source. This system is called 'Köler Illumination' and it has the advantage that any brightness of light may be used from arcs, Pointolite lamps, bunched filament projector lamps and over-run car headlamp bulbs, also plenty of spare light is available to make good losses in filters when roughly monochromatic light is required.

It has long been understood that exact focussing of the source of light upon the specimen is not demanded. There is no exact focus for a diffused source like a lamp flame, and tiny changes of focus which do not affect the working aperture of the objective do not affect resolution. Recently the matter was put to the test by Dr R. Baker of Oxford, who used a light source placed at a distance from an aperture in a screen, the aperture was imaged by a substage condenser and focussed upon the specimen in the ordinary way. The specimen was in this way illuminated normally by the Nelsonian method, the source being the aperture in the screen, but the lamp which illuminated the aperture was far out of focus.

No difference in results between this and 'critical' illumination could be demonstrated.

When using any light source it is often desirable to move certain components a little to avoid images in the field of view of the diffuser structure and dust upon lenses. This is quite permissible so long as the objective back lens remains filled evenly with light up to the aperture required and the expression 'Critical Light' should be used to mean this and should not be used in reference to the focus of the illuminant. (See 'How to Set up a Microscope', page 369.)

The Paraffin Lamp

Any steady-burning lamp with a chimney through which the flame can be seen without undue distortion is entirely satisfactory for any kind of microscopy for which it gives sufficient brightness and light, but lamps may be unsuitable in one of two ways. Many have a flame which flickers at the top edge at a rate of several flickers per second, and this often passes unnoticed until the flame appears magnified in the microscope field. The fault is due to a wrongly shaped chimney and another chimney can be tried. Some chimneys have a narrow throat at wick level, the idea being to create greater turbulence in the region of the burner to obtain more complete combustion of the gas, but this type of chimney should not be used, because the throat introduces great distortion of the flame image just where it is not wanted. A lamp with a shaded glass chimney is preferable to one with a metal chimney like those often sold in the past for microscopic work because paraffin creeps up the metal and smells greatly while being burned off when the lamp is re-lighted.

Those who prefer or have to use paraffin-burning devices are reminded that a lamp should be burned as high as possible without smoking, it will then not cause smells and it follows that a small lamp burned high is better than a large one burned low. Chemists will realize that paraffin while burning produces about its own volume of water. This rarely matters in an ordinary room where the worst result is that windows run with water condensed there, but if paraffin ovens are used the oven must be quite separate from the atmosphere of the flame. The wonderful frost patterns on windows shown by naturalists of the last century were due to the water vapour in the atmosphere of the rooms produced by the common paraffin lamps. In these days of electric lamps the frost patterns, if present, are poor examples, but the microscopist who still prefers paraffin has three dimensional frost patterns in abundance and a lesson on how much water this form of heating produces.

Some experienced microscopists believe that the large steady flame from the paraffin lamp is the best illuminant for fine, nearly transparent structure and in the Author's experience a microscope can be guaranteed to give its best performance with this illuminant though other illuminants, used with due care, can be as good.

Large Area Illuminants
The ordinary opal electric light bulb is the most usual of these and the common bulb of about 60 watts power is placed in the greatest variety of housings. It is difficult to say whether the manufacturers or the amateurs have produced the greatest variety, but nothing can be altered in the bulb itself. It is best to run it at a steady voltage and current accordingly to the maker's specification, as its brightness cannot be well controlled electrically because if this is attempted, the spectrum changes greatly. Light control is best achieved by means of neutral screens and these may be purchased or made from pieces of exposed photographic plate. A 60-watt bulb dissipates 60 W of heat, therefore it must be mounted in such a way that air flow can take away some of this heat.

The design of the lamp housing should be such that the following matters are considered:

(1) The lamp bulb should present its side to the microscope because the diffusing coating on its glass is most even at the sides.

(2) There should be present some simple device for centring the best illuminated area of the bulb to the iris diaphragm. This can usually be done by mounting an ordinary brass batten holder upon legs formed from three fixed 50 mm. screws. Finger nuts made from dry-battery terminals operating against coil springs on the screws are quite satisfactory for this adjustment (Fig. 1).

(3) An eye shade must be present to prevent direct light entering

PLATE 2

Optical Bench with Microscope and Camera with Spectrometer
Monochromator and Mercury Vapour Illuminators.

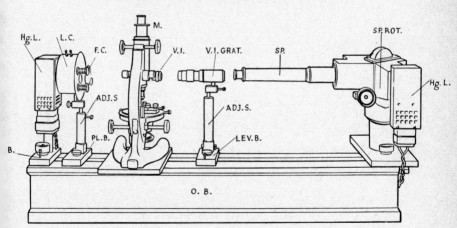

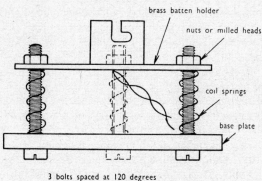

Fig. 1.

the observer's eyes and from entering the eyes of other workers.

(4) The lamp must be low on the table, i.e. its aperture should be about 100 mm. above the bench. It can easily be raised upon a stand when necessary.

(5) It should be inclinable so that the axis of the bulb and diaphragm can be placed to point at the mirror or up the microscope tube when the mirror is not in use.

(6) An iris diaphragm opening to about 25 mm. and closing to about 2 mm. must be mounted in the aperture through which the light is emitted.

(7) A holder for glass light filters should be placed on the microscope side of the diaphragm.

Lamp housings have been made satisfactorily from tin cans of all shapes and sizes. Tinplate is the easiest material to use because it can be cut and soldered easily and round holes up to 70 mm. diameter can be easily cut with a tool known in the trade as a 'Q Max' cutter. Any enamel finish will stand up to the temperature fairly well, but a properly made job should be finished in dull stove enamel. This arrangement is suitable for all visual work not requiring monochromatic light and not requiring the use of high powers above say 6 mm. objectives. The lamp will work with higher powers but when the diaphragm is reduced to limit the area of the specimen illuminated, the brightness of the small area of diffusing surface is found to be insufficient, also the structure of the diffuser becomes visible. Artificial diffusers are not good when high powers are in use mainly because their grain size is always apparent. The grain of the paraffin flame is never resolvable so, although it shows some defects similar to those of the electric bulb, it is free from this one.

Another form of diffuser is described under 'Colour Photo-micrography'.

Gas Discharge Tubes

In recent years light sources have been produced which consist of a quartz tube several centimetres long and about 8 mm. diameter evacuated and then charged with a rare gas at low pressure. An electric discharge is started in the gas by means of a kick of high voltage obtained from a choke (inductance about 30H) in series with the supply. Once this kick has ionized the gas and reduced its resistance, the ordinary steady voltage of the supply keeps the discharge going. Ionization continues to build up and the amount of light emitted by the electrically excited atoms of gas continues to increase for about 15 min. when it becomes steady. The effect may be observed in street lamps. The lamps consume a current of about $\frac{1}{2}$ amp depending upon size and a choke of the correct inductance and current rating is best purchased from the manufacturers.

When the lamps are of the mercury vapour type, small particles of metallic mercury are left in the tube and as the tube warms after striking, the mercury vapourizes and forms more gas, but the running brightnesses cannot be controlled by electrical means.

The lamps most important in microscopy are the mercury vapour and sodium types and both of these can now be made in the high pressure form, the source is then much brighter and smaller being about 25 mm. by 5 mm. in size. When working, the gas column is structureless, fairly small and bright and it forms the ideal illuminant to use with a lamp condenser and filter.

These lamps are produced complete in housing for microscopical and other technical purposes. The cost is high. However, a much cheaper form can be obtained in the shape of the GEC 'Ozira' lamp, which is quite suitable. This is obtainable with a clear glass envelope, made from glass which stops a large proportion of the ultra-violet light emitted by these sources, unfortunately the glass envelope is large in diameter and prevents a lamp condenser being brought nearer than 50 mm. from the source. This is insufficiently close but as the envelope is only a shield it may be cut off with the aid of a sharp triangular file, and the quartz tube and carbon striking resistance exposed. Some extra corrosion may be expected due to exposure of the metal connections but it appears to affect the lamp in no way at all and a good quality lamp condenser may now be brought within 15 mm. of the source, but allowance must be made for the heat generated. The quartz tube runs at a dull red heat, therefore the whole condenser assembly must be of the uncemented kind and must be allowed to warm up slowly with the lamp.

It is most important that the strong light from the lamp does not scatter about the work room. These lamps are powerful ultra-violet emitters and serious damage to the eyes will result if the light strikes them without first passing through thick ordinary glass or special thin glass, but the microscope lens system is a complete protection for the working eye.

The High Pressure Mercury Vapour Lamp

This consists of a small tube with or without a screening envelope as described above, housed in a metal box about 50 mm. square section, or of tubular form, with a window usually made of lead glass to transmit visible light and cut out ultra-violet. This part of the assembly is usually separate from the lamp condenser and is arranged to be centred to the condenser axis by some simple means such as slotted screw holes. It is best mounted upon a simple optical bench saddle as shown in Plate 2. If the lamp is made up in the laboratory it is important that a draught of air shall not pass through the housing. No ventilation is necessary other than that provided by the open light-exit hole as a flow of air causes the discharge to wander in the tube and be generally unsteady. A commercial lamp is supplied by Messrs R. and J. Beck, opticians of Watford. The life of these lamps is only about 50 hours and the life is much affected by frequent switchings on and off. The warming-up period is about 15 min. and they will not restart after switching off from full running temperature until about $\frac{1}{2}$ hour has elapsed. These factors have a considerable effect upon the economy of the laboratory.

Because the source is comparatively large, an ordinary two-lens type lantern condenser with a focal length of about 25 mm. and diameter about 35 mm. is adequate to provide an even, structureless disk of light in the microscope. The lamp, being a gaseous source, emits a line spectrum, that is, all the light energy is concentrated into narrow bands of colour clearly separated from each other and if this light is spread by means of a prism or grating the lines are clearly seen and can be compared with the continuous colour spectrum emitted by a solid source, say a tungsten filament lamp. Because the light energy is concentrated into lines of colour, the colours can easily be separated one from another by relatively crude means, e.g. by ordinary glass filters (see 'Monochromators', page 130). The colours are determined by the source and not by the characteristics of the filter, which acts as a gate only.

The important lines in the mercury vapour source are orange, green and blue with some red present as a continuous spectrum from the red-hot quartz tube and a faint red line due to neon used in manufacture, and all these colours may be filtered out with ordinary grade colour filters and all are useful in microscopy.

It is quite possible to use the source in the Nelsonian manner by focussing it directly upon the specimen by means of the substage condenser only, but the source is not really big enough for this use and the lamp condenser should be built in as part of the apparatus. A diaphragm must be mounted on the microscope side of the condenser and following this a filter carrier capable of loosely holding the filters (the whole system becomes very hot).

The mercury vapour lamp is good for almost all types of microscopy, the exceptions are projection and high power dark ground as the source is not bright enough for these applications.

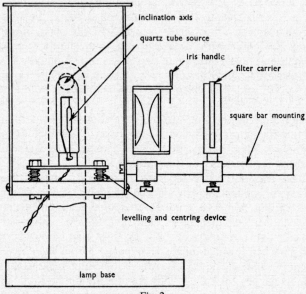

inclination axis

quartz tube source

iris handle

filter carrier

square bar mounting

levelling and centring device

lamp base

Fig. 2.

The Sodium Lamp
This is similar to the mercury vapour kind but the gas is sodium vapour. The structure of the lamp is not the same but the gas discharge takes place through a tube of the gas usually at low pressure, and a high pressure form is available from GEC of Wembley.

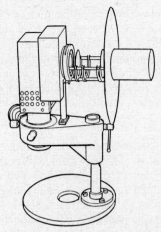

Fig. 3. Mercury arc lamp casing and filter carrier.

The advantage of the lamp is that it is nearly monochromatic, the spectrum consisting of two lines very close together in the orange. There is no particular advantage in using an orange light, it does not allow the microscope to develop its full resolving power and the eye is not particularly sensitive to it as it is to apple green, but for reasons not at present clear the eye is able to work comfortably for long periods in sodium light and is able to resolve detail much better. This effect is most apparent when fine work is being done without a microscope. The sodium lamp is not bright enough for any other than low and medium power microscopy and it is best used in the Nelsonian manner. The lamp has the same disadvantage of a 20-min. warming up time and it will not start after switching off, until it has cooled. There is no point in mounting a filter carrier on a sodium lamp because it emits only one colour. The same electrical conditions apply as for the mercury lamp but the quantity of light produced per watt of electricity is greater than that from any other source.

Future of Gas Discharge Lamps
It seems certain that gas discharge lamps will be developed very considerably during the next few years, they will be made to emit chosen spectra and their brightness will be increased. These developments will be useful to microscopists and others because they will remove the necessity of using selected filters and monochromators, and before purchasing any microscope lamps for special work the reader is recommended to discuss the matter with the lamp manufacturers. It may be that a sodium lamp as bright as the high pressure

mercury one and a mercury lamp as bright as a carbon arc will make their appearance in a year or two, and it would help considerably if the short life of this kind of lamp could be improved.

The structureless column of brightly glowing gas can be made of almost any physical size and if it can also be made to emit a chosen spectrum it will provide the perfect microscope illuminant, combining as it does the characteristics of the paraffin flame and the arc.

Point Sources

Under this heading comes all the types of lamp which have a small bright source and must be used with a well corrected condenser in the Köler manner in order to provide a large enough disk of light for the microscope field. Most require a transformer to step down the supply voltage to about 12 volts or thereabouts. All these sources are very bright and may be used with absorption filters and for darkground illumination with the highest powers. Their brightness may be controlled easily by electrical means.

There need be nothing special about the housing for the bulbs, but it must be well ventilated to allow a through draught of air. All must have condensers and must have centring movements to line up the filament with the condenser. An iris diaphragm must be placed coaxially with the lamp condenser and the whole housing must be inclinable and adjustable for height, so that the filament-lamp condenser axis may coincide exactly with the microscope axis.

The successful use of small source lamps depends upon the goodness of the lamp condenser, but very few lamp condensers are good enough for their job and this fact has caused doubt to be shed upon the usefulness of Köler illumination for critical work.

If we consider the source to be a mathematical point, the condenser must be perfect in correction for spherical aberration; if it is deficient the results will be as shown in Fig. 4.

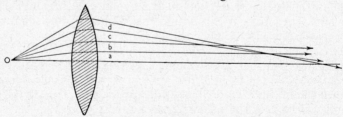

Fig. 4. Transmission of rays from a point source through a lens with spherical aberration used as a parallelizer.

The pencils of light from the source are not rendered parallel by all the zones of the lens, therefore when the lens is looked at from the right-hand side of the diagram it shows rings of dark and light areas which change in distribution as the lamp condenser is focussed. If the lens were perfect all rays passing through all zones would be made parallel, and the lens would appear as an even disk of light when viewed along its axis from the right-hand side. If the source is larger than a point, no lens can produce parallel rays from it. (Fig. 5.)

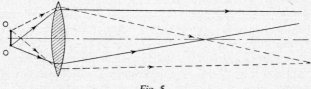

Fig. 5.

The bunched filament lamp is small enough to be considered to be a point, and the ordinary single lens and the twin lens lantern condenser are not sufficiently well corrected for spherical aberration to be used with this source (see section 'Lamp Condensers', page 137). Although the filament of a lamp is larger than a point, individual elements of it, such as each turn of the coiled filament, are very small and it cannot be considered to be an extended source in the sense used in 'Large Area Illuminants', in which a diffuser is meant. If now a poor lamp condenser is fitted, an intricate, clearly visible distorted structure of the filament will be seen in the lamp condenser instead of an even disk of light and in the best lamp condensers this cannot be eliminated entirely, but a good one will provide parallel light from the centre part of the filament and so provide an even disk of light in the lamp condenser when looked at from the direction of the microscope. The beam will be somewhat divergent when taken as a whole, as shown in Fig. 5, because of the length of the filament, this rarely matters but is a loss of light.

The accuracy of a lamp condenser can be checked very easily. It should be held above a full stop in this print, i.e. a spot about 0·5 mm. diameter, and should be raised until the spot is at its focus and exactly on the axis of the lens. If the lens is a good one a spot of this size should fill 9/10 of the aperture of a condenser of focal length about 25 mm. and diameter about 30 mm. with an even patch of blackness. If this experiment be tried with an old-fashioned bulls-eye condenser it will be seen immediately that the spot cannot fill one quarter of the lens aperture. The amount of the aperture filled by such a spot in this test is the aplanatic portion of the aperture of the condenser. The longer the focus of the lens, i.e. the lower its 'F' ratio, the easier does it satisfy this test but the less light it will gather. A bunched filament bulb requires a condenser approximately of the dimensions mentioned above.

Low Voltage Filament Lamps
In order to be able to run a tungsten filament at high temperature and so obtain a lot of light it must be small and thick to support itself, also if tight coiled, it radiates heat away from itself less quickly than if it is spread out and long. Because it is short and sturdy it must be supplied with a large current at a low voltage to raise its temperature sufficiently.

The best and cheapest bulb of this kind is a motor car headlamp. These work at 6 or 12 volts and consume 20 to 60 watts of power, the current therefore is, in the 60-watt 12-volt example,

$$\frac{\text{Watts (60)}}{\text{Voltage (12)}} = 5 \text{ amps.}$$

If the voltage were 6, the resistance of the lamp would have to be reduced so that 10 amps would be consumed to obtain the same power dissipation. The lamps will work on alternating or direct current and are usually worked from a step-down transformer, any transformer of the correct current and voltage ratings will do.

Headlamp bulbs are very robust and may be run at a few volts higher pressure than their rated amount, their life will be shortened but they are cheap, and under over-run conditions will give light of adquate brightness even for projection work.

The brightness may be controlled by means of a resistance in the primary side of the transformer. It may be assumed that the current is transformed inversely as the voltage, therefore if the lamp is consuming 5 amps at 12 volts (=60 watts) the mains is supplying about 60 W plus a little (say 10 per cent) for losses due to heat in the windings. If the mains voltage is 240, then the power in the primary winding is 240 × current = 60 watts, or the current from the mains = $\frac{1}{4}$ amp. The resistance used to control the current in the primary must be capable of carrying $\frac{1}{4}$ amp which is experienced when there is only a small amount of it in circuit. The value of the resistance may be obtained as follows:

Ignoring such matters as Power Factor (the resistance figure required is for a variable resistance),

$$\text{Current} = \frac{\text{Voltage}}{\text{Resistance}}$$

thus $\frac{1}{4}$ amp $= \frac{240}{R}$ and R = 960 ohms.

This is the apparent resistance offered by the transformer with lamp attached, to the mains. If the current through the lamp is reduced to about one half, its brightness is reduced to something like a quarter and this range of adjustment is adequate for microscopical purposes, therefore a variable resistance of about 1,000 ohms total resistance, and capable of carrying $\frac{1}{4}$ amp without undue heating is required, and will give adequate control of brightness. The resistance is connected in series with the primary of the transformer, so doubling the apparent resistance of it and approximately halving the current.

The resistance wastes (current)$^2 \times$ resistance, watts, so if we assume that at a setting of say 500 ohms, $\frac{1}{8}$ amp of current is flowing, the loss is

$$\frac{1}{64} \times 500 = 8 \text{ watts nearly.}$$

This is small compared with the power dissipated in the lamp.

No attempt has been made to incorporate all the factors in the above calculations, but in practice they will be found to be a reliable guide and quite sufficient to enable a correct choice of variable resistor to be made.

The resistance can be placed in the lamp side of the transformer but this is usually impracticable because high current low value resistances are difficult to make variable. The resistance of the lamp

referred to above is only $\dfrac{12}{5} = 2\frac{2}{5}$ ohms, therefore the resistance would

have to be variable between about 0 and 3 ohms and capable of carrying 5 amps.

Other Small Filament Lamps

A list only of these will be given because they all have the same characteristics as the headlamp described above.

The projector lamp is available up to 150 watts to work at 12 volts and its bulb is 25 mm. diameter and cylindrical. The filament is a straight coil as opposed to a coiled coil and its life is several hundred hours.

The mains projector lamp works on 250 volts supply without a transformer, its filament is long and its life about 50 hours. It is not suitable for microscopy.

Each manufacturer of lamps known as the 'Research' type provides his own bunched filament lamp and transformer, often at a high price, usually suitable for the job, but the condensers fitted to these lamps are often not good enough and should be checked with the spot test described above.

The Ribbon Filament Lamp

This consists of a filament of tungsten about 2 mm. wide and 6 mm. long and being of the strip or ribbon form it may be rigidly supported and operated at a much higher temperature without fear of disintegration. Of course, it carries a very large current of about 20 amps at a pressure of 6 to 12 volts. The filament is mounted in a cylindrical bulb about 25 mm. diameter and 100 mm. long. This lamp is the best of the small sources, it approaches the carbon arc in brightness when working under full pressure, is structureless, is large enough to give an even disk of light in the lamp condenser, and may be controlled in brightness by electrical means. The lamp has a life of several hundred hours which may be extended indefinitely by arranging the circuit so that the lamp is brought on and off gradually. A resistance in series with the transformer primary winding or a Variac transformer are the best devices for brightness control. The lamp is rarely required to work at full brightness and may be used at lower current to set up for, say photomicrography and the exposure made at full current thus reducing the exposure time and prolonging the life of the lamp.

The requirements as to lamp housing, condenser and diaphragm are as for other small source lamps but if the lamp condenser is one of short focus and therefore near the lamp it is likely that a disk of clear quartz may be required to keep some of the heat off the lens. Heat absorbing glass will not work in this position because it absorbs too much heat and will burst. Generally lamp condensers will withstand the heat of a ribbon filament lamp so long as the apparatus is warmed up gradually and all parts are in place for the whole of the warming up period. The iris diaphragm must not be oiled or it will cement solid after an hour's use. The condenser lenses must be loose in their mounts to allow plenty of space for expansion. Colour filters must be on the microscope side of the lamp condensers as they will not withstand direct heat from small sources and arc lamps when these dissipate more than about 30 watts.

The Carbon Arc

This is still the brightest artificial light and must be used for microprojection or high powers. It is rarely required for any other kind of microscopy, a possible exception being ultra-microscopy where visibility of the specimen is a function of brightness only.

Arcs are a nuisance to work. The luminous arc is struck between rods of carbon about 4 mm. diameter (they can be almost any size up to 25 mm.), the carbon now used is gas carbon, a hard product from the retorts of gas works. The voltage required is about 50 and an arc will work on a.c. and d.c. though the d.c. arc is the steadier. It may be open to the atmosphere or enclosed in a non-evacuated bulb and several amperes must pass in order to produce enough heat to sustain the discharge. The carbon rods burn away as the current flows, at a rate of about 12 mm. per hour according to the power and type of the lamp. A clockwork or automatic device must be incorporated to keep the carbons approximately the same distance from each other or the arc will elongate and finally fail. Hand control is possible but scarcely practicable.

In the d.c. arc a crater forms at the negative carbon where electrons and other materials leave and pass across the arc to the positive carbon. This crater is the steadiest source of light in the arc and the geometry of the lamp is usually arranged so that the crater faces the lamp condenser, the crater is then the source. The arc is 'struck' by allowing momentary contact of the carbons with each other.

Electrical Requirements of an Arc

The 50 V d.c. required by the arc is difficult to obtain and requires rectifying apparatus of considerable current carrying capacity. The easiest way of arranging this is to use a rotary converter, that is an a.c. motor driving a d.c. generator, or both windings may be on the same rotor making one machine. The current can be controlled by means of a resistance in the d.c. side. If d.c. mains are available the voltage can be dropped directly by means of a resistance in series with the arc. An alternative method is a bank of motor car starter batteries which can be left on trickle charge from a low current rectifier. The only rectifier which will deliver more than 5 amps necessary to work the arc direct is the mercury arc type and this is not readily available, but at least the direct current used in an arc lamp need not be well smoothed.

The a.c. arc is quite practicable but is not steady. The brightness is nearly the same as the d.c. arc but no crater forms and there is no good focal point for the condenser, but it can be used satisfactorily for microprojection when an operator is always present.

The apparatus for controlling the advance of the carbons is an integral part of the lamp. Some devices depend upon the action of the lamp current flowing through a solenoid where increasing current withdraws the rod and failing current allows them (or one) to approach by gravity, others are straightforward constant-speed clockwork. There is no particular advantage in either form and both require hand adjustment periodically.

Optical Requirements

The arc is a small source and must have a lamp condenser, and if the condenser is several centimetres away from the arc no trouble will follow, but if a high aperture lamp condenser about 50 mm. diameter is used about 25 mm. from the arc it will soon become pitted and opaque. Such a condenser is necessary in order to obtain the best results in microprojection and the difficulty can be overcome by placing a disk of clear quartz between the arc and the lamp condense. . Quartz has a very low coefficient of expansion and very high fusing temperature so it will not burst under the heat and will not pit when exposed to the products of the arc at close range.

The size of the source cannot be defined accurately but in the usual small 5 amp arc may be taken as a 3 mm. diameter disk. This is comparatively large for a small source type lamp and the ordinary twin lens lantern condenser will be filled from this source and may be used in the Köler manner.

The carbon arc emits a mixed spectrum but it is mainly of the continuous kind. A line spectrum is present but the lines cannot be isolated in the way possible in the mercury arc discharge.

Monochromators

Under this heading are treated all the useful devices for obtaining light mainly of one colour. In microscopy it is rarely necessary to obtain a particular wavelength of light entirely free from other wavelengths, for so long as the image is formed in light of mainly one colour, amounts up to say 10 per cent of other colours rarely matter, therefore monochromators for microscopy seldom need to be elaborate.

Dispersion Types

In this type of monochromator either a prism or diffraction grating may be employed. When prisms are used, a loss of light of up to 10 per cent must be expected at each glass-to-air surface due to reflection: the amount of light lost in a diffraction grating is similar, but depends upon which order of spectrum is used. When the first order spectrum is employed from which to extract the wanted colour, the overall loss of light is equal to that of about three flint glass prisms in series giving an equal dispersion. As the dispersion is increased, the loss becomes greater when the grating is used than when prisms are employed; also, diffraction gratings, either original or replicas, are expensive, while prisms may be bought cheaply as Services' surplus. Prisms of dense glass cause less loss of light because fewer air-to-glass surfaces are required for a given dispersion, but in practice any glass will do. A prism monochromator is cheaper to build and easier to maintain than the grating type, mainly because its surfaces may be cleaned easily and all parts are strong and will withstand clamping to a base-board.

The Absorption Type
These consist simply of filters, the function of which is to absorb light of unwanted colour and transmit that required. In general, several filters are required in series in order to obtain the right pass band, and the manufacturers publish graphs showing the characteristics of their filters. Usually, however, laboratory workers do not know the characteristics of the odd filters which appear in microscopical apparatus and it becomes necessary to make a test with a spectroscope. For this a bright white source of light, say a 100-watt bulb, is used as the illuminator to give a continuous spectrum, and the filters in question are placed in front of the spectroscope slit, then the pass bands of the filters appear in place of the continuous spectrum of the lamp. For use as a simple calibration, a mercury lamp may be used superimposed upon the white light, when its known lines will be seen superimposed on the pass band of the filter.

The advantages of the absorption type over the dispersion are as follows:

(1) The geometrical characteristics of the light source as seen by the microscope are not changed by the monochromator. A filter may be placed in front of any lamp and will change only the colour of the light, and any system of illumination which may be in use remains unchanged.
(2) They are simple and cheap and can be (if optically worked) mounted anywhere in the illuminating system except close to the lamp bulb where there is danger of them cracking or changing their characteristics in the heat.

The disadvantages are several:

(1) All filters pass colours other than the required pass band, e.g. all high frequency filters pass red as well.
(2) In order to obtain a narrow band of colour, a number of filters, probably four, must be used in series and this entails considerable absorption of light.
(3) It is difficult to select filters to give the exact pass band required. (See 'Colour Filters and their Uses', page 193.)

The Prism Monochromator

This apparatus is in most ways like a spectrometer but is not designed, as in the case of a spectrometer, to give good definition with a narrow slit. It is designed to transmit a large quantity of light from a wide slit and has great dispersion so that a large slit can be used without overlap of the images of the slit in various colours taking place. It follows that the lens system can be comparatively crude.

The requirements in a prism monochromator are as follows:

(1) The system must present to the microscope an image of the source which can be focussed upon the object either in the Köler or the Nelson manner; for example, a disk of light in the lamp condenser is a suitable source for the substage condenser to image upon the object, but a narrow wire filament of an electric bulb is not.

(2) The output end of the apparatus must be fixed so that when the prisms or other part of the monochromator are rotated to change the colour of the light, the microscopical set-up remains unchanged.

(3) The output image must be large enough when focussed by the substage condenser upon the specimen to illuminate at least one-third of the field of view of the microscope.

Point (1) is the most difficult to arrange and is also the most important, for unless the microscope can be set up 'critically' there is little use in employing monochromatic light. One way of obtaining the correct result is by employing large prisms as shown in Fig. 6 and a gas discharge source which produces a line spectrum.

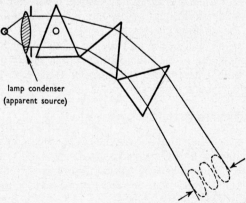

lamp condenser
(apparent source)

images of lamp condenser in colours of lines in source
(They are rendered oval by the action of the prisms)

Fig. 6.

When the dispersion is sufficient, i.e. four or more prisms with 75 mm. faces used at the position of minimum deviation, images of the lamp condenser will be obtained clearly separated from each other in the colours of the lines of the source. The high pressure mercury arc source is recommended because it is bright and has well-spaced lines of orange, green and blue, but if a solid source is used the images are merged into a continuous band and the result is useless. The prisms may be of low quality.

To present to the microscope images of the lamp condenser in the colours of the source, the whole table carrying the apparatus is rotated about on axis 0.

The system will work equally well without the lamp condenser; an ordinary image of the source will be seen in colour through the prisms and may be used in the ordinary way as Nelsonian illumination. A more refined method is shown in Fig. 7.

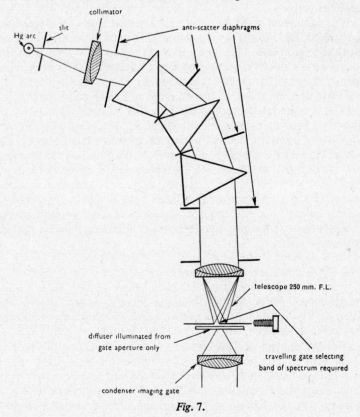

Fig. 7.

The diagram is self-explanatory, the apparatus being practically a spectroscope. If the dispersion is high, say through four prisms, a slit width of 5 mm. may be used, such a slit allowing plenty of light to enter the apparatus forming an image about 5 mm. wide on the gate. The diffuser is necessary on the prism side of the gate because were it not present, the gate condenser would form an image of some internal parts of the monochromator unsuitable as a substage condenser image for microscope illumination. The diffuser is illuminated only by pure colour which enters it through the gate; it then becomes a definite source and Köler illumination may be employed. Either the gate may be made to travel across the spectrum or the spectrum across the gate, but it is better to move the spectrum across the gate to obtain the colours required because then the gate is not moved relative to the microscope. The diffuser should be one piece of fine ground glass—500 carborundum is a suitable size particle for grinding glass screens.

When a narrow bright beam of light for, say, a vertical illuminator is required, the diffuser may be removed to permit more light to pass into the microscope. The appearance of the source as seen from the microscope is of less importance when a vertical illuminator is in use mainly because the light is scattered by the specimen to fill the objective and the shape of the source is not clearly apparent in the field. It is quite satisfactory to use a standard spectroscope set up as above to provide monochromatic light for a vertical illuminator; if it is the constant deviation kind of spectroscope, so much the better, because the prisms can be revolved on their table to transmit any part of the spectrum into the spectroscope telescope without disturbing the arrangement of telescope, eyepiece and vertical illuminator. (Plate 3.) A spectrometer is not provided with large enough lenses, prisms and dispersion to produce a large enough source image for use with transparent specimens illuminated with a normal substage condenser.

The Diffraction Grating Monochromator
A grating may be of the transmission or the reflection kind, but for microscopical work the transmission kind is easier to use. For effective use in a monochromator, a grating with lines ruled about 400 to the millimetre, about 50 mm. long, is necessary, and even as a replica this is a costly instrument. It is set up according to Fig. 8.

As with all monochromators, the concentration of light energy into narrow bands in a line source gives the best results so long as its colours are those required. The above systems will work well with gaseous or solid sources, but the amount of light is much reduced

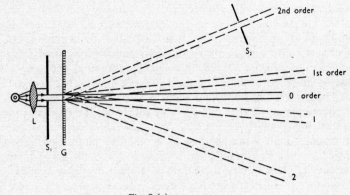

Fig. 8 (*a*).

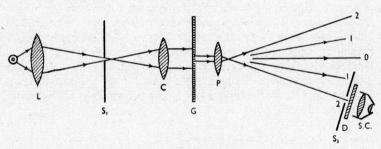

Fig. 8 (*b*).

L. Lamp condenser.	S_1. 1st slit.
C. Collimator.	G. Transmission grating.
P. Projector lens.	S_2. 2nd (selector) slit
D. Diffuser.	S.C. Substage condenser.

0, 1, 2.   Orders of spectra.

when a narrow gate is used with a continuous spectrum, naturally a wide gate transmits more light but also a band of colour.

When near-parallel light passes through a grating it is divided into a portion which passes straight through on the axis, known as the zero order component, and other orders of spectra spread on each side of the zero order, diminishing in brightness but increasing in dispersion with angle from the axis. From the grating suggested, two orders of spectra would be visible, and probably the first order would be used for microscope illumination. Having applied the light to the grating, it only remains to gate out from one of the orders, the colour required within the order, and apply it to the microscope.

In Fig. 8a a simple arrangement is shown which will work with a grating of large disperson, say 600 lines to the millimetre and about

50 mm. long, used with a line source. This system provides an image of the lamp condenser for use in the microscope. System Fig. 8b shows an arrangement for use with a grating of lower dispersion where narrower slits S are required to remove unwanted colours. To be fully effective with critical light, a diffuser is necessary after gate S_2 which, as in the prism form, then becomes a real source and may be used in the Nelsonian way.

There appears to be a considerable future for the Lipmann interference filter, which consists of layers of silver deposited photographically within the gelatine of a sensitive plate. To make the device, a very fine grain plate is chosen and a reflecting layer of mercury is placed upon the emulsion surface. Light is applied and standing waves of interference are set up between the glass and the mercury, and these affect photographically the emulsion. The plate is then developed and layers of silver remain in the emulsion where the antinodes of the interference wave pattern were situated. The whole thus becomes an interference filter of great efficiency and of resolution sufficient for monochromator purposes. If the plate is studied by taking vertical microscopical sections, they show the closely spaced layers of silver in the emulsion quite clearly, thus a satisfactory reflection interference grating is formed by the partially reflecting silver layers.

General Notes on Monochromators

It will be gathered from the text that the main difficulty with dispersion type monochromators is, that unless they have very large prisms and are of the form shown in Fig. 7, it is likely they will prevent the microscope being set up properly. It is of no use using monochromatic light to improve resolution or to increase contrast in photographs if the operator has to sacrifice accuracy of microscopical set-up. When a monochromator is in use exactly the same conditions apply as when a normal lamp is employed, i.e. an image of a source must be projected upon the specimen by the substage condenser and the objective aperture must be evenly filled with light up to the aperture required for the work. The source image must be of such a size that the object can be clearly seen against it.

If a monochromator is required for critical work, the large prism type used in either the Köler or the Nelsonian manner (Fig. 7) is strongly recommended. There is then no difficulty in focussing an image of the source upon the object in the simplest way. If large solid prisms are difficult to obtain, liquid ones can be made up with glass sides cemented with Araldite.[1]

[1] Aero Research Ltd., Duxford, Cambridge.

Lamp Condensers

Although most of the important points concerning the optical quality of lamp condensers have been covered in the sections describing various lamps, a few points remain outstanding. .

The old bulls-eye condenser which was used to increase the apparent size of the flame of a paraffin lamp is of no use at all in modern microscopy using electric lamps; it was a poor thing to use even with a paraffin flame which is nearly the size of the bulls-eye. Any single lens lamp condenser, with the possible exception of the new parabolic kind, is insufficiently well corrected for use with a modern small source lamp, while there is, of course, no point in using a lamp condenser with a large source lamp.

Several possible lens combinations can be used successfully, the first and cheapest being the standard lantern condenser, Fig. 9.

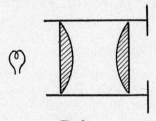

Fig. 9.

This is something after the style of a Ramsden eyepiece and consists of two plano-convex lenses of focal length about 75 mm., and diameter about the same, spaced a distance of half the focal length of either, measured from the plane surfaces. No diaphragms are required between the lenses but an iris diaphragm must be present outside the combination to limit the size of the apparent source when seen in the microscope. The lenses of this condenser can be of poor optical quality and may even be of the cast type. The combination will provide a good even disk of light for the microscope of size about 75 per cent of its physical diameter when a small source, the size of a high pressure mercury arc, is used. The best physical size of condenser is about 75 mm., which will then give a solid illuminating disk about 50 mm. in diameter. An iris diaphragm to control a beam of this size is not difficult to obtain and the outer 25 mm. of possible aperture may be sacrificed.

These combinations are comparatively common, costing little, but the test for spherical correction must always be applied (see 'Microscope Lamps: Point Sources', page 124). The size of the spot used in this test should be at most the size of the source to be used, and if the condenser will pass the test with a smaller spot, so much the better.

The lantern condenser may be built up in the laboratory from odd lenses. The spacing of the lenses is not critical and they may be made of any kind of optical glass. Spacing may be arranged by trial using the spot test referred to above.

Specially Computed Lamp Condensers

Two of the main types are shown in Fig. 10 and are comparatively

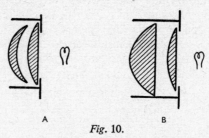

A   B

Fig. 10.

simple combinations capable of giving very good results. Both are usually about 50 mm. in diameter and with the spot test performed upon a full stop in this print will provide an even disk up to 90 per cent of their physical aperture. The usual focal length is about 25 mm. and both are obtainable from Dalmeyer and others. All good lamp condensers are asymmetrical and care must be taken to have the source parallelized on the correct side of the condenser as shown. Special lamp condensers are very expensive to purchase and there is no clearly apparent reason for this, as all except type B, Fig. 10, are spherical curves of a simple kind on common optical glass.

The parabolic kind of condenser is capable of bearing a very high aperture, i.e. it may have a diameter of 50 mm. and focal length of 12 mm. or so, and still pass the spot test. However, as there is little point in making a condenser of this focal length because it cannot be got sufficiently close to most small sources, a focal length of 25 mm. is found to be the best practical value.

Lamp Condensers: General Matters

There is no need whatsoever for a lamp condenser to be achromatic. In construction they must never be of any cemented kind because the cement will not withstand the heat of the source. For a lamp used with standard microscope substage condensers, a lamp condenser size of 35 to 50 mm. is ideal, because this allows the substage condenser (or objective in the case of vertical illumination) to project a reasonable sized image into the microscope field of view. It is convenient to place the lamp condenser near the lamp so as to collect as much light as possible and to allow an easier construction of lamp housing to be employed. The component lenses must be loosely fitted in their cells to allow for expansion with heat from the source, and all components must be accessible for cleaning on both sides. It will be found that a hot lamp condenser traps a surprising amount of foreign material on its surfaces both inside and outside.

Most mechanical considerations have been mentioned but it may be repeated that a lamp condenser used with a small source must have centring arrangements, focussing controls, and must be squared-on to the optic axis of lamp and microscope. In practice it is usually easier to centre the lamp to the condenser than vice versa.

Camera lenses do not make good lamp condensers because their apertures are low and focal lengths long.

Substage Lamps

Microscopes are now made with lamps built into the body. When these lamps are placed where they cannot heat the specimen, designed to be used in the Nelsonian and Köhler way they are satisfactory. Vertical illuminators with a built-in lamp are practical because one does not need a focussing stage. See page 465.

Microscopists employed on critical work should look to their lamp condensers because these and other kinds of faulty illumination cause many artifacts.

The direct test of the lamp condenser and substage condenser together, described under the heading 'How to set up a Microscope', should always be the final test of quality. If the observer looks carefully through a piece of really dark glass, straight at his illuminated lamp condenser, he will be able to form a fair idea of its ability to provide an even disk of light from the particular source present simply by looking at it. But it is useless to do this unless the eye is completely accommodated to the light. A dazzled eye sees all condensers as good ones.

It must be possible to focus a lamp condenser so that an image of the source to be used can be formed at all distances between, say, 2 m. and 100 mm. away from the lamp housing. The physical size of the lamp bulb often limits the range in the direction of infinity be preventing the condenser being brought sufficiently close to the luminous filament, but if the above range can be obtained without the condenser touching the bulb, the equipment will be found satisfactory for microscopy.

CHAPTER 8

The Microscope Stand

THE MICROSCOPE stand consists basically of four parts, a stage upon which to place objects, a body tube adjustable to and from the stage to carry the optical system, a tube also adjustable below the stage, to carry the illuminating apparatus, and a lamp. For various purposes the positions of these parts are altered relative to each other and are refined in detail.

It is usual to adjust the body tube to focus to and from the stage by means of a 'coarse' adjustment which is now always a diagonal rack and spiral pinion. A 'fine' adjustment is also provided and is usually a lever and micrometer screw arrangement operating either on a separate slide from the coarse adjustment, or upon the nosepiece only. The stage may be the moving part.

The short tube or carrier below the stage is called the 'substage' and this is normally fitted with only a rack and pinion coarse adjustment, but fine adjustments on substages are still to be seen.

The 'nosepiece' is the lower end of the body tube which carries the objective and any changing apparatus there may be.

The 'draw-tube' slides within the body tube and is used to adjust the length of the microscope.

The various specially-designed stands will be explained in the sections following, but there are some basic requirements common to all stands and these are listed below:

(1) The body tube should be capable of being moved along the axis of the microscope accurately, so that when a 2 mm. objective is in use, no swaying of the image from side to side is apparent when either coarse or fine adjustments are used. The test with the coarse adjustment must be applied with care because of its speed of action.

(2) When using a 2 mm. objective the image should not sway when the hands are placed normally on the coarse or fine adjustment controls. Some movement is inevitable as all structures are elastic, but it should never be so great as to be noticeable.

(3) It should be possible to operate the mechanical stage without introducing any movement except that properly transmitted

through the mechanism, i.e. the image should not move when the hands are placed normally on the stage controls. Loss of way in the rack and pinion and spiral screw should not be more than about two degrees of arc and no swaying of the image about the line of traverse should be apparent with a magnification of 1,000 times. There should be no need to re-focus a 4 mm. objective used on a flat object when the direction of movement of the stage is reversed. It was usual to make stages as good as this fifty years ago and there is no excuse for the numerous bad stages at present in use.

(4) If the microscope has a rotating stage it should be possible to rotate an object between the limits of rotation without apparent change of focus and without the object leaving the field of view when a 2 mm. objective is in use. The stage should have a separate clamping ring for locking the rotation: the clamp should not be applied to the bearing or inaccurate rotation will soon result, but it is permissible to apply the clamp to the spindle of the actuating milled head if one is fitted. The gear often causes shake in the rotation.

(5) The substage should be mounted upon a rack and pinion driven slide at least as strong as that of the coarse adjustment. The movement of the substage need not be so accurate as that of the body tube but it must be strong to withstand the changing of relatively heavy apparatus which often does not fit well. Its axial movement should not be less than 50 mm. and should be nearer 75. Whatever changer slides, etc., are mounted upon it, it should have a clear aperture of 40 mm. diameter available, as this is the standard substage gauge diameter and although much apparatus is now mounted upon changer slides, it is desirable to be able to mount standard apparatus. If condenser and objective changer slides are fitted, it is not really necessary to have milled head centring screws to the substage, but a pre-set centring device must be accessible.

(6) There is no necessity for a fine adjustment on the substage unless it is intended to adapt the instrument as an inverted microscope, when the substage carries the body tube and should have a fine adjustment.

(7) The microscope fine adjustment of focus must satisfy several conflicting requirements:

(a) It must provide the lightest and most delicate movement of the microscope to and from the specimen and be for practical purposes entirely free of backlash.

(b) It must remain stable in adjustment while photographs are being taken.

(c) It must be firm and positive in movement.

(d) It must not be resistant to movement due to grease.

In some microscopes the nosepiece only is moved by a lever, the nosepiece inner tube being lightly sprung down on to the lever, thus eliminating backlash. On the whole, this is the best system, because the body tube coarse adjustment only, carries the weight of the tubes and possibly cameras, and this is a substantial part of the microscope. The fine adjustment part is likely to remain stable because it is not loaded.

In designs where the whole body is moved, a very strong and robust fine adjustment is both permissible and necessary. This type usually gives satisfactory service but lacks the delicacy of the nosepiece type; the whole body and possibly a camera is carried by the fine adjustment, and to prevent changes of focus due to settlement of the instrument and vibration, strong control springs and heavy levers are used. It is usual to find some backlash in the fine adjustment of such instruments.

Some fine adjustments of all kinds are poor when compared with their contemporaries, but on the whole the nosepiece type is best for experienced workers and the body-slide type for students and general laboratory work. It must be remembered that the lighter springs of the nosepiece type probably will not break a slide should the objective be brought too near, but the strong springs fitted for the body type will certainly break the slide and probably push in the objective front lens as well. Some objectives now have spring loaded fronts for safety.

In the past, microscope stands were made of brass and were beautifully finished. Brass, however, is not the best material to use because it is insufficiently stiff and parts of instruments which have to stay set to within fine limits should be made of steel or cast iron because of the much greater resistance to flexure. The modulus of elasticity for brass, cast iron and steel are 10 to 12×10^6, 12 to 14×10^6 and 29×10^6 respectively, therefore it will be seen immediately that steel is nearly three times as stiff as brass for the same cross sectional area. Steel is usually used in these days for the limbs of microscopes and it is to be hoped that stainless steel will appear for use in some moving parts as soon as a satisfactory machining method is worked out.

The old-fashioned built-up microscope foot is no longer seen, and the manufacturers have standardized the jug handle limb (A) for the ordinary microscope and the jib lim (B) for the focussing stage kind. Fig. 1, A and B.

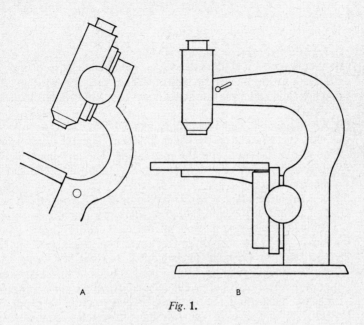

A   B

Fig. 1.

Both kinds are entirely satisfactory when made of sufficiently heavy iron or steel, and when choosing them the remarks concerning stability should be re-read. In general, the jib type limb is more stable than the jug handle kind but it must have stage focussing arrangements of the top class. With this form of stand the hands rest on the table while operating the controls, and this is a much easier position than up on the limb where the maximum amount of tremor is transmitted to the instrument.

Fortunately, the tripod foot has passed out of use for laboratory instruments. Such a device was not as stable as it appeared, because when the instrument was inclined it tended to pivot on the back leg, while if the legs were wide spread and the instrument body carried well forward to offset this trouble, it had to be placed so far away from the edge of the laboratory bench that the eye could not be applied to the eyepiece. The usual result was that the worker pulled the instrument towards him and the rear leg slipped off the bench, leaving him nursing the stand in his lap if he was lucky. Another disadvantage of the tripod foot is that the substage is inaccessible. The microscope foot should be of broad flat horse-shoe form, as indeed they usually are today.

Microscope Stages

These may be of the following kinds:

(1) A plain flat table or plain stage.

(2) A device which slides as a whole upon bearings in X and Y directions controlled by milled heads, carrying with it any object laid upon the surface, known as a 'Built-in mechanical stage'.

(3) An attachable skeleton arrangement which clamps on to a plain stage and has arms which hold and push the object slide about on the surface of the plain stage. The movement is by rack and pinion or fast screws, and is called an 'Attachable mechanical stage'.

(4) A micrometer stage which is very similar to (2) but which has a much longer travel in both directions and is more accurately controlled by slow motion micrometers (the ordinary stage moves somewhat rapidly). This stage is a special one and is not fitted to standard microscopes, its main use being the measurement of ray tracks and the positions of star images on photographic plates often carried on large apparatus. A simple design illustrating the method of construction is given in Fig. 2a, and one of the simpler commercial examples in Fig. 2b.

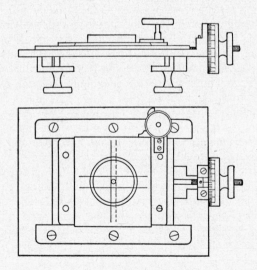

Fig. 2 (*a*).

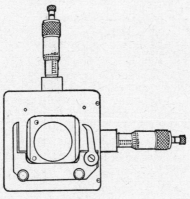

Fig. 2 (*b*).

(5) The Petri-dish stage as its name implies is large and meant to hold a Petri-dish in such a way that it can be rotated, thus allowing the whole dish to be searched whilst demanding only sufficient Y direction movement to allow the dish to be moved along a radius. Fig. 3.

The dish is searched along a circular track by rotating it while resting against the vertical rollers A. To search another track the stage is moved the required amount up or down, i.e. in the Y direction. Its structure is usually a simple plate of thin metal which can be clamped on to any microscope stage. The amount of the dish which can be searched is limited by the space between the front of the microscope limb and its optic axis, and if this distance is 75 mm. or more, any ordinary Petri-dish can be searched completely.

(6) Rotating stages may be simple or may have the addition of X and Y movements, but in modern work it is essential that the rotation be in the optic axis of the microscope and not of the old-fashioned kind where the axis of rotation of an additional top plate to the stage was moved with the X and Y adjustments. A means of centring the rotation of the stage to the microscope axis, or of the microscope axis to the stage, must be available. (See 'To Line up a Microscope for Critical Work', page 369.) It is useful to have a clamp to the rotation but this must be of a kind that does not press on the stage bearing. It may either grip a separate ring attached to the stage or may be applied to the milled head bearing if the stage has a gear drive. Any rotating stage should be sufficiently well

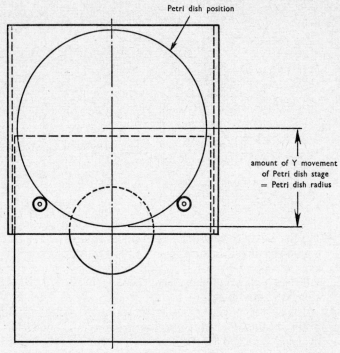

Petri dish position

amount of Y movement
of Petri dish stage
= Petri dish radius

Fig. 3.

fitted to allow an object to be rotated under a 2 mm. objective and ×10 eyepiece without leaving the field of view and without causing the object to go out of focus by more than a fraction of a turn of the fine adjustment milled head. Unless the instrument is designed to have the stage always horizontal, there should be no movement of the image away from the centre of rotation of the field when the microscope is inclined and the stage rotated.

Choice of Stage

In general, the stage and microscope stand must be taken as a whole, but the form of the stage may be the most important factor in determining the design of the rest of the instrument.

The special requirements of the metallurgical microscope are dealt with under the heading 'Metallurgical Stands' so the following remarks refer mainly to the ordinary or general purpose stage.

There are certain limits to the size of a fully mechanical built-in

stage; for instance, it must not be so large that the limb of the microscope has to lean out many centimetres to bring the axis of the tube
over the centre of the stage. Some special microscopes are built this
way, usually with plain stages, their purpose being to allow large
sections of materials like coal and brains to be studied, but such
instruments always show flexure and are unsuitable for high power
work. The stage must, however, be of such a size that a slide can be
searched over an area of 50×25 mm. without mechanical parts of the
stage fouling the substage condenser. In general, these requirements
demand a stage about 100×75 mm. in size, with an aperture 50 mm.
diameter. The top plate of any mechanical stage must be thin so
that it does not foul the condenser top lens. Fig. 4, A and B.

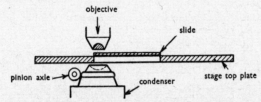

Fig. 4 (a). Bad Layout of Stage. Pinion axle or similar part fouling
condenser. Stage top plate too thick to allow condenser to come to
focus at sides of aperture of stage.

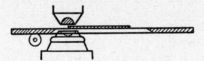

Fig. 4 (b). Correct Layout of Stage.

The state of affairs in A may be considered rather obvious but in
fact such a layout where the specimen cannot be critically illuminated
anywhere except at its centre is common.

The clearance demanded for the substage condenser usually causes
the Y motion rack and pinion to be placed at the side of the stage
aperture, but in this position it causes thrust on the dovetail bearings.
Two racks and pinions can rarely be arranged because the axle then
has to pass across the stage near the aperture, so in practice the
stage is made practically square and the thrust can be tolerated so
long as the slide bearings are good and properly lubricated with
grease.

The X motion is easy to arrange and should be of the centre pull

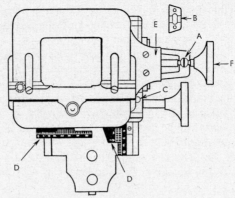

Fig. 5. A. Ball socket bearing and cap B.
         C. Tightness adjustment for Y movement spindle.
         D. Verniers.
         E. Stage top plate.
         F. X movement milled head operating top plate F with
             centre-pull screw.

type with the screw supported on an outrigger, Fig. 5. Stages with
top plates propelled in the X direction by off-set screws or pinions
should be rejected for laboratory work because they will not stand
up to the thrust from such an arrangement. (It will be noted that the
thrust produced by the Y motion mechanism is resisted by a much
heavier piece of metal than the stage top plate.)

On the whole the built-in mechanical stage is the best for the
following reasons:

(1) Any object can be secured properly to the stage by any means
    available and it will be carried in X and Y directions and
    rotated by the stage mechanism, there being no haphazard
    sliding of the glass slide upon the surface of the stage.

(2) When the underside of the slide is wet with oil from an oil
    immersion condenser, the oil does not contaminate the stage
    by the slide being pushed across it.

(3) The structure is mechanically much stronger than any attach-
    able stage and will wear much longer on microscopes which
    are continuously used.

(4) Technical instruments like stage vices can readily be fitted and
    given the stage movements.

Attachable stages are devices which clamp upon an ordinary flat
plain stage and move the object slide about, the object slide usually
resting upon the surface of the plain stage and sometimes being

sprung down upon it by means of spring clips. Such attachments are quite satisfactory and allow a very long range of movement because there is no moving stage structure to foul the condenser and no limited stage top plate aperture. The main disadvantages are that all objects must be mounted on standard slides or upon objects of similar shape, and oil from the condenser is likely to be dragged between the stage and slide, thus gumming up the works. The Author introduced a small modification, illustrated in Fig. 6, to overcome this difficulty.

An auxiliary stage carriage a little larger than a standard slide of plan view shown in Fig. 6 is attached to or held in the arms of the auxiliary stage after the manner of an ordinary slide. A light spring on the auxiliary stage presses the carriage into contact with the surface of the stage. On the under surface at each end of the carriage are strips of metal about 1 mm. thick, which space it away from the stage surface so preventing any oil which may find its way on to the underside from being spread about. Brass is a suitable material to bear on the ordinary stage surface and makes a fairly easy-running fitting. The electrical insulation material polytetrafluorethylene (PTFE) has very low coefficient of friction and can also be used for sliding fittings.

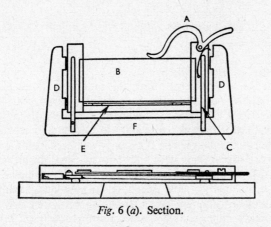

Fig. 6 (*a*). Section.

For use on rotating stages a form of attachable stage with vertically placed milled heads may be obtained, the arrangement allowing the structure to be built within a much smaller compass, thus making complete rotation of the stage possible on a microscope of normal size. In the past, the arrangement has been frowned upon but the Author has never heard any good reasons for not employing the

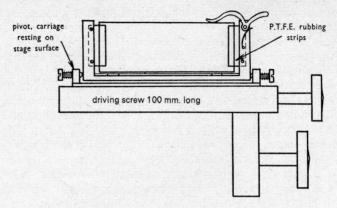

Fig. 6 (*b*). Long Range Gravity Contact Type. Useful for rough
materials (grits and carborundum).
A. Spring lever to secure glass slide to carriage.
B. Glass slide with specimen.
C. Springs fixed on auxiliary stage to press down carriage.
D. Jaws of auxiliary stage holding carriage.
E. Ledge or carriage against which rests the slide.
F. X direction slide of auxiliary stage structure.

mechanism, in fact he has found it very convenient on petrological
microscopes.

Size of Milled Heads on Stages
Some controversy exists regarding this; some workers like small
heads so that they can twiddle the controls between their fingers,
while others like the steady motion only possible with large heads.
The Author thinks that two speeds of movement are necessary on a
research microscope stage, one for low powers or finding an object,
and the other for searching with high powers. Small heads are useful
for quick traverse but are a great nuisance when high powers are in
use because sufficiently slow speed of turning and steadiness of
motion cannot be obtained. Small heads can be more easily grasped
by the soft clean finger ends of a salesman or supervisor, but not so
easily by the more usual burned, cut and hardened fingers of laboratory workers. On the whole, stage milled heads are too small and
insufficiently well knurled. Diamond knurling on stainless steel provides the most comfortable and lasting grip, but it must be well done
with sharp points. If brass is used the straight knurl will last longer
but it is not nearly so good to lay hold of. A diameter of about 25 mm.
has been found suitable for most purposes, but the effective size

250 w. compact-source, high-intensity, mercury-vapour lamp.
Approx. ×1¼.

Plate 3
Types of Microscope Lamp Bulbs.

6 v. 8-amp. spiral-
filament lamp. $\times \frac{2}{3}$.

6 v. 9-amp. ribbon-
filament lamp. $\times \frac{2}{3}$.

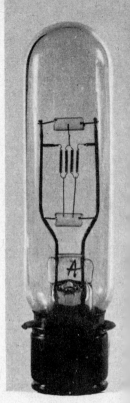

50 v. 5-amp. four-coil
grid-filament lamp. $\times \frac{2}{3}$

100 c.p. Pointolite lamp. $\times \frac{2}{3}$.

PLATE 4
Types of Microscope
Lamp Bulbs.

of the heads can be increased by boring out a large rubber bung and pressing this on to the milled heads.

A very convenient means of reducing the speed of a hand-controlled shaft and at the same time preserving its original speed for access when necessary is by employing a certain radio tuning dial reduction gear known as a Bulgin Reduction Head, obtainable from radio accessory shops. It is an epicyclic arrangement of sprung balls, about 25 mm. total diameter and about 25 mm. total length. The drive it transmits is not positive, it being a friction device, but for the same reason it is free from backlash and will transmit sufficient power to drive a stage or light substage. The fitting is a coaxial device, therefore it is necessary only to attach it to the spindle to be driven and to secure the outer ring of the structure to a solid part of the apparatus to be driven; a lug is provided for this purpose. Any knob with a 6 mm. hole will fit the input spindle of the head. A flange is left on the reduced speed side of the head to which another knob may be attached, thus presenting concentric controls for the two speeds, either of which can be used at will. In place of the second knob a drum and scale may be fitted for the purpose of measuring stage or other travel. As the drum is fixed to the side of the reduction head which is solid with the driven spindle, no inaccuracies result if the friction apparatus slips.

Adjustment of Stages

Some stages have screws placed in various positions for the purpose of tightening the dovetail slides when they become worn; others are built without this facility mainly because the adjustments are often misused and because no slide adjustment really takes the place of reconditioning the slides properly.

If a stage becomes stiff to operate it is likely that it only requires cleaning. Unscrew any stops which may be present and slide the dovetails apart, when they can be cleaned with petrol and the fit of the parts again tried. To be perfect there should be about 20 microns (just perceptible) sloppiness in the fit along the whole of the travel of the stage, and on no account must the stage plates stick during the movement. If metal parts bind when without lubrication, they will continue to bind when lubricated, and it has been found that a sloppiness of 20 microns is just sufficient to allow light silicone grease to separate the surfaces without leaving so much space that slop is still perceptible after lubrication.

If the stage slides are found to be worn, the male part must be straightened before anything else can be done because, if both mating parts are irregular, there is no standard to which to work.

Such operations are instrument makers' jobs, but if the work must be done in the laboratory, one should proceed along the following lines:

(1) If the wear is small, i.e. there is still some adjustment left on the female parts, lapping with metal polish may be attempted. Separate all parts of the stage, put metal polish between the slides to be dressed, and slide the parts together with long sweeps tightening the adjusters at intervals until an even resistance is obtained. Lapping is not a desirable process because both parts are ground and it is difficult to remove the particles from the work afterwards. However, it is an easy emergency method, and if the lapped parts are wiped clean and then polished separately with metal polish on a cloth before being lubricated and refitted, no trouble will follow. The parts may also be worked together with oil before final fitting to clear away loose matter from the rubbing surfaces. In this connection it should be remembered that the mechanical wear on a lubricated slow moving microscope is almost nothing compared with a machine.

(2) If wear is large, the male slide must be honed straight. Remove the slide, obtain a new flat India hone of medium coarseness and quality, rest the slide upon the hone and observe the angle of the dovetail. Fig. 7.

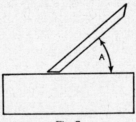

Fig. 7.

Make a block of wood to this angle A to form a guide, and either clamp the dovetail to the block or let the dovetail slide upon the block; in either method use water as a cutting collant and rub the dovetails straight. Preserve parallelism by frequent measurements with a vernier gauge. When the dovetails are straight, try the fitting again and it may be found possible to squeeze the female slide adjustments until a working fit is obtained. If this is the case, the instrument may either be left or further improved as follows:

If the gap is too large for the female adjustment to take up, a piece of copper or brass shim must be fitted, Fig. 8a, the thickness required being found by trial. It may be secured by tacking with solder,

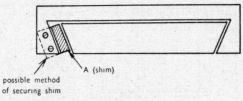

possible method
of securing shim

A (shim)

Fig. 8 (*a*).

adhesive, by set screws, or by nothing at all, depending upon the degree of permanence of the repair, but in general such packing pieces should be secured on the female member by tiny screws. It is often possible to fold the packing round the end of the slide and screw it there as indicated by the dotted lines.

When both of the slides are fitting well, the driving mechanism should be looked to. The straightness of the X motion spiral axle and the pinion axles must be checked and their bearing caps adjusted until resistance to motion is felt. Bearing caps can usually be honed and brought together to compensate for wear, but if when the bearings are tightened the spindles are found to be not round, these too can be lapped in their bearings and cleaned out as well as possible after the operation. When the axles are straight and fitted, an examination for backlash between spiral and nut, and pinion and rack must be made, after freeing these parts from grease.

If there is backlash in the rack and pinion, it will be the rack at fault unless the pinion leaves can be seen to be broken. Remove the rack, straighten it so that the tops of the teeth are in line, then pack it up with paper until, when the pinion is re-engaged, an even movement is obtained. A badly worn rack may be forced to give a little more service by securing it at one end only and bending it so that the free end and a length of rack springs up against the pinion. In this way, contact is maintained and irregularities do not cause jamming, but of course the method is a makeshift and will not allow heavy loads to be transmitted.

Little can be done about a slack spiral and nut but slackness here is rarely of importance. If the stages is being used for measuring purposes, a stout rubber band may be stretched between the outrigger and the stage structure to bias the stage one way and so eliminate the backlash. Fig. 8b.

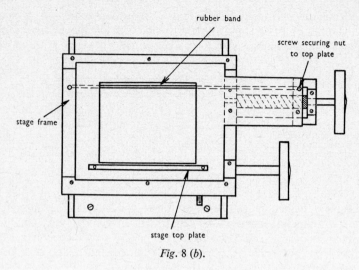

Fig. 8 (*b*).

If the rubber is placed approximately as shown, its length is such that variations in tension between stage positions maximum left and maximum right is small. Rubber is not the best material but delicate steel springs are difficult to come by; rubber is not attacked by silicone grease and will last for several months.

The rotating part of a stage may have a taper bearing and, if it has, it must be lapped to fit, there being no other way of making a sufficiently accurate job. If it has a simple straight bearing, the female part may be squeezed in a three-jaw chuck until it shows high spots, A, Fig. 9. The male part may then be trued in a lathe or lapped by hand to the smaller size required to make a new bearing. (This method of correction is not so crude as it appears. A three-point bearing is best for a device which requires primarily to be stable rather than capable of resisting great strain, and the method was employed by Andrew Ross.)

Mechanical Stages: General Matters

Whatever the kind of mechanism employed on a mechanical stage, it is exposed to a lot of use, dust, chemicals and possibly sea water, and is in addition likely to become dry or sticky where it should be well lubricated. The best lubrication for stages is silicone grease or molybdenum disulphide anti-scuffing paste, not only because they retain their viscosity over a wider range of temperatures, but also because they do not dry out and leave a sticky surface, therefore for

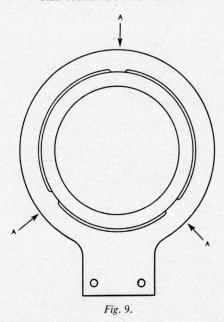

Fig. 9.

microscopical purposes they will withstand thrust as well on the moving parts. Their film-forming properties are fairly good, as conventional greases. The lubricant on all moving parts should be changed each few months because it picks up abrasive dust from the atmosphere.

Unless a rotating stage is specially made, it must not be pulled round by means of its X and Y milled heads; the dovetail slides have enough work to do without the added thrust upon them due to this action. Stage bearings have to be light for the reasons given in the paragraphs above, and if they are subjected to side thrusts they will soon start to sway when the controls are operated, so making high power work a penance.

A further hazard to which a stage is exposed is that of being driven hard against its stops by careless operators, who have not developed the necessary skill and delicacy of hand which goes with those skilled in handling instruments. In laboratories it is well worth while fitting the old-fashioned safety knobs which are secured to the milled heads by friction only and so give as soon as the control becomes stiff or solid. These knobs may easily be made in a workshop after the style of Fig. 10, or the Bulgin slow-motion heads described above may be employed.

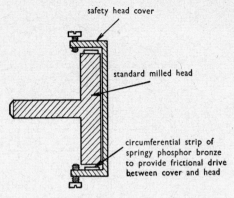

safety head cover

standard milled head

circumferential strip of
springy phosphor bronze
to provide frictional drive
between cover and head

Fig. 10.

No brass or steel mechanical device will withstand sea water and microscopes used in this work must either have all their exposed parts made of stainless steel or be protected with sheets of glass or Perspex. A fairly satisfactory arrangement is to cover the stage with a plate of glass fixed upon it with wax, about 2 mm. thick, which is fitted with sides like a shallow tray, the microscope being used with the stage horizontal. A hole is trepanned out of the centre of the plate and a thick cover glass cemented over it with Araldite, so allowing the condenser to be brought into focus through the thickness of the glass plate. Whether one uses a protecting plate or not, all parts of the instrument on and below the stage surface should be smeared with grease as a temporary protection against corrosion.

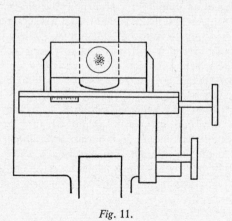

Fig. 11.

The Gapped or Horse-Show Stage

This form of stage, Fig. 11, was designed so that the slide could be grasped by the fingers easily while the microscope was being set up, the original idea being that the slide could be tilted while a high power objective was being focussed; in this way contact could be felt when the objective touched the specimen or its cover glass. The arrangement is useful mainly because it allows a condenser to be drawn out on its changer-slide without the necessity of lowering the substage to clear the stage.

Stages of this kind are not manufactured now as they have not many important advantages but some disadvantages, which are listed below:

(1) A normal built-in mechanical stage cannot be made on this pattern for reasons obvious when the stage mechanism is examined.

(2) The stage cannot be made to rotate.

The attachable type of stage or a more solid modification of it is effective on these instruments.

There is an undoubted advantage in being able clearly to see the objective-slide-condenser region, particularly when they are oiled together, and the arrangement makes more effective space available in the substage, mainly because most substages run out of their slides when a large condenser has to be lowered to clear a solid stage of normal thickness.

Although a condenser can be pulled out without lowering the substage, another condenser can seldom be put in without a change of focus. It appears then that so long as the substage has sufficient travel there is little real point in having the gapped stage. If a worker is likely to break a slide by hitting it with the objective, he is still likely to do something foolish if he can lay hold of the slide with his hands. Where there is danger of damage, either microscopes with nosepiece action fine adjustments or a safety stage should be fitted. Objectives with safety fronts are now in use.

It should be borne in mind in this connection that the nosepiece fine adjustment is the best safeguard because when a short focus condenser is oiled on to the slide before a high power objective is applied, the front lens of the condenser will arrest the downward travel of the slide on the safety stage and so limit the usefulness and safety of the device. The resulting damage to apparatus may in these circumstances reach to a pushed-in objective front lens instead of only a broken slide or cover to be expected when a long focus condenser is in use. (See 'To Line up a Microscope for Critical Work', page 369 and 'Miscellaneous Apparatus', page 507.)

How to Search and Mark a Feature on a Slide

The choice of stages for different work is dealt with in the sections dealing with the various methods of microscopy, but as a general matter the main use of mechanical stages is to systematically search a slide and to record the position of an object. Mechanical stages should be fitted with vernier scales but, if not, a satisfactory finder graduation can be made from pieces of engineers' steel rules cut to a suitable size to rest near the sliding parts of the stage and be read by a scratch or pointer fixed to the sliding part. In passing, it should be mentioned as a point to be considered when selecting a microscope that a number of commercial instruments have stage scales placed in positions which can hardly be seen from the operator's position while others have such fine markings that they cannot be read with the normal naked eye. There is no need for these scales to be very finely graduated because accuracy does not depend upon this, quite a coarse, clearly visible scale is sufficient with which to find an object, even under a high power. White filled graduation marks are always an advantage in that they are easily seen, and so long as the scales are intended only for finding and recording a specimen, the coarseness of the white filling is an advantage.

Readings on the scales of the X and Y positions of a feature on a slide taken on one microscope will not obtain on another, therefore the microscope in use must be identified on any slide which has co-ordinate figures upon it, also the position of any stops or rests on its stage and the position of the slide relative to them must also be given. It will be found in practice very useful to keep a slide marked with co-ordinates with a small specially marked feature upon it which can be easily found without the aid of scales, so that should a microscope stage be disturbed in any way after it has been used to obtain co-ordinates, it can be put back to its original position by lining it up to agree with the co-ordinates on the marked slide.

A difficulty exists when it is necessary to send a slide some distance to be examined upon a different microscope with different vernier scales. The old way of doing it was with the Maltwood Finder mentioned under 'Miscellaneous Apparatus', but this has passed out of use and probably cannot now be bought. The Author has found that the best method of overcoming this difficulty is the direct one of sending with the slide a template with a hole in it to indicate the point of interest and always to keep the template with the slide for use on any microscope. To make the template and set it up, proceed as follows:

(1) Obtain a piece of good quality postcard somewhat larger than the slide to be handled, and burn in the centre of it a minute hole with a hot needle.

(2) Set up the microscope to show the required part of the specimen, then place the card underneath it on the stage and without moving the object slide, move the card until the light passes through the hole on to the object, making quite sure that the object occupies the centre of the hole.

(3) Mark round the slide on the card with a sharp 4H pencil.

It will now be found possible to return the slide to its marked position on the card sufficiently accurately to find an object over the hole when a 4 mm. objective is in use.

To set up the object on another microscope, the instrument should be set up using critical light on any object, with a small source of light, the object is then removed and the card substituted, when it will be found easy to find the shining pin-hole in the focussed condenser light. When the position of the hole is found and centred to the axis of the microscope, the card may be temporarily stuck to the stage with adhesive tape and the slide positioned above it to lie within the pencil lines. Once the feature is found, the graduations of the new microscope may be read and recorded for subsequent use, the template being marked and preserved for use on other microscopes.

In engineering workshops and physics laboratories, the above problem often arises when vertical illumination is employed on specimens of metal and other materials which are opaque, and in these experiments the pin hole technique cannot be used.

If the specimen is a large chunk, a pointer made from a needle may be stuck to the job with plasticine, the point of the needle being placed exactly over the feature to be examined in the second instrument. A second method is to map or photograph the surface with a power sufficiently low to make the features surrounding the wanted feature recognizable in the second microscope, the wanted feature being placed accurately on the cross wires so that when the region is found, the feature is at the centre of the field. This process is in practice difficult to carry out when a specimen has many similar features and generally the direct method using a fine needle or wire as a pointer is the best.

Diamond points which mount on the objective and which can be lowered to touch the slide can be used for marking, the method being to set the point slightly off the axis of the objective, lower the objective until the point touches the slide and then rotate the stage so that a fine ring is described with the feature required at the centre.

When small bits of opaque material are being studied, they should always be cemented on to slides or squares of glass so that they can be handled. If it is necessary to mark a feature, the pin or needle may be stuck to the mounting plate in the usual way. If the microscope objectives have been properly lined up, it is possible to stick pointers on a slide or specimen under a low power and to line up the pointers by means of the eyepiece cross wires and the X and Y movements of the stage. If the wanted feature is centred accurately on the cross wires and a low power objective substituted, enough of the specimen is visible in the field for a mark to be put upon it coincident with the cross wires, well clear of the wanted feature but accurately in line with it in the X and Y directions. When these marks are set up on the cross wires of a second instrument with a low power in use, the wanted feature will be found on the cross. There is then no difficulty about the thickness of a needle coming between a high power objective and the specimen.

The actual marking of a slide is often difficult but can be done as follows:

(1) Set up the wanted feature on the cross wires under the high power, then substitute a low power to include a field of view of about 2 mm. diameter, without touching the cross wires.

(2) Obtain two tiny pieces of white paper, each with one straight edge, about 1 ×4 mm. in size, cut from an envelope flap. Moisten these and place them on the slide or specimen with the aid of a needle and move them about while wet until the straight edge of each lies along the cross wire in the X and Y direction, Fig. 12. In this way the straight edge can be made to mark the co-ordinate positions very accurately, and when the glue dries permanent marking has been achieved. To find the feature again it is only necessary to set up the straight edges of the pieces of paper along the cross wires of any eyepiece, when the feature will be found on the cross.

It may be mentioned in passing that cross wires often do not make a true 90 degree cross, this being most noticeable in the old eyepieces where wires were stretched across the eyepiece diaphragm; in the modern form it is usual to use printed cross lines on a disk of glass like an eyepiece micrometer and these are very much better and stay cleaner.

To Search a Slide

The eye is most comfortable when searching an area if it is allowed to swing from side to side instead of up and down, and so a slide

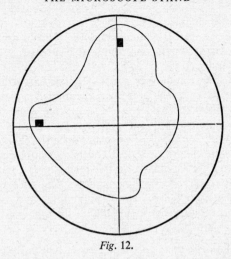

Fig. 12.

should be searched in this way. Strips of slide are examined in turn while the slide is moved in steps in the Y direction, but the steps must not be greater than about one-third of a field diameter or parts will be missed, and it is a good idea to have parallel wires placed in the searching eyepiece about this distance apart. The strip to be searched is then marked out by the wires for each step in the Y direction.

Important Accessories to the Stand: Objective Changers
No set of objectives can be made so accurately that the optical and mechanical axes of each will fall upon the same line, therefore if objectives are screwed on to the nosepiece in succession they will not produce an image of the same area of the object.

Several types of objective changer have been designed to eliminate this trouble and to make for speed in changing objectives. Some of these are as follows:

(1) The Rotating Nosepiece in which a disk suitably shaped carries two, three or four objectives, and each one may be brought under the eyepiece in turn. The apparatus is dust-proof.

    Its disadvantages are that each objective cannot be centred to the axis of the microscope and there is danger of the unused objectives fouling the stage or preparation.

(2) Sliding changers upon which the objective is centred to the optic axis, each objective being mounted and centred upon its

own male element and is set up for an indefinite time. When the objectives are off the microscope their lenses are exposed to dust but the system has no technical disadvantages and is to be recommended.

(3) Various screw-on devices where a female part is fitted to each objective in which the objective is centred, the male part being fitted to the nosepiece. It is not dust proof when the objective is off the microscope.

The disadvantages are that the objective must be lowered while being changed to clear the male portion on the nose-piece, and this means that the microscope tube has to be raised to clear the specimen safely. The turn necessary to locate the objective is considerably stiffer and more difficult to perform than the straight slide of the type described in (2).

The plain sliding type is to be preferred. It usually consists of a female slide about 25 mm. wide and 50 mm. long, secured in a mount which screws into the objective thread in the nosepiece. The objective screws into an adapter mounted upon the male part and may be centred by moving the adapter which is fitted with slotted screw holes or by a more elaborate mechanical device. With any adjustable arrangement the result is the same; so long as the slide is pushed against its stop and the objectives are centred to the microscope axis they will always return accurately to this position.

It is usual to spring-load the male slides to eliminate any possibility of looseness due to wear.

The more elaborate changer slides, particularly those intended for use in interference microscopy, have levelling screws so that the objective may be centred and squared-on to the axis of the micro-scope. The fittings are simple but difficult to set up, therefore other squaring-on devices are normally used (see Interference Microscopy', page 417). Unless there is an obvious mechanical fault in the screw cutting in the nosepiece, it is not necessary to square-on an objective, but it *is* necessary to have the objective and eyepiece axes coincident within a millimeter. An error can arise here if a sliding objective changer is badly set up (see 'Lining up a Microscope', page 369). The effect of this error is that the area of best definition is not in the centre of the field of view and may be completely out of the field.

Condenser Changers

These are usually very similar to the sliding type of objective changer but the slides are about 38 mm. wide. They may be of less fine work because centration is less important, the condenser not being an image-forming device.

Condenser changers carry various condensers into position beneath the specimen while the substage iris diaphragm remains in place in the substage on the axis of the optical system. The substage iris is centred to the optic axis of the microscope during lining up, and the condensers, each on a slide, are centred to the iris. (See 'Lining up a Microscope', page 369.)

All research microscopes should be fitted with condenser changer slides in the interests of convenience and accuracy, while the use of changers relieves the substage of much strain and makes for better mechanical working; also experimental devices may easily be mounted upon spare slides.

Any condenser changer slide arrangement should be fitted with a clamp or locking arrangement to hold the slide with its condenser in place should the microscope be inclined and the substage rotated.

Some General Matters Concerning Stands

Dust is the enemy of all fine machinery and every effort should be made to keep the room in which microscopes are used free from dust. When removing dust from any surface, first blow off as much as possible, then wipe the surface firmly once with a cloth to sweep off the particles, then finally clean or polish with another clean cloth so as not to rub the dust in the first cloth over the job. Atmospheric dust in towns is formed mainly of flue products and these are very hard vitreous and carborundum-like materials, produced by the fusion of minerals in furnaces. They will scratch the hardest surfaces and, if carelessly rubbed in on cleaning cloths, will do much damage.

At one time it was stated that microscope mechanical parts should be lubricated with clock oil and that if they would not work smoothly with this lubricant, they were wrongly made. This idea is quite wrongly based. If thin oil is used on large close fitting parts, they will adhere by capillarity and be quite unable to slide smoothly over each other. It is of course useless to bung up a bad job with heavy grease but modern dovetail slides are always an excellent fit and much better than the old ones, so crude methods of covering up bad work are not necessary. The slides must be lubricated with grease of normal consistency, preferably of the silicone type or that loaded with molybdenum disulphide which does not change in viscosity with temperature within the world climatic range.

A microscope will continue to work even though its mechanical parts are tortured, but the result will probably be sheared pinion leaves. Dovetail slides should be examined once in six months to see if they are dry, and if they appear to be, they should be run out of their guides, rubbed clean with petrol or zylol, and lubricated with plenty of silicone grease. The surplus may be wiped off after, when it is certain the slides and bearings are saturated. Should a slide have become rough through disuse, it should be polished with commercial metal polish applied on a cloth before greasing.

One should endeavour to select and mount accessory apparatus in such a way that the slides are more or less fully engaged throughout their length while the microscope is in use. If the slides are only just engaged, heavy wear of an uneven sort is set up and the exposed parts become dry and dust-laden. If the microscope has to be stored, all slides should be greased and fully engaged.

Draw-tubes sometimes become stiff and when this occurs, the danger exists that the body will be driven down upon the specimen while the draw-tube is being adjusted. Usually cleaning and greasing is all that is necessary but sometimes, particularly with a chromium plated tube, the roughness has to be removed with an engineers buffing wheel. After this treatment it will usually be found satisfactory and the waxy surface left on the metal by the buffing wax forms a good lubricant.

Microscope slides and racks and pinions are usually found to have two adjustments, one to tighten the spindle of the controlling milled head and another to adjust the fit of the dovetail slides themselves. The correct fit for a dovetail slide is as follows. When the instrument is vertical, is not loaded in any way, and has the pinion disengaged, the slide should just move under the influence of gravity. The dovetail slide adjustments should be used to secure this condition after the parts have been cleaned and lubricated. They should first be tightened to bed the parts in and take up all slack, and then slackened back to the correct fit, slackening off one screw at a time a small amount so that when all appears correct, a small tightening of any adjusting screw can be seen to affect the descent by gravity of the body; it is then certain that all the adjustable parts of the bearing are doing work.

If the male slide is badly worn it must first be straightened as described in the section 'Adjustment of Stages', page 151.

Next the stiffness of the controls should be adjusted so that the body does not run down by the influence of gravity when the pinion is engaged. They should be just stiff enough to ensure that there is no chance of creepage downward when the instrument is fully loaded with objectives, etc., and the table is thumped. The stiffness is obtained by tightening the grip of the bearing upon the pinion spindle and never by tightening the slides. The position of the various adjusting screws has to be found from inspection of the particular instrument, several modern designs have eccentric split bushes as spindle bearings.

When adjusting the large counter-balanced instruments, the same procedure should be adopted, but it will be found that there is no need to make the spindle stiff as there is no component outstanding due to gravity and a very delicate and satisfactory movement may therefore be obtained.

Should the instrument have a means of adjusting the relationship of rack to pinion, proceed as follows: set up the slides as above but do not tighten the pinion spindle bearing unduly. Inspect the rack and see that it is not badly worn and that it lies parallel with the axis of the slide; if it is not parallel, detach it and pack it until it is. Some racks are 'bridged', i.e. they bear upon their ends only, the centre portion being slightly bowed towards the pinion to form a spring and eliminate backlash. If the rack is of the bridged construction, it must be bent until it is restored to its bowed shape. Bridged racks are unsatisfactory on heavy instruments because the pinion pushes away the rack under heavy load and so does not engage properly and wears out quickly. When the rack is correct, replace the body tube and set the pinion so that there is no backlash at any point in the required travel. When the pinion is too close, the engagement of the teeth will be felt through the relatively slack pinion spindle. Finally adjust the stiffness of the pinion spindle as described above.

Several different mechanisms are used to adjust pinion distances from the rack. The usual modern one is an eccentric bush on the pinion spindle locked by means of a set screw, while older ones still in use consist of set screws tilting a bearing plate, one set to pull it in and the other to push it away, the adjustment being made one set against the other. All are obvious on inspection of the microscope, but not all instruments have all the adjustments described above because misuse has caused some manufacturers to discontinue all adjustments, but whichever others are left off, that for spindle tightness should always be present.

The Inverted Microscope

The idea of constructing a microscope designed so that the specimen may be examined from the underside, is not a new idea, but for many years its possibilities appear to have been overlooked.

In 1860, or thereabouts, the instrument was designed by chemists for the purpose of examining unpleasant liquids in such a way that the fumes did not rise on to the microscope optical system. At the present time, it is also much used by biologists who wish to insert needles into specimens without the needles coming between the specimen and objective, and there are several other advantages described below.

The optical system employed in an inverted microscope is the same as that in the conventional instrument, but an extra mirror or prism is introduced to bend the image-forming light so that it emerges in a convenient direction for the observer, see Fig. 13.

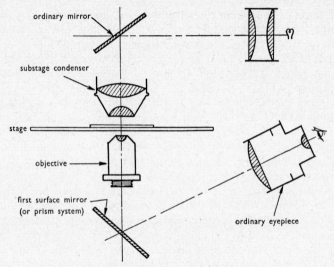

Fig. 13. Inverted Microscope.

A more complete diagram of an instrument constructed in the Author's laboratory appears in Fig. 14.

Fig. 15 shows the arrangement of the McArthur portable inverted microscope.

In construction the instruments vary a little, some having a fixed stage with focussing movements applied to the underhung body, others having a racking stage after the style of metallurgical microscopes. The fine adjustment is usually applied to the nosepiece or body tube in the usual way. The 'substage', now above the stage, is rack focussed and may be fitted with the usual mechanical and optical refinements, the mirror is normal and is usually placed upon a vertical 'tail piece'.

The form of construction has several advantages which are listed below:

(1) The main part of the microscope system, the body which sup-

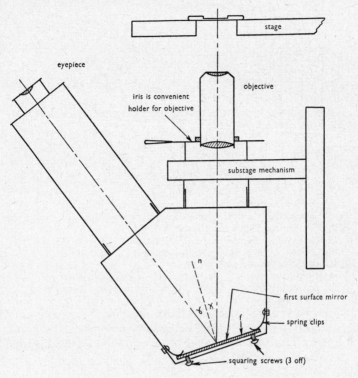

Fig. 14.

ports the magnifying lenses, is low down on the instrument, thus aiding considerably the general stability of the whole.

(2) The body focussing controls are placed near the table, so making the instrument comfortable to use.

(3) The specimens, if they are in liquid, are always observed through a definite thickness of material, i.e. the glass bottom of a trough. In the ordinary microscope, the objective has to look through the surface of the liquid which may not be optically flat, and the liquid may also be of varying depth.

(4) Any quantity of liquid may be placed in a trough on the stage to cover (say) a biological culture or chemical etching experiment, and the experiment observed through the bottom without the trouble introduced in a conventional microscope due to the limited working distance of high power objectives.

(5) Needles and micromanipulator devices may be inserted into a

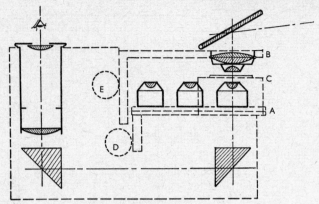

Fig. 15. Layout of the McArthur Prismatic Inverted Miscrope.
A. Sliding drawer carrying 3 objectives.
B. Bracket carrying condenser and hinging mirror.
C. 'Stage' or runners to support slide (cover downwards).
D. Fine adjustment to objective drawer.
E. Fine adjustment to condenser bracket. (It will be noted that no
   coarser adjustments are required if the objectives are par-focal.)

    trough and manipulated without any difficulty about inserting
needles in the small space between the objective and specimen.

(6) Great heat may be applied to an object on the stage without
damage to the objective.

(7) Once the objective is focussed, all objects placed face downwards on the stage must automatically appear in focus, the
thickness of the object making no difference to the position of
the face upon which it rests.

(8) Because of (7) above, such a microscope needs only a limited
travel to its coarse adjustment, only sufficient in fact to allow
for the different lengths of the mounts of the objectives. This
fact also contributes to stability.

(9) A special trouble with the modern short conventional stand is
insufficient space below the stage for accessories, but this
difficulty does not exist in the inverted microscope, where the
'substage' may be built up to the ceiling.

    There is almost everything to be said in favour of this type of
instrument for general laboratory work, combining as it does great
compactness of construction, stability in the parts most requiring it,
and space around and above the stage where it is most needed. It is
surprising to the Author that these stands are so seldom seen in use.
(See also 'Microdissection Methods', page 246.)

Adaptation of Ordinary Stand to the Inverted Form

Most microscopes may be used in the inverted manner so long as they have a solid substage provided with a good focussing adjustment.

A tube should be provided to fix into the substage and to carry a mirror or proper inclining prism after the style of Fig. 14; the objective may be held in the substage iris or, preferably, in a condenser changer slide made to take an objective thread.

An inclined-plate type of vertical illuminator is placed in the conventional body tube and an objective in the usual place is used as a condenser with a normal lamp. All the rest of the microscope with the exception of the eyepiece is used normally.

Such a set-up does not possess the stability of the properly constructed instrument because the substage slides are not rigid enough to bear a body, but it is quite satisfactory for many investigations. It must be remembered if a first-surface mirror is used in the inverted attachment that an erection takes place in the image with respect to the object in the up and down direction, but not from left to right. A proper inclined eyepiece prism allows complete reversal of the image in the normal way by introducing two reflections, thus first changing up to down, then changing down to up, there being no left to right change.

Some form of tube length extending device will certainly be required (see 'Objective Tube Length', page 61) or 250 mm. objectives may be used, for it is not practical according to the geometry of the system, to make a short tube inverted microscope of normal laboratory size. A very useful portable box arrangement $130 \times 50 \times 25$ mm. size, and sufficiently strong to be run over by a motor-car, has been demonstrated at the Royal Microscopical Society, and is shown in Fig. 15.

All techniques described in the various sections may be applied to the inverted microscope.

CHAPTER 9

Optical Benches

OPTICAL BENCHES may be divided roughly into two types; beautifully made expensive ones which can be nearly useless and improvised ones which are essential for permanent apparatus. The main functions of an optical bench are two, the first being to provide a rigid base for several pieces of apparatus, and the second to make it possible to return a piece of apparatus to a pre-arranged position. In practice, neither of these requirements is a difficult one.

Design Considerations
Obviously much depends upon local requirements but in general the apparatus required upon an optical bench used for microscopy is limited to the following:

(1) A microscope in the centre of the bench.
(2) A high power lamp at one end with room for filters, cooling trough and an auxiliary lens system between it and the microscope.
(3) A vertical camera stand near the microscope.
(4) Another type of lamp at the other end of the bench to be used either for low powers or in association with a monochromator.

The length of bench occupied by this apparatus is about 1 metre.

All apparatus used with a microscope must have its own centring devices, therefore there is no need to spend large amounts of money upon an expensive construction of bench. The bench must be heavy in order safely to bear the weight of apparatus mounted in off-axis positions, and to resist by its inertia vibrations in the bench upon which it stands. It must have such a cross-section that simple saddles can be clamped securely to it, and so long as these elementary matters are borne in mind, any structure will do.

The Author has used an old and worn lathe bed with success as an optical bench upon which much high sensitivity industrial work was done (Plate 3), and such a bed may be bought for a small sum or scrounged from a scrap yard. It weighs about 30 kilograms, has a plain flat top of section shown, is about 2 metres long, and may be allowed to stand upon its own legs if desired (Fig. 1).

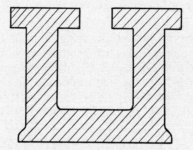

Fig. 1. Section of typical lathe bed.

The flat top of this article has the advantage that saddles may be fitted with very simple levelling and centring devices as shown below, Fig. 2.

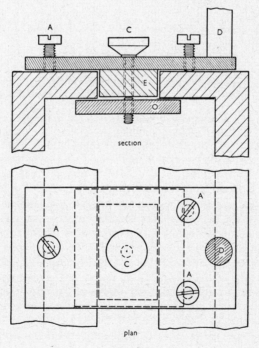

Fig. 2.

To level and centre the apparatus to an axis, the saddle clamping screw is released, the levelling screws operated and the clamping screw tightened. Locking nuts may be added to the levelling screws if the apparatus is often removed from the bench.

The saddles may be of any convenient form and a plain brass or steel sheet with tapped holes for clamping apparatus by means of bolts has been found successful. The diagrams are self-explanatory and the details vary with the shape of bench in use. The part of the saddle E is made to be a good fit in the channel of the bed but must have its corners rounded. A similar bench has been made successfully from hard wood, the parts being screwed together as shown below, naturally this is not so suitable for use when apparatus is changed about very frequently or when it must be returned carefully to the same place. The distribution of the pieces of wood is such that

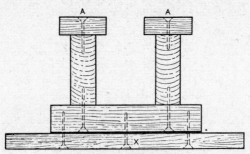

Fig. 3. A. Runners.
X. Lateral stabilizers, one each end of bench.

construction is easy and warping unlikely. Enough space should be left between the top runners A to allow a hand to be inserted to grip the saddle clamps O, Fig. 3, a space of 50 mm. is the minimum unless the bench is raised from the table.

It is useful to mount the microscope so that its inclination axis is parallel with the long axis of the bench, because when the microscope optical axis is vertical, the holes which admit light to the vertical illuminator are on the bench axis, and if the microscope has a hole at each side of the nosepiece, two vertical illumination systems may be used at will. Should it be desired to incline the microscope to ease performance of long visual observations or for any other reason, its mirror and vertical illuminator holes are not removed so far from the bench axis that the lights cannot be brought to bear. For ordinary use the microscope is left set up in the vertical position ready for photography, and when the camera is required, it is placed upon its post and is instantly ready. In the optical bench shown in Plate 3, a spectrometer is mounted at one end of the bench to provide mono-chromatic light for delicate photography, but for normal use a bunched filament lamp is mounted at the other end. If required, one lamp may be permanently directed at the position occupied by the mirror when the microscope is inclined to a particular position suitable to the observer.

It is seldom useful to mount a microscope semi-permanently in any other position than with its stage horizontal. This position makes technical work easier (see 'The Inverted Microscope', page 165), also the photographic apparatus may be simpler; objects rest upon the stage without clips, and liquids remain in place without the use of thin troughs. It must, however, be possible to place the microscope with its optical axis in a horizontal or intermediate position and to be able to illuminate it should prolonged visual work be contemplated. The Author has two saddles, one to hold the microscope in the correct position when its optical axis is vertical, and the other to hold it when in the horizontal position. The horizontal saddle has two clamping positions with fixed stops, one to place the mirror in line with the light sources, and the other to place the vertical illuminator in line with them, Intermediate clamping positions need not have stops but are useful when the lights are directed at the stage to illuminate the sides of troughs and for some forms of ultramicroscopy.

There is no point in giving measurements of this apparatus because each user has his own microscope and his own problems. All saddles may be made of hard wood or sheet metal pieces screwed together and 2 mm. thick rubber sheet stuck down with Bostik is excellent for preventing slipping of any part without undue springiness. A number of rubber door stops should be used to locate irregularly shaped apparatus, but microscopes must be clamped to optical bench saddles and this is most easily done by passing a 10 mm. diameter bolt a driven fit upwards through the saddle, the clamping pieces may then be arranged to bridge such parts as feet of microscopes and lamps. The clamps are tightened down finger-tight by the wing nut on this bolt. Bolts used for clamping instruments to saddles and those for clamping saddles to the bench should be separate from each other or difficulties will occur when the apparatus has to be rearranged.

The upper surface of an optical bench must never be painted or greased, or steady adjustment of the saddles will become impossible.

If the surface is well cleaned each week, rust and corrosion will not occur, but if it has occurred, a scouring with commercial metal polish will correct it. If the bench is of wood construction, a soaking in shellac varnish followed by a rub down with fine glass paper will be found to give a very smooth hard surface.

Devices to Aid Stability of Microscopical Apparatus

'Stability' in this sense means the absence of relative movement between parts in a system of apparatus, and it does not matter whether or not the whole system moves in space. If a lamp moves with respect to the microscope, the illumination will change, and if the camera moves with respect to the microscope the image will be blurred. If the apparatus stands upon an object of great mass, the vibrations in the whole will be greatly limited by the inertia of the mass, the rapid vibrations being attenuated to such an extent that they may be ignored. Every piece of material has a natural period of vibration which depends largely upon its mass, but it may be forced to vibrate at a different rate and power is expanded in making it do so. In general, bodies vibrate at their natural frequencies when disturbed by external shocks, while forced vibrations occur when objects are attached to machinery. The length of a time a body will continue to vibrate after a shock of given strength is mainly determined by the nature of its support and by other structures connected to it. Most relative movement will occur between object and base where a heavy object is supported upon a thin column and the next most when a heavy object is suspended in some way by a thin member. If a weight supported upon a thin rod of steel is knocked, it will continue swaying at a rate dependent upon the weight and upon the elasticity of the support, but primarily upon the weight and length of the support. If the system is inverted we have the simple pendulum where the period depends upon the physical length. The total length of time the system vibrates is also dependent upon the strength of the impulse which caused the disturbance. The problem is complex.

Microscopical problems are related to the first of the cases, i.e. heavy parts are supported upon pillars. If the pillars are light and the objects heavy, a shock will cause a considerable displacement of the base of the pillar with respect to the supported weight, also the vibration will continue for a length of time dependent upon the mechanical connections to the weight, the stiffness of the pillar and the magnitude of the disturbance. Several seconds are not unusual for the total vibration time of some microscopical structures like micrographic cameras on high stands.

A popular illustration of the effect of top weight on the stability of a structure is given by the boy who is able to balance a broom on his finger with the broom head upwards, but not with the head on his finger. The reason is that with the relatively ponderous weight of the head at the top, there is time for a large amount of relative movement to be given to the bottom by the boy before the head falls, in fact the movement is of the order of feet and if the broom contained an optical system the relative movement under vibration conditions between top and bottom would be of this order. If, now, the broom is turned over and the weight placed at the bottom, the trick cannot be performed because there is insufficient mass at the top to allow time for any relative movement to be applied between top and bottom. If this system contained an optical apparatus there would be no relative movement between the parts under conditions of vibration. It will be seen therefore that to achieve minimum relative movement between top and bottom of a structure, it must be short and light at the top. Apparatus should not be mounted on the top end of a microscope, and even a brass micrometer may be too heavy; a binocular eyepiece certainly so. The old long tube microscopes were advantageous in many ways but were always liable to tremor upon the slightest provocation from outside, mainly because of their top weight, and when they were fitted with the heavy inter-ocular adjustment of the Wenham body, they were very unstable. In the chapter 'Photomicrography' a vertical camera is recommended for various reasons, but it must be mounted upon a heavy post of at least 25 mm. diameter steel bar and be itself made of light materials.

Should the apparatus be such that shortness is not possible to obtain, the solution to vibration problems must be obtained by keeping bench and floor vibrations out of the apparatus. A heavy optical bench may be mounted upon cork or a heap of magazines. Thick rubber should not be used because its losses are not sufficiently high and movement when started is not damped out quickly. Much-used magazines are good as the air trapped between the pages provides the insulation while the friction of the paper rapidly damps out movement. The successful use of the leaf-spring in automobiles is a more familiar example of this mechanism. Cork pads are usually satisfactory for similar reasons. It is useless to bolt an optical bench to another bench upon which rough work is done because the vibrations will pass into the apparatus by force, regardless of the natural period of the optical bench.

If an optical bench is not available, an object like a paving-stone covered with cloth and resting upon cork makes a good work-table upon which to set up a microscope. A serious attempt was made by a worker in central London to obtain stability for photomicrographic apparatus by suspending a heavy timber optical bench from the ceiling by sash cords, and the system worked well. In practice it is found that whatever care is taken in mounting the bench, the apparatus itself must be properly constructed to withstand vibration without relative change of position of its parts. Anywhere within the boundary of a large town or within half a mile of a railway, the ground itself will be felt to move sufficiently to cause furniture to rattle during heavy traffic periods, therefore it is of no use depending upon massive bases. Apparatus such as cameras must be built of light materials and be supported upon stiff uprights of steel as short as possible, while microscopes should be short and have light tubes without apparatus at the eyepiece which increases top weight. Attention should be paid to the advantages of the inverted microscope in which design there is very little height above the bench, therefore easier support for a camera (see 'Inverted Microscopes', page 165).

A useful support for heavy slabs can be made from rings of drainpipe about 50 mm. long and 150 mm. inside diameter, which are laid on the table and which contain three tennis balls. A fourth tennis ball is rested between the three and does not touch the table. Three or four of these rings and balls are used to support a weighty bench or slab, the friction of the balls upon each other provides the required degree of damping. The system is very good and very stable, and has often been used in industrial establishments for supporting optical apparatus of all kinds.

CHAPTER 10

Photomicrography

I T IS always necessary when doing industrial and scientific micro-scopy to record results by means of photography. Many micro-scopical processes are destructive of the material being observed and the drawings and descriptions provided by the observer are sooner or later questioned. The emphasis upon publication in modern scientific work demands skill in several branches of the art of photography.

Choice of Apparatus for Photography at Magnifications up to Fifty Diameters

For this work a compound microscope is not necessary. The lens may be one known as a Microstigmat, which is used as a well-corrected objective. A camera lens is not designed to give a sharp image when it is magnified by a high power eyepiece because to do this the designer must sacrifice flatness of field to central definition. For very low power photography when resolution is not the primary demand, the flatness of the field of the camera lens may be used with advantage. A very low power microscope objective may also be used as these magnifications are also designed to give a flat field. The specimens are illuminated by one of the top-light methods mentioned in Chapter 11. The rest of the apparatus consists of an ordinary camera bellows and film holder, the lens being mounted in the usual place. Fig. 1.

For low power work such as this the tube length correction of objectives is not important and the camera may be used at any con-venient extension. It is best arranged vertically, should be of large size, full plate if possible, and is best mounted on a wall. The vertical arrangement allows a horizontal stage upon which may be placed powders, botanical material, liquids, heavy objects encountered in engineering work, rocks and many other things, without special supports, and a heap of plasticine makes an excellent bed upon which to orientate irregular objects. A diaphragm should be provided in the lens system in order to adjust the depth of focus of the system, but no shutter is necessary. (See later section 'High-Power Photo-micrography', page 181.) Exposures are made by obscuring the lights with a card while the plate-carrier dark slide is withdrawn, exposing

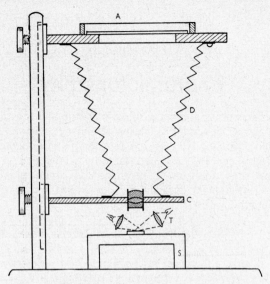

Fig. 1. A. Plate holder adjustable in height and extension.
D. Bellows Extending to about 3 feet.
C. Head holding lens.
T. Top lights.
S. Stage.

the light, obscuring again, replacing the dark slide and removing the plate-carrier from the apparatus. It will be seen in later sections, where photomicrography with a microscope is described, that all exposures of still objects should be several seconds long, the lights being adjusted accordingly. The vibration introduced by the action of a built-in shutter is undesirable but if one is present it may be tolerated in low power apparatus of this kind.

It is a considerable help to the workers to arrange a viewing mirror above the apparatus so that the ground glass screen may be seen from the front in a horizontal direction.

The focussing arrangements in this apparatus need not be of high quality. Various methods have been tried; rack and pinion control of the camera lens panel, focussing lens mount of lantern style and sliding lens panel clamping arrangements. In general, the racking lens panel usually found on plate cameras is satisfactory and gives great range of adjustment.

Mechanical stage arrangements are not necessary in low-power apparatus and an efficient substitute may be made from a sheet of polished plate glass with its edges taken off with emery paper, resting flat upon a smooth cloth-covered surface. Objects are mounted in plasticine upon this plate glass and it will be found that it can be moved easily and firmly about under the camera, with or without a guide of some kind.

A full treatment of photographic methods will be found in *Stereo Photography in Practice* by Linssen in this series.

Photomicrography Using a Microscope: Simple Apparatus

It is not always realized that a camera, as generally recognized, is not necessary in order to take photographs through a microscope. The microscope is the lens and all that is necessary in addition is a device to support a plate and make a light-tight joint between the plate and microscope. It is not necessary to use bellows or boxes to make the joint light-tight if the room is dark and the illuminating apparatus enclosed. It is, however, usually impracticable to make illuminating apparatus such that no stray light is emitted, therefore a bellows or box in which the plate is mounted, is usual.

There are many reasons given below for using carefully constructed apparatus but an ordinary camera of any kind may be supported over a microscope, with or without its lens, for the purpose of taking photomicrographs. It is necessary to be able to see the image upon a focussing screen in order to focus the microscope accurately, and if the camera is a reflex one matters are made easy. Should it be a twin lens reflex, the microscope is set up with the viewing side of the camera in operation, it is then slid along a guide until the taking lens is over the microscope when the exposure can be made with focus accurate on the film. Miniature cameras give too small a picture for the best results.

If it is practicable to remove the lens of the camera this should be done, but it is not essential to do so because the exit pupil of the microscope is only about 2 mm. in diameter and passes through the centre zone only of the camera lens. The lens therefore does not introduce appreciable aberration however crude and simple it may be; it should be placed near the eye lens, it then only adds slightly to the effective curvature of the eye lens of the eyepiece. A camera lens has a focal length of, say, 80 mm., while the eye lens of a ×10 eyepiece has a focal length of about half as much, so it appears that the effect of the camera lens in the light path is almost negligible.

When using a camera with its lens in place, it is often found that at certain distances, camera lens to eye lens, a bright circular spot of light appears in the picture in the camera. This is found to be an image of the exit pencil from the microscope reflected from components of the camera lens upon the plane upper surface of the eye lens and from there back into the camera again. The effect may usually be avoided by setting carefully the distance between camera and eyepiece while on test and a complete cure may be obtained by using an eyepiece with an eye lens polished slightly convex on its upper surface. Such a Huyghenian eyepiece is often found in terrestrial telescopes where it is employed to increase the apparent field of view. (See 'Microscope Eyepieces', page 82.)

Many arrangements for supporting an existing camera will suggest themsleves to workers. The vertical system is most convenient from the point of view of microscopy, but should it be necessary to use long camera extensions the horizontal arrangement is more convenient for the camera. It will be seen from the remarks which follow that there are very few, if any, good reasons for using long cameras and optical benches for photography with microscopes.

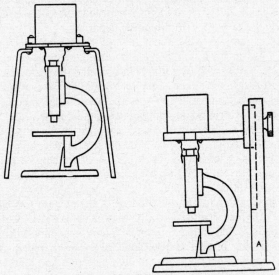

Fig. 2. Tripod stand.
A heavy timber-stem 'L' stand.

In Fig. 2 are two forms of cheap solid mounting which can be elaborated at will, the most useful of these being the 'L' stand. The disadvantage of the tripod is the difficulty of adjustment for height when objects of various thickness are under the microscope and when the microscope tube length is adjusted.

The tripod camera stand is, however, very stable and convenient and may readily be made by cutting a hole for the camera lens mount in the seat of a simple stool. So long as the work intended is not critical, that is when the objects have many planes of focus, a viewing device may be built to stand upon the camera shelf in place of the camera. Focussing may be performed upon this and the camera substituted, the advantage being that after initial calibration, with or without the camera lens in place, the camera back need not be taken off to focus each exposure. (Fig. 3.)

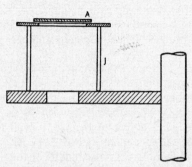

Fig. 3. A. Ground glass.
J. Viewing jig in place of camera.

A great amount of photography of seeds, cloths, powders, small irregular objects, insects, in fact any objects which do not present the difficulties mentioned in the next section may with advantage be photographed with the above apparatus and a simple roll film or plate camera.

Photography Using a Microscope: Laboratory Apparatus
It is usually necessary to construct a photomicrographic apparatus of high quality to record the results obtained during microscopical investigations. Most of the less well-known interesting features in a specimen are visible only under high resolution microscopes, and the specimens, if flat and opaque, must be focussed with great accuracy. Much interest may be centred upon fine cracks in the specimens and these cracks are lost in photographs if there is appreciable scattered light within the instruments.

Choice of Camera
A roll film camera without lens, connected to and borne by the microscope may be used with reasonable success but trouble can be experienced due to the following conditions:

(1) The camera may be too heavy for the microscope fine adjustment, even though the microscope can be focussed immediately before exposure by means of an angle attachment between the microscope and camera. After the exposure is made, a change of focus is often found and it can be assumed that a change was gradually taking place after adjustment. If the camera shutter is left open and the exposure made by moving a light stop not connected with microscope and camera, further handling of the camera can be avoided.

(2) Roll film and cassetts are a nuisance to load, unload and develop, especially when a small number of exposures is required.

(3) The focussing arrangements are not good because the operator cannot see what image is upon the camera film and focussing is only possible by observing coincidence between a deflected image and cross wires in an angle viewing device.

(4) An angle viewing device of the type which is always in the optical path scatters a quantity (say 5 per cent) of light into the camera, thus causing a small amount of fogging visible when a black object is photographed.

In a micrographic camera the following matters are important:

(5) The camera, by nature a somewhat heavy rigid device, must be held upon its own stand in the interests of stability of the apparatus.

(6) It must be so constructed that final focussing may be performed after the dark slide has been withdrawn from the plate, or accidental focus changes may occur.

(7) The focussing-cum-viewing device must show the same image as will appear for the exposure on the photographic plate.

(8) The device (7) above must not be optically connected with the image-forming light during exposure or scatter will result.

(9) The cone of light from the eyepiece should not impinge upon the inside parts of the camera. The control of size of the illuminated field, when this is less than the eyepiece maximum, should be by means of the field diaphragm on the lamp condenser (see Vertical Illumination', page 221).

(10) The camera should be short in the interests of stability. High magnification should be obtained by means of high eyepieces rather than by means of long extensions.

Photography without eyepieces should not be attempted with objectives of shorter focal length than 15 mm. approximately, nor with any apochromatic objective. No correction for tube length nor

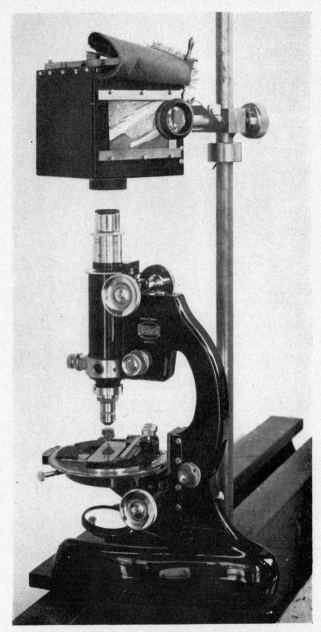

PLATE 5
Microscope and Camera Set Up for
High Power Photomicrography.

Courtesy of Engineer-in-Chief, P.O. Research Station, Dollis Hill.

for chromatic difference of magnification is obtainable without an eyepiece.

The only point above which needs extra explanation is (7).

When the microscopical image is focussed upon a screen at a given distance from the eyepiece it remains practically in focus when the screen is moved a distance of 25 mm. each side of the given position. If a reflector is arranged within the camera to deflect the image on to a viewing screen placed the same optical distance away from the eyepiece as the photographic plate, the image, when focussed accurately on the viewing screen, must be correct when allowed to fall upon the plate. No lining up is necessary and ordinary ruler accuracy only is required when building the camera. The ground screen itself makes visible a real image which can be focussed with the aid of a lens, there is then no danger of an observer's eye accommodating an image in the wrong plane, thus producing an error in focus on a photographic plate. There is danger of this error when coincidence of image and cross wire is used for focussing. The viewing screen ground glass may be rubbed with grease to make fine detail easier to see, and should be mounted in a way which allows the observer to slide it backwards and forwards in the image plane. This movement, if performed while focussing fine detail, has the effect of apparently destroying the grain of the screen while still preserving the real image seen in a magnifier, and very accurate focussing becomes possible. See Fig. 4 and Plate 1 (facing page 22) for constructional details of a camera used by the Author and built upon these principles. The viewing screen is covered with a flap of velvet during the exposure to avoid risk of light entry.

The method of construction of this camera is unimportant, the essential points being that the swinging flap carrying the mirror shall make a light tight joint with the inner box, thus protecting the plate from light when the dark slide is withdrawn and final focussing is being done. Roll film may be used by those who prefer it but the simplicity and flatness of plates makes them desirable for the best work.

When using any micrographic apparatus it is undesirable to fix the shutter to the apparatus because it is certain that vibration will be transmitted as a result of its operation, and however well constructed, a shutter is not necessary in still work if exposures are correctly arranged. The exposure should be made by shading the microscope from the illuminating system while the camera flap is operated or the dark slide withdrawn, light is then admitted to the microscope according to stop-watch timing and is obscured again while the dark slide is replaced; in this way the apparatus is not touched at all during exposure. A shutter may be separately mounted.

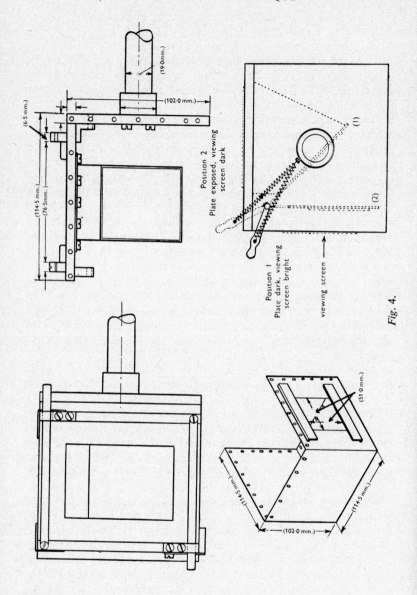

Fig. 4.

Choice of Exposure Times and Plates

In general, it is best to arrange the exposure time for still photographs to be about 10 sec. as this period of time may be measured with a stop-watch without introducing high percentage errors. Also, slight vibrations in the apparatus due to other activities in the building have a duration which is not an appreciable part of the exposure time. The periodicity of cameras when bumped varies and is about $\frac{1}{4}$ sec. in the one here described, and such a movement as this when an exposure of 1 sec. is taking place is an appreciable proportion of the total time and will cause considerable loss of definition but would have no effect upon a 10 sec. exposure. There is no point in deliberately prolonging the exposure time beyond 10 secs.

It will be found that with the apparatus described and the plates mentioned below, the amount of light required for a 7 to 10 sec. exposure is sufficient to focus with upon the viewing screen. With increase of magnifying power focussing may become difficult owing to lack of light, then colour-screens or neutral filters may be removed to admit more light during the focussing operation. (See 'Optical Apparatus associated with the Camera', page 186.)

The best guide to exposure time is experience coupled with a trial exposure, but it is possible to use the type of meter which depends for its action upon a variable brightness spot projected into the instrument field of view (SEI type). Clearly so long as light can be seen in the instrument, the projected spot can be adjusted in brightness to match the light and a figure for exposure time can then be obtained for a very low light intensity on the screen. The exposure meter must be placed a fixed distance from the camera viewing screen and a calibration drawn up of exposure meter reading against exposure required in the micrographic apparatus. The required exposure must be determined by trial and experience because it is too much to expect the commercial instrument to read direct upon its scale exposure times for a system so different from the natural one. If it is decided to adopt this method of measuring exposures it will be found successful if a particular meter is calibrated and always used, but disagreements of 200 per cent at the bottom of the low intensity range indications have been recorded by the Author in examples calibrated by the makers. The low and medium ranges easily cover ordinary photomicrographic requirements using exposure times of 5 to 10 sec.

The spot and field light colours cannot be the same when the SEI meter is used in microscopy and the user must cultivate the art of looking to the side of the spot when estimating an exposure time because this action will cause the image of the spot to fall upon the rods in the retina and not upon the cones. The rods are more sensitive to light and insensitive to colour, therefore matching of spot brightness to field brightness by this method becomes easy and will be found to give quite a definite indication of brightness equality.

Should the set-up be such that an angle viewing attachment is in use on the microscope, the exposure meter input tube may be inserted into the angle fitting eyepiece hole. For this purpose a dummy eyepiece should be made to go into the angle fitting, with a disk of ground glass resting upon its diaphragm, but with no lenses. This screen may be adjusted in position to show a real image in parallel with the image appearing in the camera. Cross wires are fitted for the purpose of focussing, for the exposure meter will not give satisfactory results unless it can see a real image. The tube of the dummy eyepiece provides a useful guide and shield for the meter input tube, the meter may then be hand held. A separate calibration must be drawn up for this. Because the angle viewing eyepiece is usually used with bright lights for photographing living objects, this application of an exposure meter is most useful.

The plates used may be as fine in grain as possible; as it is usual to have to reduce the light in order to increase exposures, the finest grain plates may be used with advantages all round. There is no particular reason for suggesting the use of Panchromatic plates except upon delicately stained specimens, but these do allow exposure times to be better estimated and standardization to be achieved. Ilford Special Fine Panchromatic plates have been found good for this work; 35 mm. camera backs with film are being used increasingly.

Optical Apparatus associated with the Camera
For details concerning the set-up of the microscope to study surfaces see 'The Study of Surfaces', page 202.

Microscope objectives are corrected for colour according to two main formulae. In the 'Achromatic' system, arrangements are made to bring two colours of the spectrum to the same focus in an image, and this means that small errors in focus will be found for the remaining colours of white light. This error, known as 'Secondary Spectrum', may be seen in the image of an achromatic system and is tolerated by the eye. It is, of course, recorded by the photographic plate and lack of definition results. There is no rule as to which colours in light are paired to obtain achromatism, the choice and result varies with manufacturers. (See 'Objectives', page 55.)

In an 'Apochromatic' system, three colours are made to form an image at the same focus but the three images may not all be of the same size except in the centre region of the field of view. This error is known as 'Chromatic Difference of Magnification'. Ignoring for the present the difference in size of the coloured images, an apochromatic objective gives a colourless image because so few colours in the spectrum remain uncorrected, and these are very nearly corrected according to the nature of achromatized systems. When such an objective is used with white light for photography with its proper eyepiece, the image will be sharp.

The difference in size of the images in the various colours of light (mentioned above) is due to the type of lens construction always used today. A solid nearly hemispherical front lens accepts all the light the system is intended to handle and performs all the magnification, the lenses above this front component correct its errors. It is not possible with this arrangement to bring all colours of light to the same focus and retain the same size of image in all colours, therefore the correction for this error is made in the eyepiece which is known as a 'compensating eyepiece'. In some combinations a compensating eyepiece helps an achromatic objective as the chromatic difference of magnification is present in achromatic systems built with single front lenses but its effects are nearly obscured in the secondary spectrum and in the outstanding spherical aberration.

Achromatic lenses used with white light usually function well only up to three-quarters of their aperture, but apochromatic objectives will function at full aperture, their spherical correction being improved together with the chromatic improvements. A compensating eyepiece may always be recognized by holding it to the light of a window, looking through it in the ordinary way, and noting the colour of light visible at the edge of the diaphragm. In a compensating eyepiece it is red-fringed, in a normal one it is blue.

Not all compensating eyepieces give the same amount of compensation so eyepieces should be obtained from the maker of the objectives in use. This fact is particularly noticeable with Zeiss objectives of 1939 make when used with oblique light.

Magnification and Resolution in Photomicrography
The resolving power of an objective is indicated by its numerical aperture, and this number will be found engraved upon the mount of modern examples. The practical maximum aperture of an objective working in air (known as a 'dry' objective) is $0 \cdot 95$, while that of one designed to work in cedar wood oil is $1 \cdot 4$. The magnification permissible before the increase in size of the Airy Disk causes definition to fall off, may be taken as 1,000 times the numerical aperture and the normal eye can see all detail in the image with this total magnification. (See 'Instruments and Use of the Eyes', page 9, and 'Magnification and Resolution', page 33.) When the image is

photographed the total magnification after enlargement may be twice this figure for optimum results because in photography the image is not presented directly to the retina and enlargement is desirable. The smallest particle or feature the shape of which can be seen with an optical microscope has dimensions of about $1/3\ \mu$ and the achromatic objective will work to this theoretical limit in monochromatic light in the green region. (See 'Interference Microscope', page 422.)

Colour of Illumination

In metallography the colour filter is not used to obtain contrast as in normal photography and in photomicrography of stained specimens, they are used to improve the performance of the microscope. The image given by an achromatic objective is greatly improved by using monochromatic light, as the objectives of this class are given the best correction for spherical aberration for the colour apple-green, this being the colour to which the eye is most sensitive. Monochromatic light also helps an apochromatic objective for, although this sytem is practically correct for all colours, the spherical correction, though better than in apochromatic systems, still functions best for one colour. Spherical correction can be made exactly right for any colour of light by adjusting the tube length of the microscope *not* the distance away of the camera from the eyepiece.

When using an apochromatic system it is rarely necessary to alter the tube length of the instrument for different colours of monochromatic light, but this alteration may have to be made when using achromatic objectives of high aperture.

It is good metallurgical practice to standardize the light filters in use because photographic exposures then become easier to estimate as experience accumulates. For consideration of the methods for obtaining monochromatic light see 'Microscope Lamps', page 130.

Setting up a Section for Photography

This section explains the photographic work only. For microscopical technique, see the appropriate chapters.

A section of a metal or other hard flat surface, when prepared by grinding, will be found sufficiently flat for photography of most of the microscope field of view to be taken in one focus. For handling, the specimen is mounted upon a glass slide by means of a piece of plasticine, then it may be pressed level by eye or by the assistance of a jig. The jig may be a plate of glass or metal $50 \times 25 \times 2$ mm. supported upon three equal legs about 10 mm. high. The specimen is set upon a piece of plasticine about 5 mm. high, the jig is placed above it and is pressed down so that the underside of the flat plate of the jig engages the surface of the specimen and carries the whole into the plasticine until the legs of the jig stop against the slide. The face of the specimen and the top of the slide are then sufficiently parallel for practical purposes. (Fig. 5.)

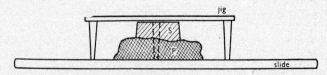

Fig. 5. S Specimen being levelled.   P Plasticine.

Very careful levelling is not necessary owing to the fact that the microscope field of view is convex and only the centre third of the field of a high power objective is useful at one focus; the higher the aperture of the objective, the more pronounced is this curvature. This effect is unavoidable if maximum definition at the centre of the field is the object of the designer. The curvature of field is of little concern during visual work when the microscope fine adjustment is used to explore the irregularities of the surface, but in a photograph a general focus must be taken and care must be exercised to select the objectives and eyepieces which give optimum results for a particular specimen. The final focus must be obtained by lifting the objective and not by lowering it, because many otherwise good fine adjustments allow the lens to settle a little and this is minimized by making lifting the last operation.

No rules can be made concerning the choice of objective focal length and aperture, but the quality and appearance of photographs is much affected thereby and also by the choice of eyepiece, it must be decided by the observer whether high definition or a general picture is required. If the eyepiece is of low power, say ×6, the field of view of the microscope as a whole is large and the curvature of the objective field appears great. Some forms of illumination (e.g. 'Contour Microscopy') do not give optimum effects over a large field and in general it is better to use eyepieces of higher power, say ×15, thus obtaining a smaller flat field and larger image upon the photographic plate. A chain of exposures along a specimen is then easier to make and easier to join up in a panoramic view.

There is rarely any need to use an eyepiece of higher power than 20 as such an eyepiece used with a 0·95 NA objective on the 160 mm. tube length microscope develops the full resolving power of the system. If higher magnifying powers are used, the photographic plate records an image of the diffraction bands and imperfections in the image which the experienced eye automatically rejects A little over-exposure and the use of high contrast printing paper makes these hazy imperfections appear real and solid in the finished photograph. Artifacts have often been described as structure, and photography can produce very convincing 'structure' if the methods are suitably arranged. Coarse grain coupled with the use of miniature cameras and over-magnification have contributed their quota to the list of artifacts which have exercised the minds of microscopists for years.

When photomicrographs of fine detail are being taken near the limit of resolution of the microscope, several photographs at slightly different focus should be taken, developed and printed. It is always difficult to decide what is real and what is artifact, but by this method the images may often be shown to pass through a distinct anti-node of condition. Different methods of illumination should be tried if possible, but common-sense is the only guide.

Before making an exposure, the eyepiece should be removed and the inside of the tube and camera inspected for stray light, as light from room lights and many unsuspected sources may enter through the illuminating holes in the nosepiece and other places. Make sure no one is about to shift machinery or bang things in the same room.

Development of Plates

It is bad in practice to try to develop plates in numbers in open dishes because it is not possible to place them all in the developer at the same time, nor is it possible to pour the chemicals in and out without the danger of the plates moving over each other and preventing the order of exposures in a series being preserved. The method recommended employs a tank with an internal removable rack in which plates may be stacked on edge and in order, as they are taken from the dark slides. When the rack is loaded it is immersed in the developer; when development is complete the loaded rack is removed, plunged in water and immersed in a tank of fixer solution. The whole is then washed and dried and the plates are numbered as they are removed. It is important in micrographical studies as in ordinary work that plates drying must have their gelatine surfaces protected from dust, and this is best achieved by inclining the racks so that the gelatine is on the underside, the ordinary room radiator with a wire gauze shelf placed about a foot above being a satisfactory plate drier It is true that the radiator causes an upward draft which may carry dust particles, but on the whole it is better to dry the plates quickly rather than expose them for a long time to the atmosphere of a typical commercial building.

Photomicrography in Colour

There is a great future for colour photomicrography because of the useful results which can be obtained from stained specimens and because the transparencies can be projected easily, this also being the most natural way to view a reproduction of the microscopical image. The trouble of specially producing lantern slides from photographs is avoided.

Basically, the methods of photography are similar to those which obtain in black and white work but some important details need attention. Exposure time is critical in colour work and in this, as in black and white, experience is the best guide. Because it is difficult to judge differing degrees of brightness of parts of an image, it is best to use as an exposure guide integrated brightness for the screen image of all subjects and adjust the illumination level to arrive at a standard figure. Any convenient device may be set up to form a comparison for adjusting the light level, and the Author has found that a sheet of ground glass illuminated from behind by means of a flashlight bulb and masked to make it the same size as the viewing screen, is convenient, adjustable and cheap. Special comparator devices are not necessary as the eye soon learns to switch from viewing screen to standard screen and to compare them accurately. The level of standard illumination is set up by adjusting the distance of the bulb from the ground glass.

If electrical means are used to alter the brightness of the microscope illumination, colour changes will be introduced, therefore graduation should be arranged through neutral wedges which may be improvised from lengths of 35 mm. film, fogged in a graduated way and drawn through a gate to keep them flat. The Author has used a method of obtaining graduation by mechanical-optical means which has the advantage that it can be calibrated. A similar arrangement has been developed independently and used with success by Mr Woollard and is described below.

The light from a filament or ribbon lamp run at full brightness is focussed by means of lamp condenser (1) upon a sheet of glass ground on both sides. A diaphragm is placed close to the microscope side of the illuminated patch on the ground glass and is used to control the apparent size of the patch. In front of the diaphragm is placed a lamp condenser (2). The intensity of the patch of light is controlled by means of the first lamp condenser (1); as this is moved to and from the lamp the patch of light on the ground glass changes in size and brightness, this condenser can be uncorrected. The change of size (if unwanted) is concealed by the diaphragm while the second, corrected, condenser (2) provides nearly parallel light and becomes the apparent source of illumination. (Fig. 6.) The distances apart of

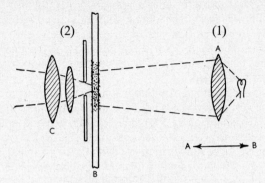

Fig. 6. A. First condenser moveable A to B.
       B. Glass diffuser ground both sides.
       C. Second (corrected) lamp condenser.

the components are measured by means of a ruler, and the current through the lamp by an ammeter. The lamp current should be always the same and intensity measurements are taken in terms of distance, screen to lamp condenser (1). By this means a good white light, obtainable from any source desired, can be changed in intensity at will without in any way disturbing the optical set up for the photograph. When a particular lamp and ground glass screen have been selected, trial experiments with the colour of the light on the film chosen can be made. It has been found worthwhile to select a few coloured specimens of blue, red and green, and to keep these for testing the apparatus or making comparisons between different sets of films. It is sometimes necessary to employ a Wratten 80A filter in the light beam to correct the preponderance of yellow in most artificial sources, but it will be found that ribbon filament lamps and arcs do not need correcting for microscopical purposes.

Having stabilized the colour of the source, the check can be made to make sure that all colours of the spectrum are present in reasonable amounts. Microscopical work is not so critical as ordinary natural colour photography but some colour correction of the source may have to be made in order to bring out all the colour differentiation that the method will give. Rules cannot be given as all sources vary, but in general the ordinary daylight filter supplied for use with oil lamps is a satisfactory correction for ordinary filament lamps. To make trials with colour, chemicals, say copper sulphate and potassium bichromate, should be crystallized upon a slide and used as objects; preferably both should be mounted in the same field of view. Accuracy of colour in the micrograph can then easily be judged against larger quantities of the chemicals comparable in amount to the size of the film image. When the colour of the light is found to be correct, a similar tint should be applied to the bulb illuminating the standard screen. The Wratten 80A filter increases exposure 4 times.

It is often found less easy to match against the standard an object lying upon a dark background than one upon a light background, but it will be found easy to integrate the picture if it be viewed through a tube of about 12 mm. diameter which has a disk of ground glass fixed somewhere near its lower end. The tube should not be greater than the suggested diameter because it is only a feature in the total picture the brightness of which is wanted. If the whole screen, object and surrounding field be integrated, the brightness indication will be too low and over-exposure of the bright parts will result.

For remarks concerning microscope illumination see chapters 'Microscope Lamps', page 115, and 'General Microscopy', page 46.

Colour Filters and Their Uses: Manufacture of Absorption Filters
Ordinary colour filters consist of a clear material usually glass or gelatine, loaded with some colouring materials chosen to have a perfectly homogeneous distribution in the base material. When glass is the base material, ordinary crown-type made up from sodium, calcium, potassium and manganese silicates is used as the base, but slight changes sometimes have to be made in this base to accommodate the idiosyncrasies of certain colours which are affected by the degree of oxidization in the base. Blue-green colours are given by adding copper oxide, blues by cobalt, greens by chromium and violets by manganese, the amount of added material being very small, usually only a trace. Tiny amounts of arsenic oxide in the melt seem to preserve in it an oxidized state and to replace the soda by potash, and the spectrum of the stain colours is usually moved towards the red by this condition. Colours in the melt, or in different melts, may be mixed thus forming selective and band pass filters. Some materials, e.g. neodymium oxide, when dissolved in the glass, produce band stop effects, and neodymium when present in amounts of about 15 per cent causes great absorption between the limits 5700A and 5900A.

The advantage of using gelatine films is that many dyes may be used to produce special characteristics when the dyes cannot be dissolved in glass. Such filters are usually made with glass outer faces which may or may not contribute something to the optical properties, and it must be remembered that these filters will not stand heat from a lamp.

Glass filters may be optically worked and the usual method of producing the basic material is to dissolve the colouring matter in the pot and then blow a cylinder about 1·0 metre long and 0·5 metre diameter on an ordinary blow-pipe. This is the same as the old way of producing window glass. The cylinder is then opened by being cut with a diamond when cold and flattened by extra heating. The thickness of the sides of these cylinders is fairly uniform and is as good optically as window glass made before 1939, and for most uses this optical quality is good enough when selected areas are used. Colour screens should always be placed in such a position in the optical system that an image of them is not formed in the microscope field. They should be placed either in the substage immediately below the condenser or about 150 mm. in front of the lamp and this applies whether or not the filters are optically worked. Any filters are likely to become fingermarked, and no more surfaces than absolutely necessary should be in focus because they tend to produce specks in the field.

Gelatine filters must not be placed near a lamp because neither they nor the cement, if of the built-up type, will withstand the heat.

If filters are to be used within image-forming apparatus such as a microscope, camera or telescope, they must be optically worked and either selected areas of blown glass described above, or cast blanks can be ground and polished by ordinary techniques. Transmission interference filters are in the optical class.

There is another class of absorption filter known as the Colloidal type, and in these, finely divided particles of gold salts or copper oxide are caused to spread through the glass. They give several varied colours but a clean cut-off is not to be obtained, and in general this kind of filter is not useful to the microscopist.

The manufacture of all these filters is empirical, the results are due to experience, and results can seldom be reproduced exactly in subsequent melts.

Ultra-Violet Filters

In any glass intended to transmit ultra-violet light, iron contamination must be avoided as an amount 0·001 per cent of iron oxide seriously interferes with ultra-violet transmission though it does not colour the glass. A large amount (70 per cent) of nickel oxide dissolved in iron-free glass allows a band of ultra-violet between about 3000A and 4000A to be transmitted with 98 per cent absorption of visible light. When ceria and neodymia are present, the glass attenuates ultra-violet light very greatly but passes visible light. Glass known as 'Crookes' (after Sir William Crookes), made with those materials incorporated, is much used for sun glasses and though it may be no more coloured than ordinary glass, stops the tiring ultra-violet radiation. The glass may, of course, be coloured as well to reduce the brightness of the visible light.

The main use of glass which transmits ultra-violet light is in fluorescent microscopy where materials are rendered fluorescent by ultra-violet illumination but are examined by the visible light they emit. For ultra-violet microscopy proper, where ultra-violet light forms the image, a more elaborate generator and quartz lenses are necessary. (See 'Ultra-Violet Microscopy', page 484.)

The only source which generates enough ultra-violet light for use in fluorescent microscopy is the mercury arc, and the modern high-pressure mercury arc lamp of the Osira type (GEC) can be bought with a dark envelope of ultra-violet transmitting glass. (See 'Fluorescent Microscopy', page 481.)

Filters may be used for obtaining monochromatic light, but unless the source be of the line type, the results are only partially successful. Examination of the maker's catalogues (Messrs Chance Bros) will show graphs of the pass bands of many filters and combinations, and in the Author's experience these may be relied upon. The amount of attenuation shown for unwanted colours should be carefully noted because often it is not very great and graphs can be drawn to any scales. It will be found that all the high frequency transmitting filters also transmit in the red. If it is decided to use transmission filters for the purpose of producing monochromatic light, a line source (say) the high-pressure mercury arc (Osira type in a clear envelope) should be used because in this source all the light energy is concentrated into a few lines instead of being spread out in a continuous spectrum as from a solid source emitter. It will be found much easier to select glasses from the catalogue to give combined characteristics suitable for rejecting widely spaced lines than to select filters with characteristics to transmit a narrow pass band only. With a line source the line is its own pass band, because the colour of the line is itself monochromatic and cannot be altered by any filter, so it is necessary only to get rid of unwanted lines which, in the mercury arc, are sufficiently far apart for this to be done with a minimum of four filters, green, blue-green, didymium 1 and didymium 2 of Messrs Chance Bros, for monochromatic green light.

Colour filters are usually used to produce increased contrast in the microscopical image, both in visual and photographic work, and true monochromatism is here not the quality sought in the filter.

When the specimen is coloured, the maximum visibility occurs when the field is illuminated with the complementary colour, the specimen then appearing black, and when seeking specially stained isolated specimens like tubercle bacilli, this is the method usually employed. In this case, the red bacilli are easiest seen by blue-green light, protoplasm is best seen in the living state with an apple-green light, but in many specimens a structure is also sought, not only contrast. In this work a filter must be chosen which aids visibility in the wanted part but does not black out internal detail nor render surrounding structures invisible, and the only rule is one of trial until the best visual effect is obtained. So long as panchromatic plates are used, a photograph will also be correct. (See section 'Photomicrography of Sections' for remarks about choice of objectives, page 188.)

The effect of colour filters should always be tried when making an observation because sometimes the smallest change in light colour makes a great difference to the visibility of detail, especially when achromatic objectives are used as the spherical correction of some achromatic systems is markedly favourable to one colour of light. The greatest advantage from colour filters may be had when stained sections of any kind are studied.

Interference Filters

These may now be purchased from the manufacturing opticians to transmit very narrow bands of monochromatic light filtered out of any spectrum containing the frequency required. The principle of their operation is as follows, Fig. 7:

Two plates of transparent material A and B are ground flat and spaced parallel with each other a distance (y) apart. A ray of light passes through A with some loss by reflection, on through B and so on its course. Some light will be lost at the surface of B as on A and this light is reflected back to A. From A some is again reflected to B and so on its course, division taking place at each surface between light reflected and light transmitted. The length of path between the plates, taking the length as three reflections, may be manufactured to be one wavelength of the light wanted. This being so, the part of the ray passing through the system without reflection will be 360° out of phase with that which experienced reflection. These components are coherent, therefore they add together arithmetically. The light input to the system may consist of many frequencies in the visible spectrum, but for only one of them will the distance (y) be equivalent to one wavelength and the output of other frequency components will therefore not add and strengthen each other to the same extent.

If a frequency a multiple of the selected frequencies is applied, it will pass through the device, the reflected and direct components being out of phase by multiples of 360° which is the same as being in phase. Fortunately in visible light it is only just possible to experience this condition with red and blue interference filters. The defect can be remedied by adding an absorption filter to cut out the multiple not wanted.

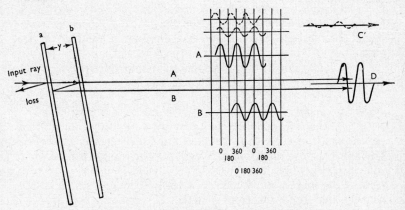

Fig. 7.

y Spacing=pathlength, 3 reflections=1 wavelength of colour wanted.
A Direct Ray.
B Reflected Ray.
C Odd phases of Rays not selected by (y).
D Direct+reflected rays.
C_1 Small amplitude products from C's.

The action of the interference filter can be seen when ordinary glass is used in its structure, but it is necessary in practice to accentuate the reflections at the inside surfaces of A and B. When this is done, two effects are more noticeable, firstly, more energy is reflected between the plates, therefore more appears finally to add to the direct beam, and secondly, multiple reflections are possible between the plates, thus accentuating the frequency discrimination of the system. If the inner surfaces are complete reflectors, no light would come out of the system; the degree of silvering usually permits about 40 per cent transmission. Interference filters may be used in cascade.

The interference filter cannot cut off suddenly and completely unwanted colours, but can exert a very powerful selective influence upon one frequency – that for which the space (y) was designed. Combined with uncritical absorption filters or line sources of light, extremely narrow bands may be selected and this type of filter is very useful for phase contrast microscopy. The essential part of the structure is that the faces A and B shall be accurately parallel with each other and it does not matter whether or not the whole assembly is flat. In practice, no two pieces of glass could be spaced the required distance apart without very special precautions being taken to preserve the assembly. Parallel sided films are obtained of the correct thickness, and coatings are built up on these by vacuum evaporation of materials of selected refractive index. Aluminium and rhodium are used to make partial reflecting surfaces. In a similar way filters can be built up by alternately depositing partial reflector, transparent film, partial reflector, and backing upon a sheet of optical glass.

For a description of the Lipmann Reflection Interference Filter see the section 'Monochromators', page 130.

Darkroom Lamps

In the Author's experience it is best to work with standardized panchromatic plates or films but there are experiments which require safe lights to be used. A coloured light may be formed from a white source in two ways, first by scattering a pigment upon a screen in front of the source, and second, by enclosing the source in a homogeneous coloured transparent material. If pigment is used in (say) the form of coloured paper, a large amount of white light leaks through the interstices of the coloured material and causes fogging of a photographic plate, though if the speckles of white are very small they are not individually visible to the eye and a false sense of security is given. If a homogeneous colour like coloured glass is used, it may still be found to transmit parts of the spectrum 'hot' to emulsions and the only sure way to test these matters is to expose half a plate for at least ten minutes to the safe light and compare the result after developing with the covered part. The original selection of a glass filter should be made with a spectroscope supplied with light from an opal domestic electric bulb, with its slit wide open. Particular attention should be paid to the amount of light at the extreme violet end of the spectrum because this is a very active region and the colour is not easily seen with the eye.

Darkroom lamps of any degree of goodness should never be placed over delicate work. They should be so arranged that diffuse light from a white wall falls upon the job. Under these conditions much more light can be used (ascertain by previous trial exposure), and the eyes obtain a better picture in the poor light. There is no point in painting a photographic darkroom black because this means that more direct light from the safe light has to be used to see the work and there is no convenient white wall against which to see the negative. If the darkroom leaks, no amount of black absorbing material on the walls will make it any good for real work.

Cine Photography

A cinematograph camera may be mounted on its own supports over the eyepiece of a microscope in exactly the same way as any other type of camera. It is not strictly necessary to remove the taking lens but, as it has no function in the microscopical set-up, it had better be removed. The camera should be driven by an electric motor because the operation of the handle is very liable to cause shake.

Films may be exposed at normal speed, that is, they will be projected at the same speed as they were taken. For this work, 16 exposures per sec. with a duration of 1/32 sec. and a bright light from an arc, ribbon lamp or over-run filament, should be used as a starting set-up. Any of the methods of illumination mentioned in the various sections may be used and it will be found that dark ground illumination of living specimens gives films of greatest entertainment value. If it is intended to take normal speed films of objects under objectives of higher power than 6 mm., a carbon arc must be used for dark ground illumination. It is nearly essential in this work to use an angle-viewing eyepiece so that the specimen can be kept under observation during a take of, say, several minutes. This is a long shot by ordinary standards of cine photography but most micro-organisms require prolonged study if the film method is to be anything other than entertainment. See Appendix 'Flash Tubes'.

Speeded-up Cine Films: Time Lapse Photography

This method of photography has yielded remarkable results in all branches of natural history. Exposures of an organism are taken at widely spaced intervals of time which may vary between one per second and one per hour. The rate of making exposures is slow when plants are being studied and relatively fast when animals are the subject. The chosen speed must be such that a smooth picture of the changes in the specimen are obtained when the completed film is projected at near normal speed; if one is looking for unknown movements in a specimen normally apparently stationary, like a sea anemone, a speed of one exposure per minute should be tried first. A projected speed of 16 frames per second lasting for one minute represents 960 exposures, and if these were taken at one per minute the total time of observation of the specimen would be forty hours, or say two days. Other taking periods may be estimated from this.

Many slow changes can take place in a specimen during two days and many of these may be cyclic and quite lost to the ordinary observer. He has no standard with which to compare the slight movements of his apparently stationary subject, also he cannot readily observe all the time. The film, by contracting the time base, appears to amplify movements and allows comparisons to be made between exposures, thus showing true movement of a specimen even though some changes may have taken place on the slide as a whole.

Systems of this sort are always made automatic and two types of control are available; first, a time clock may be made to operate a relay which switches in the light and opens the shutter at the predetermined intervals, and the second is the method which operates the light source only, the camera shutter remaining open and the film being moved through the gate by any slow-moving clockwork convenient. Several problems have to be solved in both cases. If a relay system is used to switch on the light, a delay must be arranged to make sure the lamp has reached full brightness before the shutter opens. If the lamp only is switched, a condenser discharge arrangement must be used in order to obtain sufficient shortness of exposure. Possible arrangements are as follows:

(1) A rotating disk is fitted to an ordinary clock pointer spindle. Pins in this disk are arranged to trip camera film gate, lamp switch and shutter by mechanical means. Any operation may be arranged to occur first by so positioning the pins or slots on the disk.

(2) The lamp may be left burning all the time and a camera shutter placed over it, operated by current in a magnet, or by mechanical means. Thus, lamp shutter and camera shutter can be operated with one device simultaneously.

If arc lamps are used it is practically essential to employ system (2) because these lamps take up to 15 min. to strike and settle down and their lives are shortened by a factor of 4 due to frequent switching. Ribbon and heavy filament bulbs need about 4 sec. to light and become steady so these may be switched by system (1). A useful switching system may be arranged by causing a wire to dip into and pass through a trough of mercury, the wire being attached to the seconds hand of a clock. The length of time the contact is made per revolution of the pointer is adjusted by varying the depth into the mercury that the wire dips. If a pendulum clock is used, variations of contact frequency between 4 per min. and 1 per 2 min. can be obtained by using pendulums of greatly different lengths, the actual time kept by the clock being of no importance. Contact frequencies of the order of 2 per sec. can be arranged by placing the dipping wire on the pendulum instead of on a rotating part of the clock.

Another advantage of using common clocks is that two will keep in step for many hours and can be used to switch parts of the circuits independently, thus making for easy timing.

Possible arrangements are nearly infinite but the result must be that an exposure is made of the correct length of time with a lamp at full brightness. It matters not at all where the shutter is placed or at what speed the film is moved through the camera gate between exposures.

Colour of Filter	Colour of Object	Typical Stain
Green, blue	Red	Carmine, fuchsin, eosin
Blue	Orange	Eosin, methyl orange
Blue, violet	Yellow	Safranin
Red	Green	Light green, methyl green
Orange, green	Blue	Haematoxylin, methylene blue
Yellow	Violet	Gentian violet, methyl violet

COLOUR FILTERS FOR MAXIMUM CONTRAST EFFECTS

TABLE 7

Complementary Colours	
Red	Green
Orange	Blue
Yellow	Indigo
Green	Red
Blue	Orange
Indigo	Yellow
Violet	Green-yellow

TABLE 8

Dioptres	1	2	4	5	6·5	10	20	40	50
Centimetres	100	50	25	20	15	10	5	2·5	2

$$\text{Focal length in cms.} = \frac{100}{\text{Dioptres}}$$

FOCAL LENGTH OF LENSES

TABLE 9

DEPTH OF FOCUS OF A MICROSCOPE OBJECTIVE

Depth focus in millimeters =
$$\frac{\text{Refractive Index of Medium between objective and specimen}}{10 \times \text{Numerical Aperture} \times \text{Magnification of objective}}$$

For Flash Tube illumination system see Appendix.

CHAPTER 11

The Study of Surfaces

THE EXAMINATION and photography of the surface of an opaque object has always been a matter of difficulty, yet it is one of the most important ways of studying an object and is sometimes the only way.

There is little difficulty when lower power lenses are in use because the working distance between the front lens and the object is sufficiently great to allow a beam of light to be projected from an electric lamp and condenser upon the surface to be studied. When the working distance becomes less than about 6 mm., the illuminating beam has to strike the object at an angle with the normal of about 50°. Sufficient light may be obtained in this way but if the object has an irregular surface, the shadows obscure detail, sharp edges and points cast diffraction patterns into the field, and the failings of the objective in the matter of colour correction become too apparent for serious work. When the working distance is reduced to, say, 1 mm., as is usual with most 4 mm. objectives, top light with the above arrangement becomes impossible. Several methods of vertical illumination have been tried and all are successful in some circumstances.

Vertical Illumination with Low Powers, 25 *mm. to* 100 *mm. Objectives:*
   Top Light
This consists of focussing a lamp upon the surface of the specimen, the illuminating beam passing beside the objective, Fig. 1.

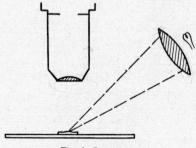

Fig. 1. Low power.

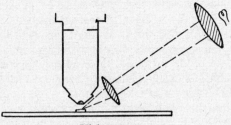

High power.

Eighty per cent of low power casual laboratory inspection is done with this sort of device and so long as the objective has a working distance of more than, say, 25 mm., there is nothing to complain about. A motor-car headlamp bulb running at 6 or 12 volts is satisfactory as a source, and the lens may be a simple plano-convex or double convex of focal length about 75 mm. and diameter about 50 mm.

If the specimen is bright like a piece of metal, this simple system is not satisfactory because the specimen reflects the illuminating beam according to the law which states that the angle of reflection equals the angle of incidence. The results are not so much misleading as they are inadequate to show the surface detail. It is impossible to obtain any scatter of light into the microscope when using a 75 mm. objective to study a bright, etched or polished metal specimen in this way. If the surface has irregularities in its structure, bright light will be reflected from some parts and not from others, depending upon the angle with the normal made by the surfaces. Should the irregularities be of a dull kind, they appear bright against a dark background as in dark ground illumination.

To overcome this difficulty either daylight may be used because this falls upon the object from all directions, or the ring illuminator may be arranged, Fig. 2.

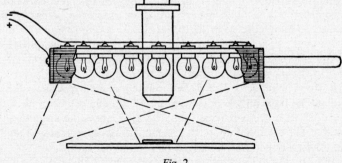

Fig. 2.

The ring illuminator is simply a dozen or so flash light bulbs mounted round a ring of insulating material about 100 mm. diameter. This ring may be clipped on to the microscope nosepiece or may be mounted upon a separate stand. A stray-light screen made of metal polished on the inside surrounds the whole. A type of bulb is now readily available in shops supplying electrical accessories called a 'lens-fronted' bulb. These bulbs take about 0·3 amps at 2V and project a roughly parallel beam of light 6 mm. diameter. A ring illuminator built with these bulbs and duly focussed at a particular working distance is very effective for all surface studies at low and medium power. The bulbs fit the standard flash lamp screw thread. The holders can be bought separately and secured by soldering stiff wire to their electrodes and to screws in the insulating ring. They may then be bent to a correct focus for a particular microscope, Fig. 3.

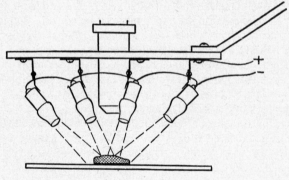

Fig. 3.

The Silverside Reflector

This device is still of considerable value as a surface illuminator when only a research type of lamp is available for all work. It consists of a hemisphere of brass with silver plating on its inside, a part being cut away to allow it to be brought near to the stage as shown. It may be mounted upon its own ball-jointed arm or upon the objective, and old microscopes had holes drilled somewhere in the stage or body for the purpose of holding this arm and other similar pieces of apparatus. The reflector should be mounted on an arm solid with the stand rather than on the objective, as the illumination may then be set up for all objectives which may be used upon the specimen. It should not be mounted upon the mechanical stage but should remain fixed while the specimen is moved or rotated beneath it.

The silverside reflector is positioned to reflect light upon the slide as shown, with its concave side nearly enclosing the object. A focussed beam from the lamp is then directed at the reflector in such a way that the reflector casts a roughly focussed image of the lamp condenser upon the specimen.

The reflector has not the disadvantage of the simple top light described above, because light falls upon the specimen from most points upon what is practically a hemisphere, so this illumination does not cause sharp and misleading shadows. The advantage of the device over the preceding ones is mainly that any bright light, say, from a high intensity filament or a mercury vapour arc lamp, may be used. The small bulbs described in the ring illuminators are not bright and cannot be usefully used with a monochromator. The silverside reflector can be used with advantage to illuminate objectives up to 12 mm. power with monochromatic light.

Difficulty might be experienced in obtaining a silverside reflector because this is one of the useful devices of pre-war times which no longer appears in manufacturers' lists. A little consideration will show that a particular curve, spherical or paraboloidal, is not called for in this device; in fact, a somewhat scattered light is better. A good reflector may be made from an old silver or silver-plated teaspoon. The handle, when filed thin and bent suitably, makes the supporting arm which may have ball joints soldered on or be simply bent to shape. Instead of moving the reflector to obtain the best illumination, the lamp may be moved, Fig. 4.

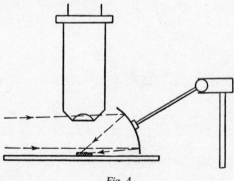

Fig. 4.

The Leiberkuhn

This is a similar device to the silverside reflector but is mounted to surround the objective like a skirt. Like the silverside reflector, it causes light to fall upon the object from many directions. It also has the advantage that any light source may be used. In this device the light must be applied from the substage around the specimen, the specimen being mounted upon an opaque stratum of dimension about 6 mm. diameter. There is little commercial use for the Leiberkuhn because a special mount has to be made and the device itself has to be fitted to its particular objective, and owing to the bulk of the objective, it cannot properly illuminate powers greater than 25 mm. unless the working distance of the objective is greater than about 10 mm. The ring illuminators now do for low power work what the Leiberkuhn did in the days before miniature electric lamps. Fig. 5.

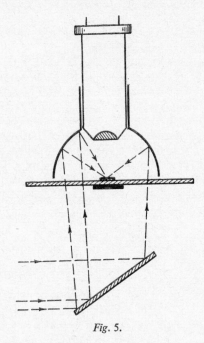

Fig. 5.

The Leiberkuhn arrangement makes an excellent illuminant for hand magnifiers and this was its original purpose as used by Descartes. A special Leiberkuhn need not be made for illuminating a hand magnifier as only the centre portion of a simple lens field is useful for image formation, therefore the outer zones may be silvered and used for illumination. Alternatively the silver spoon idea may be employed as an external Leiberkuhn. (Fig. 6.) A hand magnifier made with a large lens of steep curvature stopped down by the silvering of the reflecting part makes an excellent general inspection hand lens.

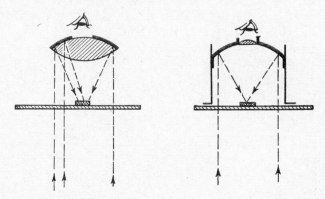

Fig. 6. Simple lens silvered on top side.   Descartes Microscope.

The Amici Illuminator

This is a simple device for obtaining light of great angle with the normal and is used for research purposes when very low contours are being studied. As an illuminator for examining ordinary specimens and for obtaining ordinary pictures, it has no value. When used under carefully controlled conditions and when in the hand of an expert, differences in levels on the surface of a specimen of a micron or two may be made visible. A slight change of height on a surface will cast a shadow even though this shadow may be confused with diffraction bands in both the light and dark areas. The shadows produced by irregularities may be interpreted if a calibration is used. It is exceptionally difficult to make a suitable calibration thickness and then put it in the correct place, therefore the method is best used in a qualitative way for such a purpose as exploring shallow etch marks which may be pits or prominences or merely colour changes only. If calibration is to be attempted, thicknesses of cleaved mica may be measured with a surface gauge and then cemented upon the slide or object being studied. A more complete treatment of the study and calibration of surface viewing devices will be found under 'Tolansky's Method of Contour Microscopy'. Fig. 7.

Brooks Reflector

This apparatus is associated with the paraboloid dark ground illuminator. The Brooks reflector consists of a flat mirror surrounding the objective and the arrangement of the two parts can be seen from the diagram. The apparatus was designed in order to obtain illumination at high angle without chromatic aberration. The paraboloid is described under that heading in the section dealing with transmitted

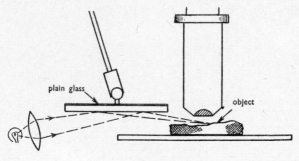

Fig. 7.

light, and it must be one which works in air and not of the immersion kind; in this application it must be raised until it touches the slide, i.e. well above its normal focus. This system shares the disadvantage with all Leiberkuhn types that the object must be specially mounted so that light can pass round it on to the reflector mounted on the objective. The ordinary paraboloid has a top which is about 25 mm. in diameter, therefore there is no serious difficulty in mounting a piece of an object if the object can be cut. There is considerable advantage to be had in a system employing only reflection to illuminate the object as tiny pits upon a surface are very apt to become artificially coloured by the illuminating light. Colouring may take place due to interference phenomena of various kinds, but this is less likely to mislead an observer when the light is applied from many directions at one time, as in this system. The Brooks reflector can be used with success to illuminate the upper surface of a mounted specimen but reflections from the cover glass and consequent loss of light upon the specimen must be tolerated.

The reflector must be specially made for a particular objective, Fig. 8. The hole through which the objective front lens receives its light must be only sufficiently large to house the lens. A piece of well-polished stainless steel on silver with hole countersunk on the objective side and stiff wires soldered to its back, will locate nicely upon the objective and be to some extent adjustable. Fig. 9.

The paths of the rays of light are shown in Fig. 8 and the best results are obtained when uncovered specimens are studied in monochromatic light. A bright source is necessary because the losses in the system are great.

The Simple Vertical Illuminator
The above two methods for obtaining illumination of opaque sur-

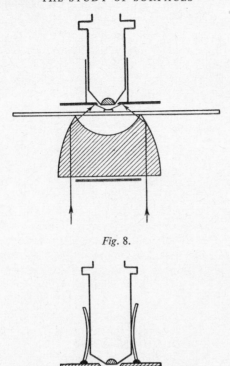

Fig. 8.

Fig. 9.

faces are mainly of academic interest and may be useful for special purposes, but the ordinary vertical illuminator described below serves best for the great majority of surface studies.

The older objectives made between, say, 1890 and 1914, which in general were very good, still did not work very well at full aperture. However, they give excellent images in transmitted light with the substage condenser diaphragm adjusted to limit the illuminating beam to three-quarters of full diameter, and it was in this way that they were mostly used. In almost all circumstances when a vertical illuminator provides the light, the full aperture of the objective is in use. If there is to be a minimum of haze fogging the image, the objective must be perfectly corrected for spherical aberration out to its periphery. The old objectives and modern ones of lower quality which are not well corrected at their edges will not bear vertical illumination. It was necessary in the past for these reasons to employ

devices which transmitted light to the surface of the specimen without the illuminating light passing through the objective on its way
to the object, thus cutting out one set of reflections and aberrations.
There must be some reflection from the surfaces in any lens system,
but the effect of these can be minimized. Modern ones are coated.

The simple vertical illuminator consists of a piece of very clean
well-polished glass mounted above the objective at an angle of 45°
with the axis of the microscope. Light from a lamp at the side of the
microscope is partially reflected from the glass plate through the
back lens of the objective to focus upon the surface of the object.
The image-forming light from the object passes up through the
objective, through the inclined glass plate and on to the eyepiece.
A loss due to partial transmission occurs when the illuminating light
first encounters the inclined glass plate and a loss due to partial
reflection occurs when the image-forming light encounters the plate
on its way to the eyepiece. Both of these losses cause trouble due to
reflected light, or due to it causing other objects to reflect light, into
the image-forming beam from the objective. Fig. 10.

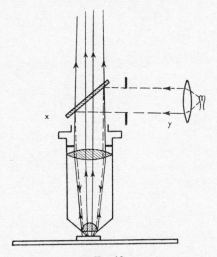

Fig. 10.

Mechanical Considerations associated with Vertical Illuminators
When a microscope is used visually, the eye can tolerate a lot of stray
light and still differentiate the detail in the object, but if there is much
scattered light, say, from bright tubes supporting the lens system,
contrast in the image deteriorates, and when a microscope is used
for photography fogging of the plate occurs even when there is only

a very small amount of scattered light present, unnoticed by the eye. This matter is dealt with in greater detail under the heading 'Photomicrography'. When a vertical illuminator is being used either for photography or visually, the effects of stray light on contrast and visibility of fine structure are much greater than in any other method of illumination.

Many modern microscopes still are built with body tubes too narrow. No surface, however well blacked, appears black when light touches it at grazing incidence, therefore microscopes habitually used for critical surface studies should have body tubes not less than 50 mm. inside diameter so that light from the objective does not touch them. The body of a research microscope must be fitted with a draw-tube which is usually suspended within the body by its bearing at the top. This arrangement prevents the draw-tube rubbing the inside of the body while in use as such rubbing causes bright areas which reflect light. This feature of the suspended draw-tube is quite adequate for most microscopical work and a check must be made that it is present on any instrument intended for serious use. The suspended eyepiece is also a feature in modern instruments and is nearly as important. Fig. 11.

When a vertical illuminator is in use, light is sent down upon the objective from the inclined glass plate and much is received by the inside of the bottom of the microscope body tube X, and by the top of the limiting diaphragm behind the objective back lens Y, while some portion of the illuminating beam passes through the inclined plate and illuminates the inside of the body tube at Z. This patch at z is in the correct position for the inclined glass plate to reflect it direct into the eyepiece. If the surfaces x and diaphragm Y are manufactured to slope slightly in the opposite direction, light falling upon them is reflected on to the body tube wall, much attenuated by its surface and stopped by the diaphragm on the bottom of the draw-tube.

The light coming from Z is best avoided by having a hole drilled through the microscope tube opposite that admitting the illumination beam to let out the waste light. Such a hole is advantageous for the additional reason that the vertical illumination may be applied from either side so long as the inclined glass plate can be moved to suit the new direction. Light will not enter through this hole unless there is a light nearby but it is desirable to cover the hole not in use with a velvet flap outside, stuck to the microscope tube with sticky tape.

The only practical way of stopping reflections from X and Y in the ordinary microscope is by sticking velvet disks over these parts. It will be found that only a pile surface appears black for angles of

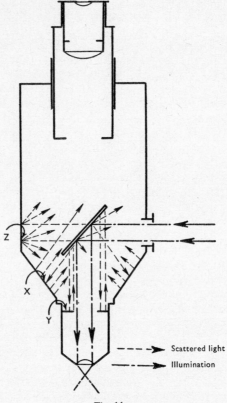

Fig. 11.

incidence 0 to near grazing angle. Velvet and similar cloths may be prepared for cutting by soaking the back with liquid glue. When this is dry, the velvet can be cut with a razor and will not fray at the edges. A type of lining material known as Paper Velvet is very useful because it has a paper back and pile front applied by spraying, and can be glued and readily cut to small shapes without trouble due to fraying; it can be obtained from photographic material manufacturers.

It is sometimes necessary to reduce with a temporary diaphragm the size of the hole at the bottom of the draw-tube because this hole is usually threaded to take an objective and the threads sometimes shine.

The choice of microscopes for various purposes will be dealt with in the following sections but it should be mentioned here that a microscope with a racking stage is most useful when a vertical illuminator is being used. The body tube then remains stationary with respect to the illuminating system and coarse focussing is done by racking the stage. The tiny movement due to the operation of the fine adjustment may be disregarded. Some modern instruments and most old ones are built with the nosepiece only moveable by the fine adjustment. If other things are equal, such an instrument is excellent for work with opaque specimens.

The fixed inclined glass plate is quite satisfactory as a vertical illuminator so long as the position of the lamp is adjustable in all directions and is inclinable, but it is much more convenient to be able to move the reflector to suit the lamp. Most commerical reflectors are mounted upon a pin which passes through a slot in the body tube or in a separate tube screwed between the objective and nosepiece for the purpose of accommodating the reflector. In addition to providing a bearing for the reflector spindle, the slot allows the reflector to be removed bodily. In an instrument made by Messrs Beck the reflector can be moved in both planes by means of concentric milled heads and this is a great advantage, especially when the microscope and illuminating apparatus are mounted upon an optical bench when they are not readily moveable with respect to each other and in fact should not be moved or the main advantage of an optical bench is lost.

The Prism Vertical Illuminator

This apparatus consists of a tiny prism large enough to cover an area about one quarter of the area of the back lens of the objective, which is usually fitted specially to a particular objective but may be removed from the optic axis at will. The advantage of this arrangement is that light may be applied to the objective without it having to pass through its central regions to reach the object. The light is therefore always oblique but many internal reflections in the objective lens assembly also pass to one side and do not interfere with contrast in the image.

On the whole the prism illuminator works well but is not so flexible in action as the adjustable glass plate. The fact that the illumination is always oblique is no great disadvantage so long as the stage of the microscope is rotatable so that a specimen can be presented at all angles to the illumination. There is usually difficulty in filling the useful part of the field of view when using low powers and all that can be done with the prism type can also be done with the inclined plate, while the inclined plate has some other advantages, namely that more light is available and a proper image of the source can be projected upon the specimen.

Concentric Systems for Vertical Illumination

When employing these arrangements the aim of the designer is always the same, that is, to cause light to arrive upon the surface of the object without having to pass through the objective lens system and there set up glare due to lens surface reflections, mount reflections and imperfections. All concentric systems must illuminate by oblique light and though this is not always a disadvantage, careful thought should be given to the job in hand before these expensive accessories are purchased. Oblique systems work well upon bright smooth specimens such as polished metal sections and minerals. For the study of materials with a surface like paper they work well and have the advantage that they are easy to set up. The light source is built on to the illuminator, therefore it is always in adjustment and the whole illuminator may be taken off the microscope as easily as an objective.

The dry type of concentric system is constructed as follows (Fig. 12):

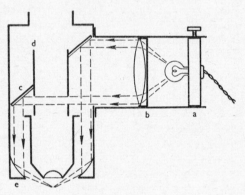

Fig. 12.

A small lamp of the flash-light kind is mounted in a tube in such a way that it can slide in and out on an axis at a right angle to the microscope axis, to give it a focussing adjustment. It may also have a pre-set centring device usually consisting of three screws working in slotted holes. The light from this small source is made parallel by means of a condenser (*b*) usually fixed within the apparatus (Fig. 12). This condenser must be of reasonably good spherical correction or uneven illumination will result and a simple lens is

A

B

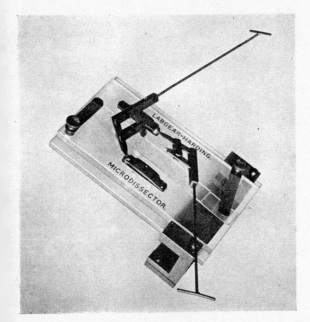

PLATE 7
The Labgear-
Harding
Microdissector.

not good enough for use in this position. Two lenses arranged as in symmetrical condenser systems or the asymmetrical arrangement will do, Fig. 13 A and B. (See section 'Microscope Illuminants', page 117.)

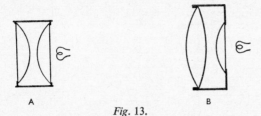

A   B

Fig. 13.

A thin tube (*d*) is held above the objective to protect the image-forming light leaving the objective from stray reflections in the illuminator. Surrounding this tube is a mirror C, the function of which is to reflect the light it receives from (*b*) down the outside of the objective lens system container. This mirror need not be of special quality nor need it be of the first surface type as it is only reflecting parallel light. An internal polished bevelled ring (*e*) diverts the light on to the specimen as shown.

It is usually necessary to use specially made objective mounts and these form part of the complete apparatus. Where size is no object, a concentric illuminator can be built around the type of objective which has a prominent front lens and a long mount. Some old 12 mm. objectives by Ross were built in this way for easier top illumination. The mirror (C) may be of any size so long as the condenser (*b*) can cover it with substantially parallel light. The reflector (*e*) may be lathe-turned in steel or brass and silvered and polished. With this type of illuminator a diffused source may be used to better scatter the light over the surface of the specimen and to destroy the shadow effects of oblique light in the image. A diffuser near the condenser (*b*) will help this lens to cover the mirror (C) with light, especially when the condenser (*b*) is not good.

A high power arrangement is possible in which light of greater obliquity and an image of smaller size is projected upon the specimen for use with high powers up to 3 mm. Fig. 14a.

The high power form is not very satisfactory because it is, by its nature, very tricky to adjust, and even when adjusted will function only upon a flat surface. The extreme obliquity of the illumination and small illumination depth allow it no advantages over the simple vertical illuminator.

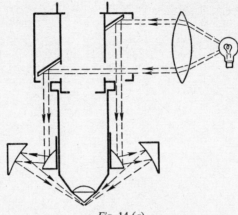

Fig. 14 (*a*).

It is possible to arrange the above systems to work in immersion oil and with immersion objectives up to a numerical aperture of 1·3. The system is then arranged as follows, Fig 14b:

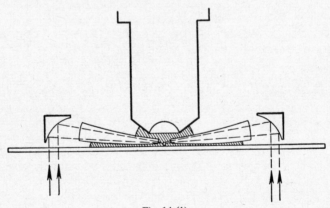

Fig. 14 (*b*).

The mechanical support for this optical system may be arranged in several ways at the discretion of the designer. The working distance of most 2 mm. oil immersion objectives is insufficient to permit a glass annulus to surround the front lens, but lower power immersion objectives do allow room for such an annulus although there is seldom any point in reducing the working distance of such objectives as a long working distance is one of their chief properties. A special glass front lens mount has been used in place of the usual plated metal one and this is undoubtedly the most satisfactory method but is very expensive.

When such an illumination system has been set up with care, it is sometimes possible to show certain objects clearly resolved, but on the whole the results are not worth the considerable expense of the illuminator. Similar results can be obtained by applying the beam of light from a research type lamp direct to the side of the oil drop when an immersion objective is in use. Two lamps may be used from opposite sides to offset the worst effects of shadow and diffraction bands inseparable from the use of extremely oblique light. The immersion objective has more than double the working distance for the same aperture as a dry system and this makes top illumination by direct methods much easier than it might appear. As with dry concentric systems, most information for research purposes is to be had through the use of the simple vertical illuminator properly managed.

The concentric systems are most useful for routine inspection purposes requiring a power not higher than 6 mm. They are easy to set up, are self-contained, dust-free and portable. There is no reason why a powerful external source of parallel light should not be applied through the hole normally occupied by the lamp but if this were done the condenser (*b*) (dry systems) would have to be removed.

Vertical Illumination Systems with Built-in Lamps
Mention has been made of the use of a racking stage so that the body of the microscope with the illumination system in line need not be disturbed when the specimen is focussed. Concentric systems usually have built-in lamps which do not introduce focussing difficulties and these have been described in principle. The ordinary vertical illum- inator can be obtained with a lamp housing attached, which may be screwed to the microscope body or to an attachable piece which fits between the objective and body tube. It takes either of the follow- ing forms shown in Figs. 15 and 16.

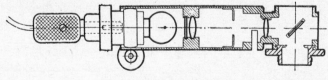

Fig. 15.

In the vertical type (Fig. 16) the illumination assembly is screwed rigidly to the side of the microscope body. The part marked (a) is rotatable upon the axis of the microscope assembly, so providing an input of oblique light to the inclined plate within the microscope, thus when the two inclining parts (a) and the reflector (b) are operated together, light of any degree of obliquity can be applied through the objective to the specimen.

Of these two systems, the vertical type is to be preferred for the following reasons:

(1) It is not liable to vibration and consequent disturbance of the focal adjustment of the microscope;

(2) it is not likely to be knocked by the hand;

(3) the rotation of the head (a) combined with the movement of the inclined plate allows light of any angle to be used.

Both of these devices have the disadvantage that the observer is tied to one light source, one optical input system, and usually one colour of light, that of the small filament lamp. There are many industrial tasks, particularly in the inspection departments, where this form of instrument is useful, but it should not be acquired for research purposes because it is not sufficiently flexible in its uses.

Adaptation of Vertical Illuminators to Microscopes without Racking Stages: Auxiliary Stages

One way of overcoming the difficulty of the vertical illuminator moving with the coarse adjustment is to use a conventional vertical illuminator and mount the specimen on the substage. This is arranged by using a block of hard wood turned to the shape of a flat-topped mushroom with a stem a few inches long, the mushroom stem being inserted through the stage aperture into the substage ring where it is tightly located. The flat top of the mushroom then becomes the stage and can be racked up and down by the substage gear. A small amount of rectilinear motion is also available from the substage centring screws.

The External Reflector

This is a simple mirror or prism attachment as shown in the diagram, Fig. 17, which may be added to any elementary vertical illuminator system.

Small changes of distance between the inclined plate (a) and the lamp due to focussing do not have a serious effect upon the images as the racking of the microscope does not affect the direction of the incoming light. If a long optical train is in use another mirror may be placed at (b), when light coming in the direction of the arrow may

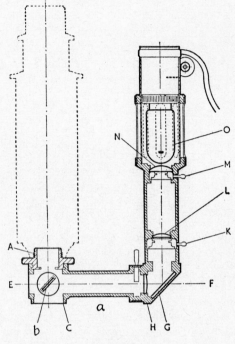

Fig. 16.

A.   Standard objective thread attachment to nosepiece.
b.   Vertical illuminator reflector.
C.   Female objective thread to take microscope objective.
E.F.   Horizontal axis of illuminator.
G.   First-surface mirror changing vertical axis to horizontal one E.F.
H.   Illuminator body casing detachable at this point.
K.M.   Field and source size limiting diaphragms.
L.   Illuminator field lens.
N.   Lamp condenser.
O.   Small electric filament lamp.

come, as in the usual racking stage arrangement, along an optical
bench or other device. Mirror (*b*) may also be placed above the
vertical illuminator and (*a*) arranged to receive downcoming light.
If the microscope has a large stage aperture, mirror (*a*) only may be
used in conjunction with the microscope tailpiece mirror, this
arrangement having the advantage that the normal microscope lamp
may be used for vertical illumination without the necessity of changing
its position. The mirrors (*a*) and (*b*) need not be optically worked

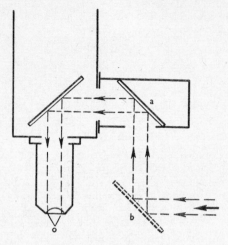

Fig. 17.

and need not be of the first surface type. Right-angled prisms can be used in both positions and will eliminate multiple reflections, but the mirror is best as it can accept light from any direction without introducing distortion. Double images of the light source are not serious in an illumination system.

With the vertical illuminator, polarized light may be applied to the top of the specimen, and particularly when metals and chemicals are studied much information may be gleaned by this means. It is necessary that the microscope stage be rotatable when polarized light is used, for purposes of analysis rather than for reducing glare and unwanted reflections.

(See 'Vertical Illumination with Covered Specimens', page 221, 'Metallurgy', page 472, and 'Polarized Light', page 95.)

To set up a Simple Vertical Illuminator
Once the simple vertical illuminator system is understood, all others will be found easy to handle.

To set up the simple inclined glass plate vertical illuminator, proceed as follows:

(1) Remove and clean the glass plate with a well-washed old cotton handkerchief.
(2) Set up the microscope upon a dense object (a piece of scratched first-surface mirror is good), using a 12 mm. objective and any

light available, the idea being to focus the objective upon a specimen of any kind.

(3) Select a high power lamp fitted with a lamp condenser and mounted upon a stand, and project from it a beam of light into the input hole to the vertical illuminator.

(4) Look through the microscope and adjust the inclined plane until the light is seen to illuminate the specimen. If there is no adjustment to the glass plane, then the lamp must be adjusted in position until the specimen is seen illuminated.

(5) Adjust the size of the illuminated area to be as small as possible compatible with seeing what is required. A small illuminated area means less glare.

(6) Check that the illuminated area shows an image of the lamp condenser controlled by its diaphragm. If it is not controlled, either the lamp condenser is not focussed upon the vertical illuminator input hole or the lamp source and condenser are not on an axis which intersects that of the microscope at the vertical illuminator plate. The remedy is to adjust the lamp in position until the required disk of light is seen. The distance between lamp and microscope may require adjustment until the lamp condenser image is sharp on the specimen.

(7) Having set up the system for one objective, it will be found correct for all.

(8) Do not use the microscope coarse adjustment for focussing or the illumination will be disturbed. (See section 'Vertical Illuminators', page 218.)

The Application of Vertical Illumination to Covered Specimens
In general, the presence of a cover glass over a mounted specimen makes vertical illumination unsatisfactory for the following reasons:

(1) Light is reflected from the top surface of the cover glass, thus drowning the light from any reflecting object below.

(2) Light is also reflected from the underside of the slide, causing mixed effects in the image such as partial transmitted light illumination due to the reflected light passing through the object into the objective.

(3) Unless the object has a refractive index or opacity considerably different from the mounting medium, it cannot reflect any light from its position in the medium.

In some circumstances something can be done to enable the observer to obtain a look at the structure even though the picture he obtains may be unsatisfactory as a demonstration of a favourite object. There are several advantages to be obtained by looking at the surface of an object by vertical illumination. Firstly, the light by which the specimen is seen has not passed through the other, deeper parts of the specimen, for should the specimen be of regular structure like a diatom, the use of transmitted light may cause false images to appear in the plane of the part of the object being studied due to diffraction from the lower regions. A surface view always gives a better idea of the relative densities of parts of the object and better general idea of the structure. Also only one plane, that illuminated, is visible at a particular objective focus.

If it is assumed that the structure to be examined is relatively dense, like a metal, sections of wood, a large diatom, wool fibres and so on, which can reflect light out of the surrounding medium, reflection at the cover glass may be avoided to some extent by polarizing the input light and examining the image through an analyser. Adjustment of polarizer and analyser must be made to suit the specimen, i.e. render it most clear against the background of haze.

It must be remembered that, owing to reflection of the polarized input light by the bottom of the slide, a proportion of polarized light passes upwards through the specimen, so should the specimen be birefringent, allowance must be made for differences in brightness of parts of the specimen not due to density differences but to polarization effects. (See 'Polarized Light', page 95.)

A considerable improvement may be obtained by oiling the slide on to a rough, black dark-well of some kind. Such a dark-well was often provided for use with Leiberkuhns but as these are seldom available today a satisfactory alternative has to be made from a piece of acid-blackened brass. This or some other similar material about 10 mm. or so diameter is caused to adhere in optical contact with the under-surface of the slide by means of a drop of immersion oil.

Little can be done when the object and mountant are nearly similar in the optical sense, but the following method allows a small part of an object to be seen with the inclined plate type of vertical illuminator.

The condenser should be removed from the vertical illuminator lamp exposing the small bright filament. The lamp should be placed the same distance away from the vertical illuminator input hole on the microscope as the distance from this hole to the eyepiece flange. Centring of the lamp is of little importance because only the objective now remains to focus the light upon the specimen and the objective takes a very wide view. A dense specimen must be placed upon the stage, and as the objective is carefully lowered a small bright illuminated patch will be seen upon the specimen in which the structure will be seen. It is necessary to set up in the first instance with a dense specimen because the illumination is carried down with the objective and is sharply focussed; there is therefore little illumination present until the object comes into focus. Were the object a semi-transparent one, it would be missed.

When an image is obtained, the distance of the lamp from the microscope may be adjusted to make the image of the filament fall exactly at the point of focus of the objective. This last adjustment is important because with care only the stratum containing the specimen can be illuminated strongly to the exclusion of other layers.

The use of polarized light and the dark-well may be necessary at the same time in order to obtain contrast with the background.

It is usually necessary to employ an oil immersion objective to remove cover glass reflections, and a water immersion objective may also be made to yield satisfactory results in biological studies of organisms on surfaces. When the vertical illuminator lamp distance and position are once set up, and if the object is semi-transparent, it will be found easier to find the specimen and to set it up roughly with transmitted light. The vertical illuminator need not be disturbed while this is being done but, if a dark-well is in use, this must be removed. As soon as the specimen is in the focus of the objective, the transmitted light should be attenuated, when probably the bright spot of the vertical illuminator lamp filament will be seen on the specimen; if so, observations can start with the top light, and the transmitted light source be extinguished. Often considerable care is required to see the specimen illuminated, and if the setting-up procedure is not followed there is considerable chance that a reflection of the filament from either the top or bottom of the cover glass or slide will be studied and the object not be found at all.

The Study of Surface by Means of Replicas

The technique of reproducing diffraction gratings consists in pressing a plastic material upon the grating in such a way that an impression of the grating remains upon the material when it hardens. The impression may be silvered to produce a successful reflection grating or the whole may be transparent, it then being a transmission grating. This kind of replica is somewhat different from that described in the following paragraphs because it must be a substantial object capable of retaining the dimensions and shape of the original. Details of the production of this type will not be given here because it is not a microscopical matter, the replica used for recording and studying microscopical surface detail need not be optically clear nor need it be stiff; a mere film is quite satisfactory and is usually used.

A replica is a very accurate representation of the surface undulations of the object from which it was taken, so long as due regard is paid to cleanliness. The study of replicas used in electron microscopy shows that structures of atomic dimensions are faithfully represented when the replica medium is used in the liquid form and when solid, stripped off. The advantages to be obtained by using the replica technique are several:

(1) Replicas may be studied as transparent objects.

(2) They may be filed away in books as a permanent record of the structure.

(3) They may be taken from large and heavy objects which may be impossible to move to a microscope, and in such positions that a microscope cannot be got to them.

(4) They may be shadowed by the gold evaporating process to show up undulating structure clearly when examined by vertical illumination, contour methods and oblique light. Whereas the replica of a diffraction grating must be thick and rigid over an area of several square centimetres, the micro-replica need not be thick or rigid. The area under examination is always small and the relative position of one feature to another several centimetres away does not matter. In general, the micro-replica may be a thin, flexible film like a piece of cellophane.

Replica Media

The requirements are that the material shall be easy-flowing at room temperatures, quick drying, free from contraction and expansion during and after drying, free from properties which cause adhesion to the surface, permanent when stripped off and transparent.

Several replica media have been published at various times and no doubt each worker has found one which suits his particular job best. In the experience of the Author all the materials of the cellulose acetate class may be used. A thick solution of celluloid in acetone is satisfactory.

A good replica media is made from 0·5 per cent volume of Formvar in ethylene dichloride. This is first poured into the job and allowed to dry, then is backed up by a coating of nitrocellulose in butyl acetate. The backing coating allows the replica to be stripped off easily and may later be dissolved off (if required) in amyl acetate. The job should be washed and soaked in toluene before taking the replica.

Method

The surface to be studied must be cleaned in the ordinary way by washing with the appropriate chemical. Clearly the treatment must depend upon the nature of the surface and in general washing with a camel-hair brush is better than rubbing with cloths, but no amount of cleaning of this sort is so good in the final stages as taking several 'waste' replicas before the required one. The first two films taken off the surface will carry away with them all the loose material present, and this should always be the final cleaning process before the wanted replica is taken. Should it be required to take a sample of a loose surface for permanent reference, a more sticky medium must be used. The new plastic varnishes and paints made from synthetic resins adhere well to most surfaces and may be removed as a film by mechanical means, but this medium has not been recommended for normal replicas because it is apt to adhere too well to a solid surface. Each problem requires modification of the basic methods.

After the surface has been cleaned, the medium is poured on to make a liquid layer about 1 mm. thick. This is left to dry naturally and completely. When it is dry the transparent film will appear to be stuck to the job and it will in fact be in optical contact with it, but should not be adherent. If a razor blade is applied at a waste edge, the film will be found to lift without trouble and may be stripped off the job and examined. Several replicas should be taken, one may be mounted on a slide, and one or two in a note-book with a description of the job and process.

Microscopical Examination

When replicas are studied as ordinary transparencies, the structure shown looks very like the magnified view of the real surface by vertical illumination. High spots are covered by a thin layer of medium, while deeper holes are filled with a thicker layer, and when these differences appear as an absorption pattern, the effect is realistic and may be interpreted directly.

Another way is to coat the replica with a metal such as gold in an evaporating chamber, the gold being applied in a direction normal to the surface. This produces a fine reflecting surface which may be studied with a vertical illuminator in the microscope. Alternatively, the gold may be applied at a great angle with the normal so that shadows of high spots on the replica are shown by spaces in the film of gold. This preparation is studied as a transparency and the effect is most revealing. A high degree of resolution in depth can be obtained by this process because the lengths of permanent shadows can easily be measured by means of a standard eyepiece micrometer;

the calibration is obtained by observing the height of a large prominence which can be measured by means of a calibrated fine adjustment. The shadow length from this can be measured micrometrically, and other heights obtained by measuring shadows and applying geometrical methods of interpolation, Fig. 18.

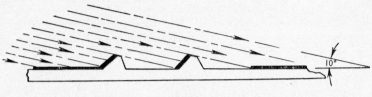

Fig. 18.

The general appearance of a shadowed replica is still similar to that of the surface of the specimen because although the replica is vertically inverted with respect to the object, the shadowing leaves bright spaces representing shadows of depressions in the original object, Fig. 19. The results obtained from examination of replicas viewed as transparent objects are entirely satisfactory for the study

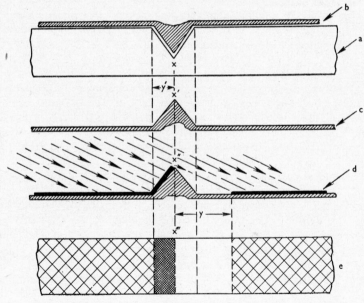

Fig. 19.

of most surfaces. Specimens such as etched metals, ground surfaces of all kinds, rocks and minerals, are well shown by the method.

In any replica work which is shadowed, the angle of shadowing must be arranged to give the length of shadows best suited to the job in hand. The most realistic effects are obtained when the shadowing is at about 45°, but small differences in surface levels can be shown up with shadowing at grazing incidence. If a specimen has high spots as well as low level changes, high angle shadowing cannot be employed owing to the great length of the high spot shadows. In general, fine detail requires a limited amount of shadowing material to be applied at an angle of about 70° with the normal. The amount of material used should not be greater than is sufficient to absorb about 50 per cent of the light transmitted through the microscope from the heaviest shadowed parts. For photographic reproduction of known structure, the amount of material can be much increased but delicate features will be lost by swamping out.

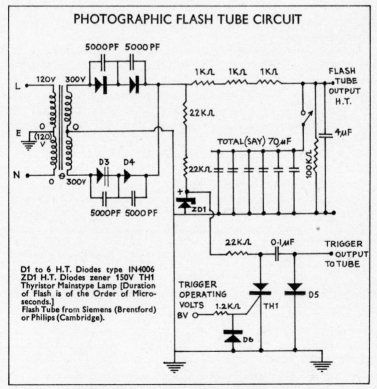

PHOTOGRAPHIC FLASH TUBE CIRCUIT

D1 to 6 H.T. Diodes type IN4006
ZD1 H.T. Diodes zener 150V TH1
Thyristor Mainstype Lamp [Duration
of Flash is of the Order of Micro-
seconds.]
Flash Tube from Siemens (Brentford)
or Philips (Cambridge).

Fig. 20. Circuit of a 5 Joule Flash Tube Power Unit.

CHAPTER 12

Methods for Handling Small Amounts of Material

Pipettes

THE SIMPLE pipette consists of a glass tube about 100 mm. long with its upper end cut off at a right angle with the axis of the tube. To use the instrument, the finger is placed over the upper end while the lower is inserted into the liquid to be gathered up. When the finger is lifted from the upper end, the liquid rises in the tube. The finger is then replaced and a quantity of the liquid which entered the tube is lifted clear. This instrument may be used to catch small objects in water and may be calibrated after the style of a burette or measuring cylinder.

All kinds of modifications may be made to the instrument, i.e. its shape may be varied, its diameter changed in various places and it may be operated by means of a rubber teat instead of the finger. The simple pipette is useful when measurements are not required, but if its diameter is greater than 8 mm., it is not possible to hold water against gravity and the collected liquid and specimens will run out.

If the simple pipette is to be used to catch such objects as animalcules it must have a comparatively wide mouth, about 4 mm. diameter, for if it is narrower the surface tension about its mouth is so great that the wanted specimens do not enter with the stream of water, and the pipette does not take a representative collection of suspended particles unless its diameter is about fifty times that of the largest particle in suspension.

For measuring and transferring small quantities of liquids, one of the specially designed pipettes described below should be used.

(1) The thermometer tube pipette, mouth operated, which can be made to any degree of fineness and calibrated as required.

(2) The Micro-pipette

This is usually teat operated, and may be easily filled up to the calibration marks. Liquid is drawn into the small tube until it overflows into the larger so when the liquid is expelled only that quantity in the small tube is delivered. The capacity of the small tube is adjusted by breaking pieces off the external end, or the capacity of a long piece of this tube is predetermined and is then divided geometrically. The instrument is best fitted together with Araldite but may be made up with glass tubing and a rubber cork.

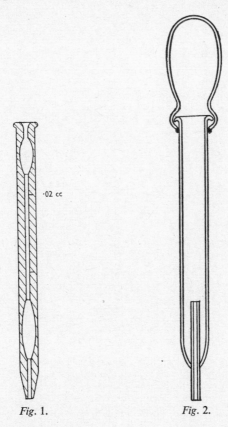

Fig. 1.   Fig. 2.

(3) The 'Limited Flow' Pipette

This is usually teat operated but the constriction prevents a rush of liquid into the calibrated part of the tube. Quantities of $1\mu l$ to $100\ \mu l$ may be handled without encountering difficulty in setting up the liquid level to the calibration marks.

(4) The Screw Pipette

It is usual to arrange the nut A in which the micrometer screw works to slide in the upper part of the pipette so that the pipette can be filled by plain suction. When liquid has entered the capillary tube

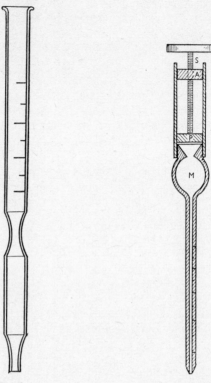

Fig. 3.

Fig. 4. A. (See text).
      P. Piston.
      M. Mixing chamber.
      S.   Micrometer screw.

the screw motion is used accurately to adjust the quantity to some calibration mark. The whole may be washed out after use by removing the micrometer part.

(5) The Piston Pipette

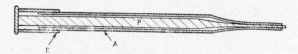

Fig. 5. A. Outer tube.
      P. Plunger.
      R. Soft rubber control spring.

This instrument is difficult to use but has some uses for collecting single specimens of pond living organisms.

The outer tube is of ordinary glass drawn to as fine a point as required. The inner rod is selected from stock to slide easily within the undrawn part of the outer tube. The rod is then drawn fine to a smaller size than the bore of the drawn part of the outer tube, and the rod is then inserted into the outer tube and pushed in gently until it sticks, so that its drawn end protrudes through that of the outer tube; the drawn end of the rod is then broken off so that the outer tube and inner rod end at the same place. Because the rod was pushed into the tapering tube until it stuck, the fit between rod and tube at the capillary end is extremely close. Remove the rod and flatten the butt end by heating it to redness and pressing upon a hard wood surface. Insert the rod again and connect rod and tube at the butt end by means of soft rubber tubing as shown and adjust the position of the tubing so that the drawn end of the rod is about 3 mm. inside the tube. When light pressure with the finger is applied it will cause the inner rod to move in a piston-like manner under the control of the rubber tube acting as a spring.

When the pipette is in use immersed in a liquid, the inner rod may be slightly protruded to act as a prod, and when a specimen is required the rod is allowed to nearly touch it, at which instant the plunger is released. As the plunger flies back, the object follows into the tube with the minimum of liquid. As indicated above, easily obtained results should not be expected, mainly because of the difficulty of manipulating the contrivance under the microscope, but if the apparatus were mounted in a micromanipulator and a better system arranged for withdrawing the piston at the right instant, greater success could be expected. A pneumatic withdrawal device would work well.

(6) Pipette for Lifting Large Objects

The instrument is made of thermometer tube about 100 mm. long and the object is held in the bell-mouth part while liquid is drawn up the capillary shaft, without taking with it the specimen. Any amount of liquid may be taken in without affecting the position of the specimen. The device is used mainly for transferring entomological specimens between bowls and tubes of liquid and is commercially available.

Micro Balances

The instruments described below are not intended to take the place of the elaborate chemical apparatus often found in laboratories today, but are examples of apparatus which can be built up in a laboratory and calibrated to do delicate work.

Fig. 6.

(1) Torsion Balance (Fig. 7)
For very light loads the quantity to be measured can be hung at
A or B with only the torsion resistance of the wire to control the
apparatus. For heavier loads known balance weights may be placed
on B and the card used only as an indicator of balance in the usual
way. The range of a balance of this kind is about 0 to 100 mg. with
an accuracy of $\pm 0 \cdot 1 \ \mu g$. The whole should be enclosed in a glass box.

(2) The Spring Balance (Fig. 8)
B is a weight into which is fitted a piece of fine steel piano wire,
phosphor bronze strip or quartz fibre A of the approximate shape
shown, C is a square glass box or jar. The deflection of the wire A
can be read against the scale as shown. When the balance is constructed of very fine wire to obtain maximum sensitivity, considerable swaying about of the wire is experienced and in this case the
deflection is best read through a reading telescope or cathetometer.
The sloping balance wire is set up against a convenient part of the
scale by sliding the weighted block B along the base of the glass case.

   All these microscopical measuring devices must be directly calibrated over the whole of their ranges and if difficulty is experienced in obtaining tiny weights, lengths of bare copper wire may be
divided geometrically, as modern wire production ensures a very
regular product. Obtain a length of bare wire of such a gauge that a
length a few centimetres less than one metre weighs 1 gramme, then

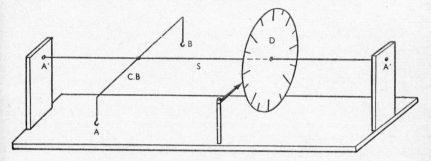

Fig. 7.

A₁   Fixed supports, one or both of which can be rotated for pre-
       setting torsion bar.
S   Thin steel piano wire torsion bar.
C.B. Cross beam.
D   Graduated cardboard disk fixed on wire.

stretch the wire until its length is exactly one metre, taking care to
hold the wire at the extreme ends in square ended tweezers. The
main function of the stretching is to straighten the wire, but if greater
accuracy is required a longer length can be stretched until a round
figure of weight is obtained for any metre of it. Glue the straight,

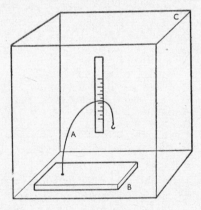

Fig. 8.

weighed wire on to a metre stick with water-soluble glue, divide up
the length with a razor blade into (say) millimetres, giving 1/1,000 gm.
weights or any multiple required, and detach the pieces by dissolving
them off with a paint brush and water. It is quite easy to obtain

milligram weights in this way with an accuracy dependent upon the skill of the worker, and if smaller weights are required finer wire can be used, but if the wire is finer than 40 S.W.G. an average figure for several weights so made must be used because the accuracy of drawing of the wire cannot be guaranteed to show less variation than 10 per cent on diameter and 5 per cent in weight per 5 mm. length. As a guide it may be remembered that 1 metre of 34 S.W.G. tinned copper wire weighs 390 milligrams approximately, therefore 1 mm. of this weighs 390 μ grams and there is no difficulty in weighing and cutting wire 1/10 of this diameter. Weights obtained in this way are still coarse compared with the possible accuracy of the balances described, but the deflection of the balance over the range required can be considered proportional to load, therefore a fine scale can be employed and divided between 0 and full scale deflection obtained with a relatively heavy weight. The average value of several calibrating weights can be obtained by observing the deflection while setting up the scale and taking the mean position.

Micro Physical Processes
(1) The Micro Filter (Fig. 9)

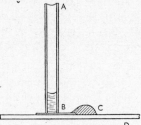

Fig. 9. Micro filter.
A. Square ended pipette.
B. Filter paper.
C. Drop from which filtrate is required.
D. Glass plate.

The filter paper and pipette are arranged as shown and are moved across the glass plate until the filter paper touches the drop. The filtrate rises in the pipette and may be lifted clear in the usual way. The piece of filter paper must be small or loss of liquid into its structure will result, and the lower end of the pipette must be ground square with its long axis.

(2) Determination of Boiling Point

A capillary tube several centimetres long is prepared and a drop of the liquid to be tested is drawn into this, and the end sealed off in a flame. The capillary tube is tied to a thermometer and the whole immersed in a beaker of high boiling point oil. When the drop under test boils, it rises quickly in the capillary tube, thus boiling points can be determined to within 1° C. by this method.

(3) Determination of Refractive Index of a Solid

In microscopical work this is always done by comparison with a liquid of known refractive index. If a body is immersed in a liquid of the same refractive index as itself, light experiences no deviation when passing between the substances; should deviation occur, it is due to a difference in refractive index between the materials, and a dark band or fringe is apparent around the object in the microscope when oblique light is employed. When the substance to be tested has a higher refractive index than the surrounding medium the dark fringe is outside the substance and if its refractive index is lower, it is inside, but in order to get used to the appearances it is recommended that workers make the observation with drops of oil in water and with air bubbles in balsam mounted as ordinary microscopical objects.

In all laboratories where microscopical analytical work is done, a stock of liquids of known refractive index should be kept, and below is a list of suitable materials. To make the test, place the unknown material on the slide, use a 20 mm. objective and transmitted light, obtain oblique light as required by passing a piece of card below the substage condenser, thus cutting off a part of its aperture, and place upon the object various known liquids blotting off each after the test, until a liquid is found of the same refractive index as the specimen. The more oblique the light, the more sensitive is the test and monochromatic light should be used.

From the books of Physical Constants many more refractive index measurements may be obtained for special purposes in addition to those common ones and mixtures given in the table.

Mixtures of liquids may be made up to have the same refractive index as a solid under test; the refractive index of the larger quantity of the mixture may then be determined with a spectroscope in the usual way.

All mixtures and liquids of measured or known refractive index, however obtained, should be well corked and kept aside, for in this way a useful battery of test liquids can quickly be built up.

(4) Refractive Index of Liquids

These may be obtained with or without the aid of the microscope, depending upon the size of the material and the accuracy required, by reversing the above process. A sample of the liquid is placed in a well-slide (which should have a flat bottom) and crystals of a solid

of known refractive index are added until one is found which shows
no fringes under oblique light.

It should be noted in connection with methods (3) and (4) that the
shape of the object to be tested does not affect the test when the
refractive index of liquid and solid are the same but lenticular effects
are apparent until the refractive indices are matched.

(5) Chemical Apparatus

Almost all the usual laboratory chemical apparatus may be built in
micro sizes and used in the ordinary way with due regard for the
tiny quantities involved.

(a) Distillation

In Fig. 10, A, B and C are test tubes of any size required, D is a
thermometer. In B is placed a high boiling point liquid to heat the
specimen in A by conduction. The cork in B must not be tight unless
a vent tube is provided. Transformer oil is suitable.

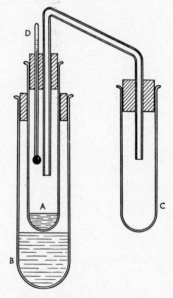

Fig. 10.

When tiny quantities are involved, the container A can be blown
in the delivery tube in the form of a bulb, the thermometer may then
be tied to the bulb as for boiling point determination.

(b) *Vacuum Distillation*

This may be performed with the same apparatus. The vacuum pump must be connected to C, and the end of the delivery tube in A should be drawn to capillary dimensions to restrict the entry of liquid if bumping occurs, or a glass bulb blown in a capillary tube slightly smaller than the inside diameter of A can be placed in A and will stop bumping taking place. Fig. 11.

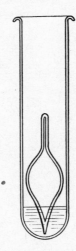

Fig. 11.

(c) *Steam Distillation*

This process may be undertaken with the apparatus shown in Fig. 12 which is self-explanatory, but care must be taken to cut off tube A with a stop-cock when the process is finished or there is a danger that the substances under test will be sucked back into the steam delivery tube.

(6) Micro Gas Apparatus

In Fig. 13 is a thermomenter tube with an enlarged end B into which a liquid may be poured. C is a rubber sleeve with cork D and clamp A. A specimen of gas may be drawn into the capillary tube B and rubber tube C may be filled with a liquid through the plug D. When the clamp A is open the liquid C acts as a piston to move the column of gas in thermometer tube B, the column being graduated and calibrated. In the enlarged end of B can be placed any other liquid or solid which is to react with the measured quantity of gas in the capillary portion. The gas may be drawn backwards and forwards

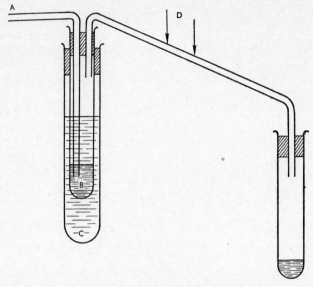

Fig. 12. A. Steam input.
　　　　　B. Substance under test.
　　　　　C. Bath.
　　　　　D. Flow of tap water.

along the capillary by means of the liquid piston in C and any measurements of absorption etc. of the gas by the material in the enlarged part of B made.

(7) Determination of Fusing Point of Metals
A tiny filament of the metal to be tested is heated electrically in an

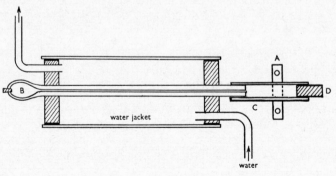

Fig. 13.

evaporation chamber and allowed to condense as a continuous film on a glass slide. The film must form a continuous electrical circuit and be connected in series with a battery and galvanometer. The film is raised in temperature in a furnace fitted with a temperature measuring device and when fusion of the film takes place, the metal runs into globules and breaks the electrical circuit, thus indicating the moment of fusion.

(8) Determination of Fusing Point of Crystals

If a material shows any activity under polarized light, this activity disappears instantly when fusion takes place. The crystals should be obtained either on a slide or in a capillary tube, placed on a covered hot stage fitted with a thermometer and studied under polarized light with a low power objective. A hot stage may be made and filled with transformer oil which can be raised in temperature greatly above that of boiling water without much vaporization. The temperature of such a stage is readily observable with a thermometer and a great range of temperature is available for melting point determination.

Separation Techniques

(1) Gravity Separation

In microscopical work it is often necessary to separate particles of dust or other small materials from a mass of substance. If the materials are powders, the best general method is that of flotation. One must have to hand a collection of non-corrosive liquids, the actual density of which need not be known, with which tests can be made. Small amounts of the materials to be separated are shaken up in the liquids until it is noticed that a liquid is found in which one of the components of the mixture behaves differently from the rest, i.e. it either floats or sinks. It is not possible to give a comprehensive table of liquids because each separation experiment is a problem of its own, but the following materials have been used with success.

Liquid	*Materials Separated*
2 vols. glycerine + 2 vols. water	Coffee and chicory
Chloroform + bromoform (SG 1·8)	Starch and flour float, other organic matter sinks
Potassium iodide 40 per cent + water	Viscose wool, cotton sink; cellulose acetate silk, true silk and cotton float
Carbon tetrachloride	Foraminifer float, sandy materials sink

TABLE 10

The usual method used when the materials to be separated are near each other in specific gravity, is to perform the flotation in large watch-glasses so that the floating component may be poured off easily.

The following list of liquids has been found useful for making up separation media:

	Specific Gravity
Chloroform	1·526
Bromoform	2·884
o. toluidine	1·003
m. toluidine	1·998
Alcohol	0·800
Carbon tetrachloride	1·595
Water	1·000
Glycerine	1·270
Water	1·000
Pyridine	1·965

TABLE 11

(2) Hand Picking

When the materials to be separated are spread upon a clean side, particles can be picked out from the mass by hand. The method is as follows.

The powder is washed clean by any method which does not affect the wanted material. A drop of the mixture in liquid is taken in a pipette and dropped from a height of (say) 100 mm. on to a clean slide. The liquid is allowed to dry off, leaving the objects spread upon, but not stuck to, the slide. If it is possible to use distilled water or carbon tetrachloride for the last washing, matters will be found easier. The suspension of solid matter must be thin as when the strewn slide is examined microscopically there must be plenty of space between the particles in which to manipulate a needle.

A sliding carriage is arranged upon the stage of an erecting low power stereoscopic microscope in such a way that the strewn slide and a clean slide can each in turn be brought into the field of view of the microscope while the needle is still visible. The needle may be hand operated or controlled by a micromanipulator, but most microscopists can do better by hand under magnifications less than 200 times. (See 'Micromanipulators', page 246.)

We are at present considering needles used on dry subjects, and several substances mentioned below are satisfactory. The holder for hand operation may be any convenient wooden handle of the artists' brush or pen-holder type. If the objects to be lifted are tiny, like diatoms, the best needle is made from a pig's eyelash; the natural free end should be used, the needle being about half an inch long. A number of other bifurcating bristles will do, and such may be selected from a shaving brush, the back of a fox terrier or the under-neck of a cow. The bristle should be passed over a piece of damp cotton-wool before a particle is touched, then it will be found to adhere easily to the bristle. No modern substitute is known.

The particles should be deposited upon a slide which has been covered with a dilute solution of gum arabic or one of the cements described under the heading 'Mounting', which has been allowed to dry. When such a slide is exposed to the breath (not blown at) the surface of the gum becomes sticky and readily detaches the particle from the bristle as soon as it touches. The gum film should be so thin as to be invisible and the bristle should not be allowed to come into contact with the cement or the particles will not come off. For delicate work it is a great help to support the hand on a bag of lead shot with another bag on top of the wrist. The finest work of picking out particles and arranging them such as the wonderfully arranged slides of diatoms made in the last century, were done by this method.

(3) Separation of Organisms or of a Few Particles by Volume Division

A drop of material is placed upon a slide which may be considered clean when the drop spreads evenly over the surface. From this drop is drawn by means of a needle a track of liquid along the slide, and microscopical observation will show whether or not a wanted particle is somewhere in this track. If too many particles are present, cross tracks are drawn, and when a wanted particle is seen to be sufficiently isolated the rest of the drop is wiped away, distilled water (or other liquid) is added, and the particle washed clean in this and mounted.

For swimming organisms, a drop of the water containing them is placed upon the slide and another slide is allowed to touch this, thus removing a part of the water from the drop. Examine each slide and continue the volume division until an organism of the required kind is found on one slide, when it can be treated as required. An advantage of this method is that none of the liquid need be lost.

To obtain an organism free from debris, either employ the above method, adding clean liquid at each division until a satisfactory dilution of the debris is obtained, or place a drop containing the organism at one end of a slide and a drop of clean medium about say 10 mm. away. Connect the drops by a track drawn from the clean to the dirty drop, shade the dirty drop from the light and illuminate the clean drop, when, after a time, the organisms will swim into the clean lighted medium. After this, wipe out the track. This process may be applied to bacteria but light is then not necessary.

To concentrate organisms in water, set up the following apparatus:

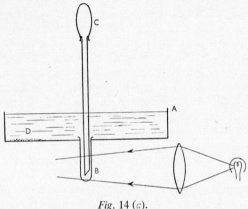

Fig. 14 (*a*).

A is a dish of any material (except metal if sea-water is to be used) and B is a tube about 10 mm. diameter and 25 mm. long. The muddy water is poured into A in such a way that most of the sediment collects round the sides, say, at D and it is usually possible to tilt A to keep most of the sediment back, but if mud settles in B it should be pipetted out as soon as possible. Darken container A but illuminate strongly B, then after about one hour a representative collection of organisms will be found in B. These may be pipetted out (C) and further separated by the methods described above. It is important that tube B is not unduly heated by the light source or it will tend to repel the organisms. As an artificial source for use in the dark, a flash-lamp or cycle lamp is about right, and ordinary daylight concentrated by means of a concave mirror, for field work.

(4) Separation by Means of the Centrifuge

The ordinary hand driven centrifuge may be used to separate emulsions and to concentrate suspensions of any kind. When small quantities of material are being handled, small test tubes or capillary tubes are placed within the normal holders of the centrifuge.

(5) Separation of Liquids by Capillarity

Small quantities of emulsions may often be separated by allowing the mixture to spread along a strip of filter paper. The components pass along the filter paper at different speeds and so, if there is sufficient emulsion available, separation into zones takes place and microscopic amounts of liquid can be lifted from the filter paper by means of a pipette pressed upon it according to the method of micro filtration.

Sorting Dust Particles

It is often necessary to apply some processes of elimination before particles are set out for selection by means of a micromanipulator or other device. In general, the laws of chemistry apply on the microscopical scale and the following examples of methods may indicate ways of approaching other problems.

A sample of dust may be cleaned of carbonaceous matter by applying concentrated hydrochloric acid. A cleaned weighed slide should be selected and a sample of the dust suspended in distilled water allowed to drop upon the slide from a pipette held, say, 150 mm. above. The spread sample should now be dried, when it will be found that the particles adhere slightly to the glass and remain in place. Apply a drop of hydrochloric acid at one end of the slide and incline the slide so that the drop flows along it. Several applications of acid may be necessary to dissolve all carbonates. The loss in weight of the material indicates the amount of carbonate which was present, and the residue on the slide contains well-spread and clean silicates, felspars, etc., not affected by hydrochloric acid.

In a similar way hydrofluosilicic acid can be used to dissolve silicates, but a temperature of less than 70° C. must be used or decomposition of the acid will result. A glass slide must not be used as it is attacked vigorously by the acid, while materials like polytetrafluorethylene which are polythene-like solids are soft and are not transparent. Platinum foil is the best material upon which to perform this process.

When particles of silica only are being sought, a smear or spread may be taken on a Pyrex glass slide, dried and raised to red heat for a few minutes; the silica will then be found to have a clearly recognizable appearance under top light, looking like polished beads. This method may also be used to fix particles to a cover glass for mounting.

Having obtained a spread slide of material, the liquids mentioned on page 252 may be applied and the refractive index test used. When a liquid of the same refractive index as the particle is found, a further test by polarized light can be applied. For the polariscopic examination of quartz occurring as particles of any size, it may be immersed in a bath of benzyl alcohol as this has practically the same refractive index as quartz, is not harmful to the hands or to lenses, but will strip paint at an alarming rate. When it is necessary to examine large lumps of quartz before cutting for optical or radio components, a glass tank containing benzyl alcohol is illuminated from below through a Polaroid screen, the quartz being manipulated in the tank by hand and examined with an analyser.

Method of Detecting Radioactive Dust Particles

The samples of dust are collected by any convenient method and spread upon filter paper, clean white paper, or glass plates lightly coated with gum arabic. A photographic film is then pressed against the dust layer and left in the dark for several hours when, after development, dark spots will be seen where a particle was radiating energy. Having in this way located accurately the position of the particles of interest, they can be picked out with a micromanipulator or by hand.

It is important that the film be in contact with the dust film because if the glass sheet or the paper comes between them a large amount of the beta and all the alpha radiation will be stopped by the thickness of glass or paper, leaving only the gamma radiation to affect the photographic film. If it is required to separate the radiations, a screen of 20 microns thick aluminium foil between the dust and the film will transmit beta and gamma radiation but stop alpha, while a foil about 0·6 of a millimetre thick will stop alpha and beta, and pass gamma radiation. Several photographs taken of the same dust sample with these screens in use will give a good idea of the kind of activity in the dust in question, but it must be remembered that no allowance can be made for the different energies of the particles of the same radiation, and some betas of high energy will pass the 0·6 mm. thick foil yet others are stopped by the 20 micron alpha screen.

Method of Handling Small Objects during Processing

With practice and dishes of the right colour, most tiny pieces of material can be handled by the ordinary method of rubber-nosed tweezers and wire mesh section lifters, but some objects become very transparent under treatment and for convenience should be attached to small pieces of material which will not be affected by the reagents.

Insects and similar parts are easiest handled in little silk or gauze bags made from stockings, etc., and smaller objects may be stuck with (say) albumen on to scraps of paper marked in some way to identify the specimen. For this, filter paper may be soaked in albumen and cut up into squares or other shapes of about 2 mm. side and fragments of material are attached to them by means of a drop of albumen, the shape of the paper being a better guide to identification than ink marks. The pieces of paper then pass through all the processes with their passengers hanging upside-down in the liquids. If sectioning is intended, the paper must be got rid of before the microtime knife is used and it is usually possible to float the paper on the wax, allow the specimen below to be embedded in the surface, then pull off the paper.

Method for Removing Loose Materials from Surfaces[1]

It is often useful to be able to take a specimen off a surface without disturbing its arrangement, an example is a mould growing upon a building. To do this a liquid should be made up consisting of ether 75 ml., ethyl alcohol 25 ml. and cellodin about 10 grams. This liquid is thin and can be applied to a complicated surface but will dry into a substantial film, preserving the arrangement of the particles. It can then be stripped off in a way similar to a replica. In order to avoid air bubbles and to clean the specimen before applying the above mixture, it should be washed in alcohol or ether or a mixture of both first, but it is usually satisfactory to gently pour on the cellodin solution without previous treatment of the specimen. When the surface is nearly horizontal, a rubber ring pressed down forms a convenient fence to control the liquid, but if the surface is vertical a rubber or cork trough must be made which is tied or stuck to the surface with wax to make up the irregularities. As a makeshift, clay and plasticine can be moulded into a pouch as shown in Fig. 14b.

For rougher work the material known as Scotch tape can be used; this is a transparent, sticky tape about 25 mm. wide and a few hundreds of microns thick, with an adhesive on one side only which does not dry. If this tape is applied to a dusty surface or to paper it will hold anything which is loose on the surface and will take with it the surface of the paper as well. Samples of all kinds can be collected in this way but the pieces of tape must be well separated or they will all stick together like the Clans and attach themselves to everything else besides.

Micromanipulators

These devices are designed to transmit to a needle or other microtool, motion in two or more directions at a much reduced speed

---

[1] See also 'The Study of Surfaces'.

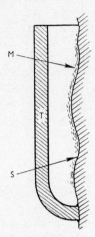

Fig. 14 (*b*). S. Surface of wall.
              M. Material wanted.
              T. Trough or pouch to hold cellodin.

suitable to be observed under a microscope magnifying, perhaps 1,000 diameters.

The first kind of instrument consists of a holder for the tool mounted solidly upon the microscope, capable of being preset in X and Y directions but having a micrometer control in the Z direction. To use the instrument a needle is fitted, the movements of the holder released from the clamps and the needle moved about until it appears nearly in the centre of the microscope field of view, then it is clamped in this position. The micrometer Z direction control is then employed to lift the needle to and from the specimen in the axis of the objective, the specimens to be operated upon being moved under the needle by means of a good mechanical stage. The advantage of the system is that the needle is always in view regardless of the limited field of the objective; the system is easy and cheap to arrange and any simple clamping device may be built on to the microscope, Fig. 15. The ordinary mechanical stage will move objects sufficiently steadily for dissection under a 4 mm. objective, and should a very slow movement be required the stage may be fitted with Bulgin slow motion heads as sold for use on wireless-set tuning dials. This, the 'fixed needle' kind of micromanipulator, is the best for selection and mounting of particles, and it is often called a 'mechanical finger'.

The second kind of manipulator is that where motions in all directions are transmitted to the needle in sympathy with the move-

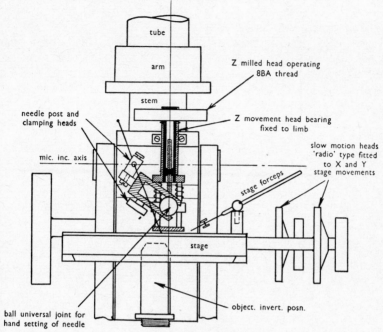

Fig. 15. Front elevation.

ment of the hand, and such apparatus is always mounted on a separate stand. Some apparatus of this kind is entirely mechanical, while others are hydraulic or pneumatic. The mechanical kind consists of a form of wobbling lever directly coupled to the needle, Fig. 16. The pneumatic and hydraulic kinds have no bearings but depend upon the action of metal bellows connected to a joy-stick control lever, which are separately connected to a similar set of bellows to which the needle is attached. Fig. 17.

Almost any arrangement employing hydraulics or levers may be pressed into service to make a micromanipulator. The important points are that either all motions transmitted to the needle are reversed with respect to the hand or all are in sympathy with the hand. Some mechanical instruments have the disadvantage that one direction of movement is reversed while the others are not. Such an apparatus is a great nuisance because one direction of movement is not natural when either an erecting or an ordinary microscope is used. Should one have to use such an instrument, the microscope can be made to accommodate it by making an inclining

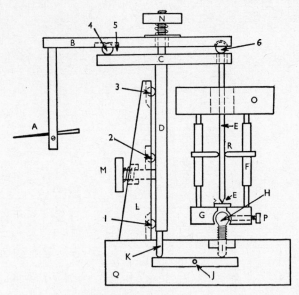

Fig. 16.

A.   Adjustable needle carrier.

B.   Cross beam mounted on bearings 4, 5, 6 (ball-groove, ball-plane, ball-socket, respectively).

C.   T piece supporting B solid with D and bearing on 1, 2, 3 (ball-groove, ball-plane, ball-groove respectively) and retained in contact by spring and milled head M.

D.   Vertical bearing plate resting on upright L.

E.   Push rod terminating in ball and socket (6) and on point-plane at bottom.

F.   Sliding carrier of 3 point bearing R.

G.   Solid base supporting structure G F O on ball head of screw H and adjustable in stiffness by P.

H.   Ball head of screw HI bearing in base plate Q by which height of structure GFOEB is adjusted by rotating GFOE by ring O.

I.   Round end of screw HI.

J.   1 : 1 heavy lever, fulcrum at J.

K.   Push-rod actuating DCBA in vertical plane.

L.   Stout upright.

M.N.   Milled heads acting through springs to locate D and B respectively.

O.   Handle ring for rotating structure GFO and for moving it in horizontal plane on ball joint H.

P.   Clamp screw by which GFO can be freed from screw HI.

Q.   Massive base containing coarse mechanical adjustments for setting up position of needle.

R.   Three point bearing on E sliding with carriage F to adjust throw of E (minimum throw when R is lowest).

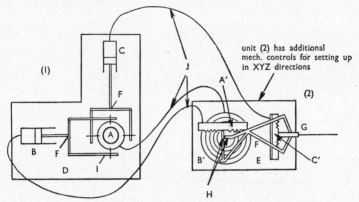

Fig. 17. Diagram of Beaudouin Micromanipulator. Typical Pneumatic type.

A, B, C.	Pneumatic cylinders, vol. approx. 25×10 mm. Ball jointed to base D.
A′, B′, C′.	Pneumatic bellows, vol. approx. 30×30 mm. Supported rigidly by their back faces.
D and E.	Heavy bases, hand control inst. (1) fitted with support for hand and screw actuated piston in A.
F.	Stiff mechanical linkages.
G.	Needle holder (stiff with respect to C′).
H.	Spindle linkage with flexible steel wire connections to F and A′ also from B′ to F.
I.	Forks fitted by bearings to jacket of cylinder A.
J.	Air tubes about 1 metre long, 3 mm. diam. int. (Velocity ratio 30 : 1 with needle 100 mm. long).

eyepiece employing a mirror instead of a roof prism, as a mirror so placed inverts the image of the microscope in one direction only and so compensates the micromanipulator. Fig. 18.

Micromanipulators may be used with any form of microscope and when low powers are involved most workers appear to prefer to have a conventional microscope with the needles operating between the objective and specimen, but this is certainly not the best arrangement. The inverted microscope used with the needles above the specimen is much the better plan and the only one which can be made to work with high powers as the tools do not get in the way of the objective. When specimens are in liquid the inverted instrument is almost essential.

It is possible to use micromanipulators in hanging drop preparations, Fig. 19, but there is usually trouble due to the liquid wetting on to the needles and running down them. If the drop is kept very small this trouble is largely avoided, and for coarse work a spine from a cactus plant will be found to make a very good needle; it is

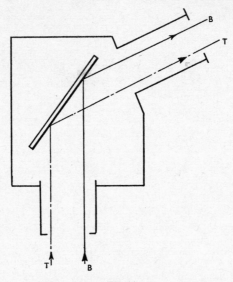

Fig. 18.
Single reflection type inclined eyepiece showing inversion up-to-down
but none left-to-right.

both fine and stiff and it will not wet with water, and it is for these
reasons excellent for sorting out things in water.

The following apparatus has been devised for isolating micro-
organisms and can be used for other jobs in micromanipulation. To
an ordinary glass slide are cemented pieces of glass to form a box,

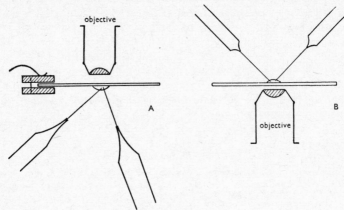

Fig. 19. Hanging drop. A .
Inverted microscope, lying drop. **B**

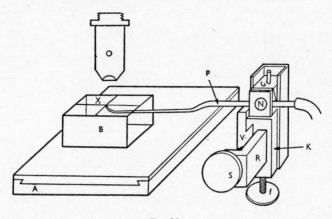

Fig. 20.

open at the top and at one end (Fig. 20), the dimensions of the box being 40 by 25 by 18 mm.; a cover-slip 25 by 40 mm., cleaned and sterilized, is placed on the upper edges of the box, previously vase-lined. On the under surface of the cover is placed a drop of nutrient fluid, and near to it a drop of culture containing the organisms to be isolated; the whole is then placed on a microscope stage and a fine capillary pipette with a curved tip and a brass holder is clamped to the microscope. The box, with its open end towards the pipette, is adjusted so that cross lines (x) on the cover are in the centre of the field, the pipette is then adjusted by moving it in the groove at the side of G, and by turning the screws that move the parts R and G of the holder, until the point is nearly in the centre of the field; the pipette, with the parts G, K and N holding it, is raised or lowered by the screw f, the part V being clamped to the microscope.

The portion of the cover bearing the sterile drop of medium is now brought into the field, the tip of the pipette is raised into it, and partially filled, the pipette is then lowered and the culture-drop is brought into the field; the pipette is then again raised until it comes into contact with the micro-organism to be isolated and this at once enters the pipette (often in company with other cells). The cover is then moved by the mechanical stage until the tip of the pipette can be brought into contact with an unoccupied part of the cover, when its contents are discharged, being blown out gently by means of the rubber tube. The process may be repeated as often as desired with fresh drops of culture media until a single organism in a single drop is obtained, when the culture may be left to grow or be transferred to a well slide.

Alum (sodium)	1·439		Potassium bromide	1·559
Alum (potassium)	1·450		Strontium nitrate	1·566
Alum (ammonium)	1·459		Barium nitrate	1·571
Alum (potassium chrome)	1·481		Ammonium chloride	1·640
Alum (ammonium iron)	1·485		Caesium chloride	1·645
Potassium chloride	1·490		Rubidium iodide	1·650
Rubidium chloride	1·494		Potassium stannochloride	1·657
Sodium chlorate	1·515		Potassium iodide	1·667
Sodium chloride	1·544		Caesium bromide	1·698
Quartz	1·544		Ammonium iodide	1·700
Rubidium bromide	1·553		Arsenous oxide	1·755
Quartz	1·553		Silver chloride	2·071

TABLE 12
REFRACTIVE INDICES OF SOLIDS

Methyl alcohol	1·32		Bromoform	1·58
Ethyl ether	1·36		Carbon disulphide	1·62
Ethyl alcohol	1·37		Water	1·336
Amyl alcohol	1·4		Quinoline	1·622
Chloroform	1·44		Bi-iodomethane	1·74
Carbon tetrachloride	1·46		Phosphorus and sulphur	
Glycerine	1·47		in methyline iodide	2·06
Benzene	1·49		Nitric acid	1·410
Chlorobenzene	1·53		Castor oil	1·490
Fennel seed oil	1·54		Canada balsam (raw)	1·549
Mononitrobenzene	1·55		Oil of cassia	1·641

TABLE 13
REFRACTIVE INDICES OF LIQUIDS

The Labgear-Harding Micro-dissector

This is an inexpensive instrument suitable for manipulating objects under the highest powers of a binocular dissecting microscope (Plate A), that is, with magnifications up to about 100 diameters. It is not intended to take the place of elaborate and expensive instruments which are necessary for cytological work, but which are unsuitable for those intermediate-sized objects which are too small to be manipulated by hand except by unusually gifted individuals. The complete instrument consists of a pair of dissectors each with a very fine tungsten needle. The dissectors are mounted on a transparent base which clamps to the stage of the microscope, and the object to be manipulated is placed on a movable depression slide between them.

The principle of the dissector is that of a pantograph. Instead of a pencil copying on a different scale the movement of a pointer over a drawing, the point of the needle follows on a greatly diminished scale, every movement of the control held in the fingers. The drawing instrument works only in one plane, but by using a ball-and-socket joint for the fulcrum of the pantograph the dissector may be operated in three dimensions.

The Base and the Method of Clamping it to the Stage of the Microscope
The pair of dissectors are placed in the correct position relative to each other with the needles in the field of operation, by mounting the housings for the ball-and-socket joint in suitable positions on a transparent and easily cleaned base (Plate 7, A and B). The clamp is so designed that the base can slide in it and be centred for a wide range of sizes of microscope stage. Tightening a pair of small screws enables the base to be locked in any position on the clamp, so that upon removal of the complete instrument from the microscope stage, after slackening the thumbscrews, it may be subsequently returned to the same predetermined position.

Ball-and-socket Joint
The principal movement of the pantograph takes place at a ball-and-socket joint. The ball is located at the end of the terminal member and is about 6 mm. in diameter, thus making it easy to adjust precisely the tension of the spring which maintains its position in the housing. The spring consists of a curved phosphor-bronze strip which holds the steel ball in its brass housing by a single adjusting screw located at its centre. The simple construction facilitates the convenient removal of the pantograph movements for the replacement or repointing of needles, etc. If the spring is turned through a right angle with the fingers, the ball is released and the pantograph can be lifted off the base, Plate 7B.

Freedom from Backlash with Smoothness of Movement
Reference to Fig. 21 will show that if three plane surfaces are separated by two steel ball-bearings located in holes passing through the members (the diameter of the holes being somewhat less than that of the balls), the centre member may be firmly clasped by the outer members. In this position the bearing area will be only a few hundreds of microns wide at the point of contact with the ball. The centre member will then be quite free of backlash and will have but a small though constant resistance to movement. This form of construction is applied to all four elbows of each of the pantographs of the micro-manipulator. A countersunk screw connecting the outer members enables the pressure on the balls to be precisely adjusted to the requisite degree, this will also take up any slackness due to wear.

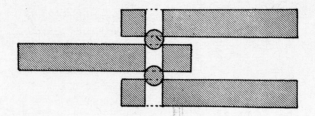

Fig. 21. Freedom from backlash with smoothness of movement.

The Needles and their Attachment to the Instrument

The glass needles for micromanipulators are too brittle to be of much use for the type of work for which the present instrument is intended. The tungsten needles described by Cannon (1941) are more suitable. The needles are made by dipping the tips of 200 and 100 μ tungsten wires into boiling sodium nitrite. The tip is eroded away to leave a finely tapered point. These needles are mounted in nickel tubes by crimping and are fitted to the pantograph mechanism by a simple clamp which consists of a channel which passes through the bevelled hollow into which a countersunk screw-head fits. The nickel shank of the needle is inserted into the channel with a pair of forceps and is firmly gripped by tightening the screw. This arrangement is free from any projecting parts, so that no matter what the position of the pantograph the view of the needle is not obstructed. It is advantageous to remove the pantographs from their respective ball housings on the base to facilitate the fitting of the needles.

The Knee-Focussing Attachment

So that both hands are free for needle manipulation a knee-focussing attachment is arranged. It is normally attached to the right-hand focus control of the microscope by means of the rubber-covered V slot and tensioning thumb screw, with the weighted section free to hang over the front edge of the bench to rest upon the thigh of the operator. The telescopic member will allow for adjustment to accommodate all normal bench heights and leg lengths. If the ball of the foot is raised, say, 100 mm. from the floor upon some object, it will be seen that raising the heel will result in a smooth adjustment of the microscope focussing. The universal joint offers a further fine adjustment of the focussing by a sidewards movement of the knee. The weight of the attachment tends automatically to rotate the focus control in an anti-clockwise direction, whilst raising the knee governs the clockwise rotation.

Where it is considered inconvenient to have the attachment protruding over the front of the bench the entire device may be reversed on the same control knob, after detaching the weighted member which is subsequently connected again through a 10 mm. hole in the bench. A rather larger degree of control is effected by this method since the weighted extremity rests upon the knee instead of the thigh, thus giving a greater range of movement.

Adjustments

The smoothness of working and the precision of the instrument are dependent upon the following adjustments:

(1) Centre the base accurately on the stage of the microscope so that the 10 mm. hole between the housings is concentric with the field of view.

(2) If home-made needles are used, see that the point of the needle is in line with the T-shaped control and the ball. If it is too long, lateral movements will be impossible, as the point will remain stationary for movements of the control towards or away from the ball.

(3) Adjust the screw holding the curved spring down on to the ball so that it is just sufficiently tight to support the weight of the pantograph. It is important not to increase the tension on the housing beyond this point as a tendency to twist may result, giving backlash to angular movements.

(4) Adjust the two screws on the pantograph itself so that it will support its own weight with the pantograph in a vertical plane, and the handle horizontal.

When correctly adjusted the instrument will respond to the slightest touch on the control, but will remain stationary for an indefinite time in whatever position it is left.

REFERENCES

CANNON, H. G. (1941). 'A Note on Fine Needles for Dissection', *J. R. micr. Soc.*, 61 (3), 58–9, 1 text-fig.

HARDING, J. P. (1939). 'A Simple Instrument for Dissecting Minute Organisms'. *J. R. micr. Soc.*, 59 (3), 19–25, 3 text-figs.

CHAPTER 13

Methods of Studying Objects
in Liquids

MOST OF the notes which follow refer to the study of living animals in water but as this is the most difficult study it covers all other techniques employing liquids.

In general, it is necessary to hold a liquid between two sheets of glass in order to form the combination into a flat slab of transparent material through which light can pass and form an image without undue distortion. This arrangement also holds the liquid still so that it is less liable to distort the image of objects immersed in it by movement due to vibration.

The simplest way of mounting a liquid is to stick three supports made from chips of cover glass, each a few millimetres square, on to a slide, place a drop of liquid between them and drop on the cover, capillarity holds the lot together. Any material between the cover glass and slide is clearly visible and colouring matter and reagents may be added at one side and pulled through by means of filter paper at the other. If living organisms are present they will live on this slide for a long time (weeks) so long as the water is kept from evaporating and the temperature is kept down. (See 'Objects in Liquid', page 259.) Slides with a continuous wall surrounding the drop should never be used because living organisms suffocate and liquids cannot be drawn under the cover.

Numerous live boxes have been designed and consist of two thin glass plates mounted in various designs of brass holders which allow the plates to be set at various distances apart, Fig. 1 shows a

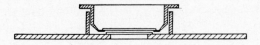

Fig. 1. Typical, though unsatisfactory live box.

typical arrangement. Generally the devices are of little industrial use, mainly because they are cumbersome and most designs are such that the brass plates foul the condenser and objective, so preventing the field being explored completely and often preventing critical illumination conditions being satisfied. If live boxes are to be used (they are sometimes useful as compressors) the design should be as follows (Fig. 2):

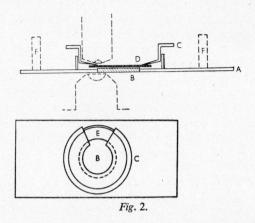

Fig. 2.

The brass slide A must carry the lower glass tablet B within its thickness and the whole should not be thicker than 2 mm., which is the working distance of a typical high power condenser. The compressor ring or barrel C must be of thick brass and wide so that an objective (shown dotted) will view the edge of the drop without fouling the brass work. The cover glass D must be secured to C with an easily soluble cement so that it can easily be replaced after breakage. A gap E should be cut in C so that the edge of the cover glass is visible and water can be added to the preparation without disturbing it. C should slide, not screw, into its tube, for if a screw is used with or without slot E, damage to the specimen due to the turning of the cover always results. In some forms of live box there is a pin which prevents the glass plate revolving with the barrel, and so long as the whole device is kept well lubricated it works well but is no real advance on the sensitive sliding barrel type. This simple arrangement does all that a live box can do and it should not be elaborated.

For use with sea-water the structure should be of stainless teel, but as this material is most difficult to work, commercial chromium plating has been employed which works satisfactorily and resists most liquids. It is useful to fit a live box with three detachable legs (F) of such a length that the box can be used on the stage upside down, and there is no reason why the tablet of glass B should not also be a cover glass.

In Fig. 3 can be seen several arrangements for examining living creatures or any other material in liquid.

Preservation of Life in Liquids

Microscopical animals and plants are very sensitive to changes in their surroundings and though it is true that cultures will thrive under different conditions, creatures from a particular culture, natural or artificial, do not tolerate changes of temperature of more than a few degrees, changes of oxygen content of the media in which they are growing, nor changes in salinity. It follows that animals must be specially catered for when they are under microscopical examination. It does not seem to be generally realized that the temperature of a pond or river varies very little from month to month and negligibly from day to day.

Slides must be designed so that air can enter the drop of liquid freely at least at its edges as in A and C, Fig. 3. The microscope must

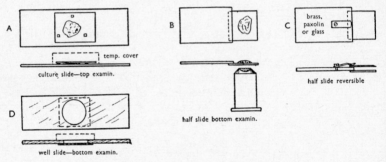

Fig. 3. Half slides are secured by one stage clip and project over stage aperture.

be kept cool and this is best achieved by always using a water bath or a piece of Chance heat-resisting glass in front of the lamp. Microscopes with lamps under the stage should always be avoided for biological work and work with chemicals.

Salinity changes usually follow evaporation of the liquid, therefore this must be prevented if possible because the addition of extra liquid does not rectify the situation. Many forms of growing slide have been designed and a list of diagrams with brief explanations will be given, but for use with active living creatures the Author has found that most growing slides of the type which depends upon an external supply of water are either mechanically unreliable or introduce chemical changes in the liquid. It is preferable when an observation can be interrupted to remove the slide from the micro-scope and store it in a humidity chamber.

A bell glass about 200 mm. diameter is inverted into a trough of water so that a water seal is formed all round its mouth. Standing in the water is a rack made of wire or any convenient material upon which slides are rested, Fig. 4. Slides are examined on the microscope for (say) fifteen minutes at a time and are then returned to the humidity chamber. It is found that active creatures like rotifers will live satis-factorily on such slides containing a single drop of water if those slides are kept in the chamber. Water will not be lost by evaporation and a few drops of water are likely to condense on the slide. Should topping-up or water change be necessary, this condensed water, picked up in a pipette, may be used. A convenient food for most organisms in this kind of captivity is pulverized pond-weed added to the slide in minute quantities such that some is always present; micro-organisms feed during twenty-four hours of each day and when food is not immediately available they are likely to encyst.

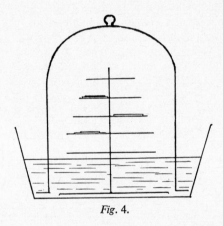

Fig. 4.

The humidity chambers must be kept at a temperature suited to the organisms and out of direct sunlight, and to arrange this the water trap of the chamber may be stood in a sink and cooled by a steady, slow flow of mains water; a cellar is usually at about the right temperature for pond life. For some reason a typical house bedroom is about right for culturing unicellular animals like amoeba and paramecium (an uncomfortable, unheated English bedroom, of course).

It is a fair guide to the success of culturing methods for pond living organisms to observe whether or not diatoms will live on the slides. If they will do, then the culture is existing in about the right conditions of light and oxygen content. In a similar way the purity of the air in a district may be assessed by observing the presence of lichen growth on the barn roofs and fences: lichen will not grow where there is the slightest amount of air pollution and they are never found in towns. Exposure to cold wet or dry conditions appears to make little difference to the amount of growth though the species are affected, so we have, provided by nature, an excellent air tester.

Difficulties with Humidity Chambers

If a normal slide with a cover glass is placed in a humidity chamber, moisture condenses upon it, and usually it will clear by evaporation after a few minutes and allow the objective to see through to the preparation. If, however, an oil immersion objective is in use, the oil cannot be wiped off without disturbing the cover, and if it is left on it goes milky due to the dampness of the chamber. A solution is to use water immersion or dry objectives, but a better solution is to use an inverted microscope and slides like B or D, Fig. 3. The liquid under examination then rests upon the solid base of the cell and the base can be wiped clean of oil without disturbing the specimen. A second method for use with the conventional microscope is the hanging drop technique, Fig. 5.

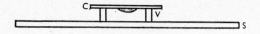

Fig. 5. C. Cover glass.
V. Vaseline seal.
S. Slide.

The diagram is self-explanatory. The method is useful for growing bacteria and moulds but the drop, when once placed, is inaccessible; the apparatus can, however, be made sterile.

Many modifications to suit particular experiments can be made to this technique and an important one is to make the cell, Fig. 5, of rubber with aluminium rings to clamp the cover glasses in place. Two covers making top and bottom of the cell are used, sterile fluid is placed in the cell, and the whole is sterilized and then inoculated by means of a hyperdermic needle thrust through the rubber. The capsules may then be treated in incubators, etc., and will remain sterile indefinitely.

Of growing slides, like lamps, there is no end. Several are shown in Fig. 6.

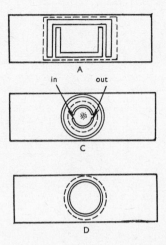

Fig. 6. C,   Liquid taken in and out with fine pipette.
A, C, D, Dotted lines show position of covers.

The moist slide used by Dallinger and Drysdale in their study of flagellates is still a useful piece of apparatus which may be adapted and modified in many ways for a variety of studies. The Author has used it constructed as follows (Fig. 7):
The glass plate D is about 100×40 mm. size and 1 mm. **thick.** C is a piece of often-washed cotton material with a hole 25 mm. diameter cut at F, and a tail G 100 mm. or so long. The cover glass E and the liquid rest within but not touching the hole in the cotton material F. A is a glass ring about 35 mm. diameter and 5 mm. high. B is a piece of toy balloon rubber which has been softened by being blown up for several days. The parts are secured by means of elastic bands. The rubber diaphragm can be secured to the objective in several ways and the Author found that a connection made with a rubber band is best, the amount of rubber sheet present being adequate to allow the objective to explore the slide without disturbing the glass ring.
The cotton tail G is immersed in the liquid with which the chamber is to be kept wet, usually it is water, in which case rain or distilled water should be used as tap water in most districts deposits lime, etc, in the cotton which interferes with the flow of water.
The chamber is operated as follows (Fig. 7):

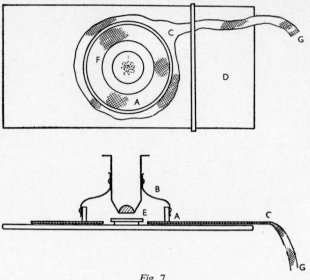

Fig. 7.

(1) The object to be studied is placed on the glass base-plate D and a cover glass, less in size than the hole F is placed over it.
(2) The cotton sheet C is wetted in the liquid required and wrung out. It is then placed on the base-plate as shown.
(3) The rubber sheet is attached to the glass ring with an elastic band. It is then attached to the objective as shown.
(4) The objective is fitted to the microscope.
(5) The glass ring is placed on the moist cloth as shown so that a humidity chamber is formed between the side of the ring, the rubber sheet and the base-plate with a ring of the cotton cloth inside.
(6) The tail G of the cotton cloth is dipped in a pot of water, say, 100 mm. away. The length of the tail determines the amount of water reaching the chamber.
(7) When the specimens have to be exposed, the objective and glass ring are lifted off without disturbing the set-up.
(8) The apparatus may be adapted for the inverted microscope, the only changes being that a hole must be drilled in the base-plate to carry a cover glass, and a glass plate used in place of the rubber sheet.

This apparatus and modifications of it are excellent when combined with a warm stage, for growing fungi and for studying the germination of spores and seeds. The only trouble likely to be encountered is due to inconstancy of flow along the cotton thread because this changes with external humidity and temperature. If the cotton tail is kept away from objects and is suspended by cotton threads, the apparatus becomes quite workable and will operate for several days in a building with normal heating arrangements.

Troughs

These are usually made up from thin slides with a spacer between them. They are usually used on the microscope with the stage in the near-vertical plane so that the narrow surface of water is uppermost. In this position the vibration due to working the microscope is least transmitted to the water and object in the trough.

For exhibition purposes the trough should have a transparent spacer so that light can be injected from the sides, Fig. 8. The trough can be of any size. (See 'Miscellaneous Apparatus: Henderson's Trough', page 513.)

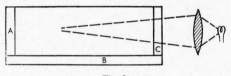

Fig. 8.

The safest and easiest way of building these troughs up to almost any size, say $300 \times 300 \times 50$ mm., is to grind strips of 5 mm. thick plate-glass and cut them up into blocks A, B, C, so that they are a good fit at the corners. The whole is then united with thermal setting Araldite (Aero Research Products, Cambridge). The strip to form the sides of each tank should be made in one piece, then, when cut, it may be assembled so that the flat polished surfaces of the glass are presented to the sides of the tank for cementing. If they are made separately, the thicknesses will be incorrect unless an expensive job is done, and cementing then becomes difficult. Araldite has virtually no solvent but boiling acetic acid will break it down. (See 'Laboratory Methods for Working Glass', page 491.)

If it is essential to use the microscope stage in a horizontal position with a trough in place, the inverted microscope is the only satisfactory instrument because except with the lowest powers it is impracticable to look into a free surface of water and obtain a steady image of what is below. The procedure for making a suitable trough is similar to that above but a circular piece of glass cracked off a tube about 50 mm. diameter forms the walls, Fig. 9.

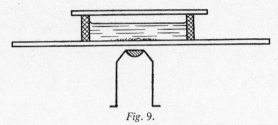

Fig. 9.

No good comes of making troughs with cements such as shellac, Bostik, marine glue, Chatterton's Compound, Canada balsam, and many others, because these will all crack sooner rather than later, whereas Araldite does all the jobs and resists most chemicals.

Troughs like those shown for use on the conventional microscope are also used for spectrometry. For ultra-violet work they can be made of Perspex cemented together with chloroform.

A type of trough designed by the Author and used by a number of outdoor workers consists of two plates of glass typically about 150 mm. square between which is placed a curved length of laboratory rubber tubing of diameter equal to the thickness of the trough required. The structure is held together by means of ordinary spring clips of the kind used to clip tools to boards in workshops. About four clips per trough are required to make a water-tight job and it helps to rub the tubing over with glass paper before use to remove bumps and glaze from its surface. Smaller additions of the trough may be made in the same way but by employing jam jar rings and electrical resistor clips.

If required, the troughs can be considered to be of permanent construction but their main advantage is that they can be dismantled for cleaning easily, can be assembled on site and if necessary out of junk materials. Being springy they are much less liable to breakage than is a rigid glass trough. It is a real advantage to be able to clean properly the inside of a trough, and it becomes easy with demountable apparatus.

The largest trough of this kind used in the Author's laboratory was 450 mm. square and 50 mm. thick, employing a piece of hose pipe as the rubber ring and wide-reach paper clips as clamps. The trough worked perfectly with sea water for several months until the investigation was finished.

Collection of Animals and Plants

The naturalist's side of this matter cannot be dealt with here, but a few notes concerning the collection of samples of water and effluents are included.

The expression 'animalcules' was invented to describe the enormous numbers of microscopical creatures of sizes up to a few hundred microns in diameter which inhabit smelly water and in fact any water which has organic matter dissolved or suspended in it. But actually these creatures are not the crude, indestructible lovers of filth that laymen think. They are really delicate creatures and they live their lives during a particular phase only of a putrefaction. If when samples of water are collected, the temperature is allowed to change more than a few degrees or the oxygen content of the water altered by a few per cent, the creatures die and may or may not make way for another species. In either case, the results of examination would not be an accurate estimate of the state of the water in the natural situation.

When specimens of water, mud, algae and weed are taken they must be separately stored in bottles of about half a litre capacity, this size being considerably greater than that often recommended. It has been the Author's experience that bottles any smaller cause the contents to experience changes within a few hours. Bottles must not be carried on the person or over-heating of the contents results, a naturalist's vasculum is the best carrying device.

When the bottles arrive in the laboratory, they should be opened within a few hours of collection and the contents poured into flat vessels about 50 mm. deep and 100 mm. diameter. The ratio of volume of liquid to surface area must be small or lack of natural aeration results causing changes among the creatures. If the depth of the liquid is only about a quarter of its diameter in the vessel, all will be well and this rule applies strongly to aquaria which are almost always of the wrong shape. Aquaria of the wrong shape can be aerated mechanically but microscopical creatures cannot stand such changes in the natural water and cannot be artificially aerated.

The flat vessels should be set aside for an hour or more after which time the mud will have settled and life will have regained composure. Samples of the mud, weed, plants, etc. as required are removed with a pipette or tweezers and placed on a slide such as that shown in Fig. 3. Several slides should be made at one time. The slides should be put in the humidity chamber, Fig. 4, and left for a further hour or more again to settle down, and when they are taken out and studied a fair representation of life in the sample will be obtained.

Plankton

It is not possible to study animals in sea-water so easily as in fresh water for the reason that the level of oxygen in sea-water is so great due to the great turbulence of the sea that sea-water taken out and made stationary cannot support the life in it. The water is not replaced naturally by highly oxygenated water from other places and an artificial flow cannot be maintained on a microscopical scale. If larger marine creatures like sea urchins, larvae and the younger stages of crabs and jelly-fish are being studied, it is possible to aerate the tanks by means of Perspex disks drawn up and down in the water about once per minute, dash-pot fashion, and by forced air.

Microscopical creatures in sea-water must be separated from the collection with a centrifuge and studied immediately. Plankton contains larger creatures which may be 25 mm. long and these are usually collected by towing a net made of nylon or bolting silk behind a boat. The net must have a bottle fastened at its apex so that delicate creatures pass down the net and find refuge from the abrasive action of the water flow, in the bottle. Periodically the bottle is emptied of its great concentration of animalcules into another container because the creatures cannot live more than a few minutes in such a concentration. If counts and preservations are being made the creatures are killed and fixed while fresh by adding 50 per cent by volume of commercial 40 per cent formalin. After killing they may be preserved for years. When they are studied they are usually taken out of storage with a pipette and spread in Petri dishes, from which they are sorted by hand.

Fresh water plankton can also be treated in this way but the creatures are usually smaller and consist mainly of rotifers. The net should have its threads spaced about 100 μ apart or many animals will be lost. A marine plankton net can be of double this spacing but in any case two nets of different fineness should always be used.

If mounting the collected freshwater creatures is intended for a work of reference, they must not be killed with formalin but must first be narcotized. (See 'Mounting: Hanley's Method of Mounting Rotifera', page 300.)

The smaller creatures like typical single-celled animals of the paramecium kind, either marine or freshwater, cannot usually be collected in nets and so must be centifuged out of large volume water samples but one of the best nets for general collecting purposes is made from the section of a standard nylon stocking between the knee and ankle. The knee end is secured by rolling a quantity of it round a smooth brass supporting ring and a boiling tube is secured by elastic at the ankle end. The denier of the stocking should be of the coarser kind.

Microscopic marine creatures are narcotized by Hanley's method while in their salt water environment, but when doping is complete they must be immersed in fresh water by gradual replacement of the salt before killing. The salt water can be drawn off with filter paper while the fresh distilled water is added with a fine pipette.

Culture of Animalcules

For some purposes it is necessary to culture single-celled animals and the basic methods are here described. For details of special work, medical and biological literature must be consulted.

Almost all free living flagellates, ciliates, rotifers and amoeba live on bacteria and the bacterial zooglea produced by them. It is clearly necessary when attempting a culture to establish a colony of bacteria in a neutral sample of water before an attempt is made to grow animals in it. The bacteria may be started in many ways, for example, hay may be boiled in water to release the nourishment from it, grains like wheat, oats, etc. may be used in the same way and a pinch of mustard in water grows bacteria very well after a week or two. Inoculation is performed by immersing hay, soil or small pieces of old dung in the liquid or it may be left to inoculate itself from the spores in the air. Nearly any organic material suspended in water will grow bacteria and consequently there is no end to varieties of culture methods. All single-celled animals do not feed on the same bacteria and several media should always be tried in an experiment, also the bacterial content must not be so high that the medium stinks with decay products. For producing a lively gathering of ciliates for other experiments, the Author has used extract of horse dung made in the proportion of one spheroid of dung to a litre of water boiled for about ten minutes to evaporate off any ammonia there may be present. A similar extract of fresh cow dung is not good for ciliates, but, when, old grows flagellates of the euglena kind very well. The mixture can also be recommended for causing protococcus and other unicellular algae and lichens to grow on new roofs and give them a weathered appearance, but for this the mixture needs to be much stronger owing to the leaching action of rain so a concentration of about ten cow pats (dry) to a dustbin-full of water is right. If the pats are fresh, boiling is not necessary but the infusion should be left to stand for two days before use. (See *The Study of Microfungi* for other culture methods.)

If less haphazard methods are required, the following, used by Mottram, may be employed.

An agar plate was inoculated with *Staphylococcus albus* and after a few days the culture was scraped off and washed in water in a centrifuge. It was then diluted until it looked like milk diluted 100 times with water. This solution of bacteria was then placed 1 cc. at a time in bottles and sterilized, as dead bacteria are good food for ciliates and less trouble than living ones. Bakers' yeast may be used in place of bacteria.

The creatures required for culture are separated from the collected material by one of the methods described under 'Handling Small Amounts of Material'. They are then grown in a 50 per cent dilution of the above material in well slides and are kept in a humidity chamber. A culture of small size such as one on a well slide must be sub-cultured about twice per week, that is, as soon as the animals have used up all the food in the drop.

Should it be necessary to start a new culture from one which has become loaded with unwanted living bacteria, the following method, also due to Mottram, has been found successful in many experiments:

(1) Obtain a capillary tube about 100 mm. long and fill it with sterile culture media.

(2) Place some dilute contaminated culture in a tube and centrifuge it so that all unicellular animals and bacteria are thrown to the bottom.

(3) Allow about 10 min. before the next move so that the ciliates and more active animals can swim up from the bottom.

(4) Take up a few of the wanted swimming forms with a fine pipette and introduce them into the capillary tube.

(5) Centrifuge the capillary tube slowly until the larger animalcules have reached the bottom.

(6) Sterilize the outside of the tube by dipping in absolute alcohol and dry it in acetone.

(7) Break off the end containing animalcules but not bacteria, and place it in a bath of fresh sterilized growing media.

(8) The animalcules swim out and start a fresh pure culture.

For keeping alive rhizopods and animalcules of all kinds the Author uses boiled whole wheat grains, about three grains to a Petri dish of water, almost exclusively.

In all experiments involving the growing of microscopical water living creatures, it is important to start with reasonably clean pond or river water and this should be boiled or filtered to clean it up. Tap water is usually so chlorinated or otherwise treated that many culture methods fail. Pond water is better than distilled because it contains more naturally occurring trace elements which appear to be essential to life.

The Study of Micro-organisms in Soil and Similar Materials

If soil is mixed with a drop of water and spread upon a microscope slide it is unlikely that any organisms except perhaps a flagellate or two will be seen even after the most careful search. But if the same sample of soil is mixed into a culture medium and incubated at normal temperatures it is soon clear that an enormous number of organisms, tens of thousands to the cubic centimetre, were present in the sample, many of these being the larger single-celled animals. It may be argued that many of these were in the spore or the encysted form before being brought out by the culture methods and this is quite true, but if the soil sample is treated with a 3 per cent solution of hydrochloric acid for 8 hours, all free living forms are killed; subsequent culture of the treated sample will show the proportion of organisms which were in the spore or encysted condition and survived the acid treatment. It is usually found that in garden soil about two-thirds of the organisms are in the free-living state but this varies greatly with the degree of dryness and amount of chemical and other manuring which has taken place.

Details of soil microbiology cannot be included here but the basic methods used in this kind of work are given below.

As indicated above, it is useless to attempt to wash organisms out of dust, soil, debris and other similar materials and expect to get an accurate idea of what is living there. Some larger forms may be washed into suspension but many others will not. It appears that the organisms are able to live very close to the particles of solid matter and remain in a way stuck to the particles even though they be vigorously washed. Most minute creatures are individually transparent and so cannot be seen on the solid particles. It is therefore essential that a culture method be employed to cause the occupants of a sample of material to grow and reproduce and so increase in numbers in a suitable medium that they can be found and seen.

A suitable culture medium is ordinary agar jelly doped with an extract of the medium being studied. If the medium is soil, a sample of the soil should be boiled for an hour in water, filtered and the water used to make up the agar mass, for by employing this method the trace materials so necessary for normal growth are present. If the special interest is fungi, an extract of rabbit dung should be made as above and added to the agar.

It is always an advantage to use a solid growing medium where possible because then the colonies of the organisms do not readily mix up with each other as they do in a liquid. If there are too many colonies the sample used to inoculate the agar must be diluted a known amount until the colonies, when they develop, are well spread out. In connection with culture media it should be remembered that ordinary tinned commercial foods are excellent sterile materials and, if of suitable salinity, may be used directly.

It is usual to pour the sterile agar into a Petri dish, allow to solidify and then smear the surface with a dilution of the material under examination. Incubation temperature must be arranged to suit the organisms in question or trials must be made. Usually the natural temperature of the material under examination is used for one plate.

When soil or other protozoa are being studied, it is necessary to get going on the plates, a culture of a typical soil or other food bacterium before seeking the larger organisms. Bacteria and fungi can grow on doped agar directly but the single-celled animals cannot, and must have a food organism. Some fungi are predaceous and catch eel-worms, while most amoebae and flagellates live on bacteria.

It is clear that when an investigation is about to take place, a general culture must be made, preferably on agar doped with different materials, say meat broth, soil extract, yeast extract, etc., depending upon the subject. Books dealing with bacteriological techniques give endless numbers of agar media and these specialist works must be consulted. As stated above, all flagellates and single-celled animals must have bacteria to feed upon, and the reason why a common hay infusion (clean hay boiled in soft water, cooled, and inoculated with another piece of unboiled hay) is so good for growing infusoria of all kinds is because it first develops a large clean colony of the hay bacillus surrounded by plenty of bacterial zooglea. This infusion does not readily decay into stinking products.

It is a mistake to make culture media strong in organic content, about 1 or 2 per cent of food material to the water or agar is sufficient, or a swarming growth of bacteria results and suffocates all other creatures. The material to be tested should be spread thickly in or on the culture medium so that all forms present have an opportunity to grow and live on each other. Ordinary microscopical examination will then show the variety of creatures involved and methods may be developed to study each. A suitable culture of soil bacteria may be taken from one of these random plates and propagated for use when growing protozoa.

It is convenient to place glass rings about 25 mm. diameter upon the surface of the agar in order to keep a particular protozoan in one region of the plate, and in this way an inoculated agar plate may be used to culture several species of protozoa. It is also a great advantage to be able to locate the small area defined by the ring for subsequent microscopical study.

To discover the kind of organisms living in a place, an ordinary microscope slide may be covered with agar and placed in the location studied for a length of time such that it experiences all influences there, then after a day or two of incubation colonies will be seen on the plate.

For testing reservoirs and sediments, several clean slides may be immersed in the liquid in various positions and examined after varying intervals of time. Organisms will be found attached to the slides and alga may be rooted or attached in their growing positions. Such slides must be studied in a trough of the liquid concerned.

An adequate idea of the activity of mycetozoa and other wood-living forms cannot be obtained by a single observation, but if sections of the wood are cut and stained according to the methods given in *The Study of Microfungi* the hyphae may be seen but little information of value will result. The wood in question must be observed over a period of a year during which most moulds will show characteristic fruiting bodies, but the mycetozoa (slime fungi) usually adopt the plasmodium (creeping slime) form once per year, the phase lasting about one week, before the fruiting bodies are produced and the plasmodium dries into a skin. If a piece of wood or a structure is thought to have slime fungi within it, it should be preserved warm and damp, say under a bell-glass with water present, at room temperature. Sooner or later, perhaps several months later, the slime will creep out of the wood, exist on the surface as a plasmodium for a few days and then dry up, leaving coloured fruiting bodies erect upon the dried film. Although the plasmodium may be 25 mm. square or so in size, it is very difficult to see upon the wet surface of wood. Should a particularly slimy patch be visible it is possible to see the streaming movement of the large number of nuclei present in these organisms by means of vertical illumination with a 4 mm. objective used to look at the block of wood itself. A specimen part of the plasmodium can rarely be removed from the wood because it is so thin that it cannot be picked up and if separated from the mass of protoplasm the streaming ceases. The plasmodium is usually in the form of a network of protoplasmic threads and the movement suggests that of cyclosis in the cells of plants. Mycetozoa are recognized by the shape, form and colour of the fruiting bodies and these may be stripped off a piece of wood and mounted as dry objects in deep cells.

The mycetozoa, though probably the most interesting group of plant-animals, appear to have little economic significance and exist in timber which is already far from fit for structures due to the action of water and other fungi. Tan pits are always covered with specimens where they are known as 'flowers of tan' because of the bright orange colours of many species. The mycetozoa must be considered to belong doubtfully to the vegetable kingdom because they must have built-up food material to live on and, though so far as is known they have never been seen to imbibe particles in an amoeboid manner, they are, during the plasmodium stage, very amoeba-like. Their reproduction however, is certainly fungoid, so under present classification it is to the plant kingdom under saprophytes they must go.

The dry and wet rot fungi are, of course, of the greatest economic importance, and it will surprise many to learn that both are most difficult to study in culture. The best way of conducting experiments with dry rot is to seek out a well infected place where temperature and humidity are such that a strong growth of fungi has occurred, and conduct the experiments there, bringing sections of timber and other materials to the laboratory for microscopic preparations only. A quick test for whether wood is experiencing active rot attack or is only soft and wet is to insert the blade of a screwdriver into the wood and feel whether or not the wood tends to resist the withdrawal of the instrument. If it does, the wood is probably only wet and not attacked, as it retains its elasticity. (See *The Study of Microfungi*.)

CHAPTER 14

Principles of Staining

WHEN STRUCTURES are transparent it has been common practice since the days of Leowenhoek to try to stain or dye them so that they become more visible. Processes of staining have often been rediscovered and improved and at the present time some approach to standardization has been achieved in medical and biological circles. The manufacture of microscopical stains is now an important and exact industry.

Under the heading 'Staining' are several processes not similar to each other and not designed to produce the same results; they are as follows:

(1) INJECTION. This is a means of extending or distending tissue or other substances by pumping them full of a coloured mass like gelatine by means of a syringe, thus showing the tracks of tubules and vessels within the structure. Injection is performed before section cutting. The injected mass is usually compounded of the jelly substance which hardens and retains the shape of the extended tissues when it cools, together with one of the dyes described below, the function of which is to stain the tissue as well as to colour the mass.

The injection technique is widely used in biology and medicine, and beautiful preparations of stomach, kidney and intestine tissue are produced showing the course of tubules and capillary blood vessels.

(2) PRECIPITATION METHODS. These consist of injecting a structure with a thin and probably clear fluid which is caused to deposit some opaque precipitate upon the tissue within the specimen. This method is usually more delicate and refined than mass injection as the thinner fluid can be caused to penetrate into capillary vessels and air ways over a long period of time and can be caused to precipitate the required opaque material right in the smallest spaces where it is most wanted. It is usual to deposit an insoluble residue like a metal because other processes may then be carried out without fear of dislodging the precipitate.

(3) PHAGOCYTOSIS. This consists of feeding to a living organism like paramecium a finely divided material like cochineal which

can be ingested by the organism, thus showing through its transparent skin the course of its digestive tract. But the method is not limited to digestive organs; coloured material can be selected to be taken up by many organs and structures, and a modern variation of the technique is to feed or inject into a body or system a dose of artificially radioactive material. The radioactivity must be such that it decays rapidly, usually after a few hours, and so after it has indicated where the concentration, leak, etc. takes place, becomes harmless to the tissue. The Atomic Energy Research Establishment at Harwell, Isotope Division, offers a service providing suitable radioactive isotopes for this purpose, and publishes a catalogue describing the characteristics of all isotopes available. The method is not strictly microscopical but is much more sensitive than any staining method for tracking a material. The radioacitivity is indicated by means of Geiger counters, scintillation screens composed of sodium iodide (gamma radiation) or zinc sulphide (alpha and beta particles) and by small photographic plates surrounding the job. When handling animal bodies it is usual to clamp dental photographic films to various parts of the structure. When a result on a large scale is obtained, the subsequent microscopical work is much reduced and can be concentrated at the points indicated.

Colour Tests

Again, these are not designed to differentiate parts of tissue or other materials for detailed study but are designed to indicate the presence of a material by means of an empirically derived colour test. Examples of such tests are as follows:

(1) Starch turns blue when treated with iodine.
(2) Plant cell walls treated with dilute sulphuric acid causes them to turn blue when subsequently treated with iodine.
(3) Resins and essential oils are coloured by Alcanna.
(4) Fat is coloured by Sudan III.
(5) Bismarck Brown colours fats in living cells.
(6) The presence of nitrites is indicated by a pink colouration when a mixture of $0 \cdot 5$ gm. sulphanilic acid plus 150 cc. of s.g. $1 \cdot 04$ acetic acid is applied.
(7) For protein decomposition by bacteria (Indol test): Add a few drops of $NaNO_2$, 1 per cent solution to the specimen solution in a test tube, shake and allow to settle, then run in steadily concentrated hydrochloric acid down the inclined side of the test tube. Raise the tube to the vertical and observe a red ring at the junction of the liquids. Allow some minutes for the test.

(8) To detect cellulose: (*a*) Suspend the material in a small quantity of water, add a solution of iodine in potassium iodide until the water is pale straw coloured. Add a drop of concentrated sulphuric acid and cellulose will turn blue.

(*b*) Schultze's solution: Dissolve 50 grams of zinc chloride and 16 grams of potassium iodide in 17 cc. water. Add iodine to excess. Decant the solution after several days and store in a dark bottle. Cellulose turns violet in presence of the solution. It may also be used as a permanent stain for cellulose cell walls of specimens.

(9) To detect iron: Pass hydrogen sulphide through water. Use the water to treat the specimens when, if iron is present, the structure will turn black.

(10) To indicate pectin: Pectin is coloured red by a 1 in 10,000 solution of ruthenium red. The dilution is not important.

(11) To indicate fats and oils: Sudan black or Sudan blue colours fats and oils black or blue but requires about fifteen minutes to act. Osmic acid stains black.

(12) Tannin is made dark green in the presence of a dilute solution of ferric chloride.

These tests are really analytical ones but are useful in making preparations designed to show the presence of certain bodies and materials and so are to some extent staining methods.

Vital and Non-vital Staining

This process is 'staining' in the normally accepted sense of the word. Non-vital staining is an expression reserved for the artificial colouration of dead tissues. The colours probably do not enter composition with the material to be coloured but appear to be adsorbed only. The process is undoubtedly a combination of chemical and physical processes and is not fundamentally well understood.

Vital staining consists in adding a non-toxic dye to living tissue and with this method the process of life in a cell can be watched, but it should be noted that placing a living plant stem in a coloured liquid does not constitute vital staining but comes under the heading phagocytosis. The stain passes into and around the plant in the sap but it does not stain the living protoplasm of the cells.

Examples of vital stains are methylene blue, neutral red, and congo red. The method is to use a 1 per cent solution of the dye and allow the specimens to live in the solution. Many vital stains are toxic but those mentioned work well upon water-living infusoria.

Preparation of Stains

Staining a section of tissue or a smear of microbes is more than simply dunking the specimen in a bath of liquid like a maid dyeing curtains. The preparation of the staining solutions themselves requires much care of detail. The industrial worker can be quite sure that the manufacturers of stains take all possible care in their production and the failures are likely to start in the histological laboratory, and the water used to make up the solutions is the usual offender, even distilled water is not the simple substance it appears as the stain sees it. If the distilling condenser is very good and delivers the water nearly cold, the concentration of hydrogen ions is such that an acid reaction can be detected by the brown thymol blue test. The water should be delivered from the still quite hot when it will be found to be neutral. Stale distilled water also shows acid reaction quite sufficient to spoil several staining operations, particularly those involving eosin.

A further point is that stain mixing is apt to be too casual and the aim should be constant strength in all stains. When diluting or making a solution, it is best to mix a few cc. only of the liquid with a quantity of powder sufficient to make a stiff cream, and this should be ground in a pestle and mortar. However great the tendency of the powder to clump and cake in a normally made solution, it will not do so when it is placed in a deficiency of water. The creamy mass is then dissolved without trouble in larger quantities of liquid. Some stains are so difficult to dissolve that heating is demanded, and some may be dissolved only by mixing them with materials like silver sand and grinding the whole together in a deficiency of water. All stains must be filtered through filter paper before use and most benefit from being left to stand for a few days, after which they are decanted and again filtered.

Many materials are added to the usual aqueous solution of stains for various purposes and a list of these follows by way of indication of method :

(1) Acid added to an acid dye generally increases both its speed and intensity of action.
(2) Alcohol restrains the action of a watery solution of stain. (This may be necessary to prevent over-staining and consequent loss of differentiation.)
(3) Aniline and phenol cause many stains to differentiate tissues more readily and may be added to acid and neutral solutions. No one knows how they work but they are useful ingredients.
(4) Camphor and thymol are harmless materials as stains see

them, but are useful preservatives and it may be mentioned in passing that fungi of some kind will grow in any bottle of stain if given time, particularly if it has a natural cork fitted.

Dyes and staining mixtures are legion but they all do basically one job, that is, stain the substance it is required to see and not stain to the same degree the surrounding mass of material. The correct stain for a job is usually discovered by trial and error and below is a list of raw dyes which are useful:

Methylene blue, aniline blue, eosin (all the aniline dyes); Bismarck brown, phloxine, erythrosin, bengale, iodine cotton blue, carmine, gentian violet, carbol fuchsine, night blue and haematoxylin.

A list of compounded standardized stains is given at the end of this chapter.

In the past dyestuffs were not so precisely characterized as at present—and even now there is too much scope for error, the result is that in early papers a dye is named which it is impossible to identify with certainty at the present time. A case in point is the magdala red formerly recommended for staining algae.

These facts are sufficient to demonstrate the necessity for a proper system of nomenclature and classification of dyestuffs, together with some certain means of identification. This need very early manifested itself to chemists and technologists in the dyeing industry. The only rational system of classification is according to the chemical constitution of the dye—azo-dyes, triphenyl-methane dyes, quinone-imide dyes and so forth. The first index of dyestuffs was published by G. Schultz (1902), and this has gone through many editions and revisions. The standard index in this country is the Colour Index of the Society of Dyers and Colourists (1924). In this each individual duestuff is listed under the name given to it by the manufacturers and also carries a systematic number—The colour Index Number (C.I. No.)—so that given either the name or the number, any dye can be identified immediately by the reader. Thus these publications, by the introduction of a rational system of classification of dyestuffs, have helped to remove many of the complexities inherent in the earlier literature of the chemist and the dyer.

The user of dyes for purposes of biological staining is faced with other difficulties, of which some, however, are of his own making. One is the retention by biologists of obsolete and ambiguous names such as gentian violet; another is the failure of the author of a new technique to specify precisely in his description the nature of the colour employed. This he could do quite simply by giving the Colour Index Number of the dye, but since this is not stated on the labels of his bottle he may not know it, and quite probably as a biologist he is not sufficiently familiar with dyestuffs chemistry to be aware of the confusing synonomy which exists. A sentence from Conn's *Biological Stains* may be quoted in this connection: 'The manufacturers and dealers in stains have sometimes encouraged this confusion by their practice of taking care to have the label on the bottle agree with the name used in the customer's order, regardless as to what the usual name of the dye may be.'

There are, however, other factors for which no responsibility can be attached to the biological user. Most of the dyes available are prepared commercially for use in the textile industries and they are seldom or never chemically pure individual compounds; indeed, the nature and proportions of the 'impurities' may be of fundamental importance as regards the 'shade' actually produced on the fibre. This is indicated by the various letters or numbers attached to the names of the dyes—Methyl violet B, 2B, 6B, R, etc.; Eosin Y, W, WS, G, etc., and many others. This fact is of great significance to the biologist, for sometimes a crude dye is found to give an excellent result, whilst a highly purified product used in the same technique may fail to stain at all. The nature, number, and even position of alkyl groups or the presence of sulphonyl groups may profoundly modify the behaviour of a dye with respect to a given tissue element. Again, the strength of the dye solution may affect the type of staining; it is difficult to adjust this to any particular value unless the dye-content of the sample of dye used is known. The composition of many dyestuffs is difficult to assess by chemical analysis, since closely related compounds may be practically incapable of separation and accurate estimation. The use of the spectrophotometer may assist here, for many of these substances have very characteristic absorption spectra, and suitable experiments will reveal the presence of different colouring matters and often indeed may enable an approximate estimate to be made of the proportions in which they are present. All these factors greatly complicate the biological application of dyes. The chemical and photometric data are of limited utility, and these criteria can seldom take the place of a biological test using the actual technique in which the dye is to be employed.

Granting, then, that these potentialities for confusion exist, the question arises as to what steps can be taken to eliminate them. The first person to attempt a solution of this problem was apparently Gruebler of Berlin, who is reported to have tested various batches of dye until he found one which gave satisfactory results and then bought up the stock of it to retail to his customers. This appears to be the practice of our most reliable English suppliers at the present time. Gruebler certainly built up a great reputation for his products, as have some of our own retailers since the reorganization of the British industry, but this method seems to leave the scientific worker very much in the hands of the dealers.

A different approach to the problem was made by H. J. Conn in 1922 in America, when the supply of stains from Germany, hitherto the main producer, had been disorganized by the war of 1914–1918. The significant step was the formation of the American Commission on Standardization of Biological Stains. This body is a co-ordinating committee organized under the auspices of the National Research Council and representing all the principal societies and associations whose members might be concerned with, or interested in, the preparation and use of stains, including chemists, zoologists, botanists, bacteriologists, pathologists and others. Its activities are directed by an executive committee of five members, under the chairmanship of H. J. Conn. The work of the Commission was, in the first place, to assemble all the available information concerning stains and staining techniques; and secondly, to enlist the co-operation of the manufacturers in the production of reliable grades of materials. In connection with the former of these it has published *Biological Stains* (H. J. Conn, 1925) and the periodical *Stain Technology*. In the other direction the policy of the Commission has been to encourage manufactures to submit samples of each batch of their products for examination by experts approved by the Committee. The necessary chemical and biological tests are carried out, and samples proving satisfactory are certified as such by attaching a special label to the container. Only a sufficient number of these labels is supplied to cover the particular batch of stain. Altogether, the Commission has drawn up specifications for about fifty dyes, details of which, and the numbers of certified batches, are published in *Stain Technology*. A small extra charge for these certified stains goes to the Commission to defray expenses, but the Commission is not a profit-making concern and most of the testing is done gratuitously by members of the Commission itself or their associates. Conn himself is most emphatic on the need for keeping the Commission free from entanglement with commercial interests.

Notwithstanding the fact that the first synthetic dye was prepared in this country, and despite the warnings of the more far-sighted of our sientists and politicians, the British manufacturers and the British Government allowed the dye industry—and indeed the fine chemical industry as a whole—to be captured by Germany; and German it remained until 1914. Then, faced with the appalling reality of chemical warfare, even those who did not wish to see were forced to admit the folly of this policy, or, perhaps rather, lack of policy. The foundation of the British Dyestuffs Corporation with State and legislative backing subsequently gave the British industry its chance. Industrially, it has completely justified the confidence thus placed in it. But from the point of view of biological stains we would now have been in little better position than was America in 1922 had not some of our larger distributing houses adopted Gruebler's policy and themselves tested their materials. The confusion of nomenclature is still with us; the consumer has no means of knowing, apart from possibly unpleasant experience, whether any sample of dye is actually suitable for the technique he wishes to use. It must not be assumed, however, that no attempts have been made to solve these problems. In the early 'thirties the Royal Microscopical Society drew up specifications for several stains, including methylene blue, acid fuchshine, neutral red and others. There is no means of ascertaining whether any particular sample conforms to these specifications other than by testing it oneself. Not unnaturally, many workers regard the position as unsatisfactory, and have turned their eyes to the work of Conn in America. An invitation was sent to all learned societies likely to be interested to appoint delegates to discuss the formation of a British Stains Commission. A meeting was convened in the rooms of the Royal Society of Medicine, London, on June 10, 1949. The delegates were unanimous as to the desirability of forming such an organization, and were especially interested to learn that, in a private communication to one of their number, Dr Conn had signified his support, emphasized the mutual benefits which would ensue if the British specifications corresponded with the American ones, and assured the British Commission that he would be willing to supply any information required as to the organization, secretarial work, system of code numbers, or other features of the work of the American Commission. The title 'The British Biological Stains Commission' was adopted for the organization.

The first step has therefore been taken, but it must be admitted that it is only a very small one and that much more remains to be done before an effective organization can come into existence. The collaboration and confidence of the producers and distributors as well as of the users of stains in this country has to be won. This may not be easy, since the concept of standardization may very well be thought to cut across the interests of some of the retailers, especially perhaps the more reliable of these. The manufacturers—of whom there are not a great number making biological stains—might be expected to support standardization since it might readily lead to an improvement in foreign and especially American sales. The Commission must persuade both makers and users to employ only standard and precise descriptions and to make full use of the Colour Index Numbers of the dyes; it must draw up the necessary specifications and devise the machinery for the chemical and biological testing of samples; it must then ensure that these tested dyes are clearly distinguishable on the market from other non-certified brands. In its early stages the Commission will have to face this complex of scientific and economic problems, and, most probably, will not find its course by any means plain sailing.[1]

A typical example of microscopical differential staining technique is given by the Romanowsky method described below.

If a smear of blood is placed upon a slide and examined either wet under a cover glass or dry, very little can be seen except dark blobs or transparent ones, depending upon the mounting media. If, however, the film is fixed to the slide, washed and stained, different stains affect different parts of the blood cells and make many structures and parasites clearly visible in a way that is both useful and beautiful.

Romanowsky used as his stains methylene blue and eosin, the blue dye being one which seeks and stains nuclear bodies, while eosin stains red, ordinary dead protoplasm. A further point was observed in the experiment which demonstrated the extreme sensitivity of microscopical stains to traces of material in the liquids. Romanowsky found that when he used old and consequently slightly oxidized methylene blue solution, he obtained three colours in his blood film, the nucleus of the white cells stained purple, the red cells (not nucleated) red and cytoplasm blue.

Romanowsky's stain is still a standard material and is made up from dilute aqueous solutions of the two dyes added one to the other until a precipitate begins to appear. The mixture is then filtered and left for some weeks to oxidize. When manufactured, the methylene blue is now oxidized ready for use.

There are several methods of applying stains and the method must often be chosen to suit the shape, bulk and type of the specimen. Bacteria are easy to show by the process known as dry staining, where a film of bacteria is spread on a chemically clean slide, allowed to dry, passed organisms downward once or twice through a bunsen flame to kill them and cement them to the slide, and then subjected to the action of a drop of 1 per cent solution of any of the dyes mentioned above for a few minutes. When washed with a stream of water smoothly and gently flowing from a tap, dried and mounted in balsam, the organisms if stainable in the dye used will be seen clearly.

[1] For much of the historical material given here I am indebted to C. F. Ferns.

If it is desired to watch the action of stains, the specimen should be mounted on the cover glass of a slide shown in Fig. 3, page 258. Drops of various stains can then be drawn under the cover by means of filter paper and when the desired effect is observed the cover may be removed and mounted.

The above method may be employed with living animals but it will be found that few materials besides 0·001 per cent solution of quinoleine blue or Bismarck brown will stain living protoplasm, that is, without killing the organism. It used to be a test for life whether or not an organism would stain in aniline dyes and the test is still a useful indicator, but the dyes must be applied in aqueous solution or the living cell is killed almost instantly and misleading results obtained.

Sections are usually stained after they are cut. They can be placed in zylol to dissolve away the embedding wax, placed on the slide and fixed down with some insoluble cement, then dipped in stain, but it is not always possible to stick down a section as much depends upon its fragility. Sections are usually cut thick enough to hold together while they are washed in zylol, alcohol, dipped in stain and washed in water, being borne to and fro on a fine mesh section lifter. If non-aqueous solutions of stain are used, egg albumen makes a good invisible cement for sections, while for aqueous staining solutions a thin solution in acetone of nail varnish makes a useful cement. Both of these materials are used in traces only and are quite invisible when the mount is complete.

Objects may be bulk stained by washing out the object and immersing it in stain for a time which must be determined by experiment and is usually an hour or so. Textile materials are best bulk stained for microscopical study even if this is not the best manufacturers' dyeing technique.

In conclusion it should be remembered that there are no rules in staining technique. Each object presents its own problems and so long as a supply of dye is available experiments must be made. Stains are made up by the manufacturer on a trial and experiment basis, and this must also be the rule for the user. Following is a list of stains and staining methods useful as background knowledge.

Examples of Staining Techniques

1. Test for Acidity of Distilled Water

Draw about 80 ml. of the water in a tall jar, stand it on white paper and add a few drops of brom thymol blue. If the colour becomes yellow, the water is acid and may be corrected by adding a few drops of tap water until the colour becomes greenish. It is then neutral and may be used in staining techniques.

2. To Stain a Milk Smear (Shutt's method)
Smear a sample of milk on the slide and allow it to dry naturally in a warm room. Mix the following stain:

> Ether 100 ml., Abs. Methyl Alcohol 100ml.
> Methylene Blue 1 grm., Hydrochloric Acid 0·7 ml.

Mix the ether and alcohol, and add the methylene blue. When the solution is complete, add the hydrochloric acid. Keep the prepared stain cold and the bottle well-stoppered. Apply to the smear for 15 seconds.

3. Quadruple Stain for Animal Tissues (Groverman's method)
Paraffin embedded sections are cut and brought down to 70 per cent alcohol. They are then stained in a solution (a) composed of 0·5 grm. saphranine O, 50 ml. of 50 per cent ethyl alcohol, 50 ml. ethylene glycol monomethyl ether, 40 ml. formalin, 0·5 grm. sodium acetate for 24 hours, then washed in tap water. Stain again with 0·1 per cent gentian violet solution for 1½ minutes and wash. Rinse for 30 seconds in equal parts of 95 per cent ethyl alcohol and ethylene glycol monomethyl ether and stain in a saturated solution of equal parts clove oil and ethylene glycol monomethyl ether, one part of which is added to 2 parts of 95 per cent ethyl alcohol. Wash in a mixture of 2 parts 95 per cent ethyl alcohol and 1 part ethylene glycol monomethyl ether. Stain in gold orange prepared as for solution (a). Wash in a mixture of equal parts of clove oil, abs. ethyl alcohol oxylol, rinse in a mixture of zylol and abs. alcohol, rinse in zylol and mount in balsam. Cartilage stains bright red, nuclei purple, connective tissue green and muscle light red. In intestines, submucosa is orange, goblet cells green and other mucosa cells gentian violet. This process has been given as an example of superimposed staining techniques, any one of which may be useful by itself.

4. The Smear Method of Demonstrating Chromosome in Plants
Dissect out the anthers from a young flower, place them on a clean slide, crush them with a clean smooth scalpel and smear them across the slide with the blade. Fix the cells with a saturated solution of mercuric chloride in 1 per cent acetic acid for an hour or more. Wash well in water and stain with a solution of either gentian violet and iodine, alum haematoxylin or suphranin. Wash in zylol and mount in balsam. If it is desired to show mitosis, the anthers or a root tip

should be taken from the plant at about five o'clock in the morning as most cells will be found in mitosis at that time of day.

5. Method of Obtaining Sections of Fresh Plant Tissue

Mount the leaf etc. to be sectioned between two blocks of paraffin and press together for a few minutes with a force of about 1 kilogram weight. Weld the edges of the blocks with a hot iron, mount in the microtome and cut in the usual way. For bulky objects use elder pith as embedding medium, and clamp it together with the specimen in a divided brass tube for cutting.

6. To Stain Radulae and Light Chitinous Materials

Extract the radula or other organ and wash in sodium hydroxide. Oxidize the part in potassium permanganate solution until it appears black, then decolourize in a saturated solution of oxalic acid. Stain in 1 per cent solution of Hofmann's violet for 30 minutes, dehydrate in alcohol, clear in zylol and mount in balsam.

7. Method of Staining Yeasts: From *Stain Technology*, 24, pp. 85–6

Make a smear of yeast on a clean slide and fix with heat gently applied. Stain with a mixture of crystal violet 5 grm., ethyl alcohol 10 mil., aniline 2 ml., distilled water 20 ml., poured on the slide and warmed (not boiled) for three minutes. Wash in tap water and de-stain with a mixture of 95 per cent ethyl alcohol and 3 per cent hydrochloric acid. Wash in water and stain in a mixture of 2·5 per cent saphranin O, in 95 per cent alcohol and distilled water 100 ml., then wash in tap water and drain dry.

When used on ascosporogeneous yeasts the spores are stained violet and the vegetative cells red.

8. Method for Staining Nerve-Cells; Tomaselli's Method

Pieces of nervous tissue are immersed in ammoniacal alcohol (absolute alcohol 100, ammonia 4–5 drops) for 6–7 hours. They are then immersed in pure pyridine and kept at a temperature of 36–37° C. for two days, the pyridine being very frequently changed, especially at first. The pieces are then washed in running water for 2–3 hours. The material is treated with an acid solution of molybdate of ammonia for 12 hours, imbedded in paraffin, and the sections stained with thionin (1 : 10,000).

9. Method of Staining Spirilla Bacteria

These bacteria, and particularly the tiny ones like *Treponema pallidum*, are likely to escape detection by means of the ordinary stains and a good test material upon which to perfect the techniques is sputum taken before a meal. A method due to Follet is as follows:

Glycerine 40 grm., acid-fuchsin 2 grm., pure carbolic acid $\frac{1}{2}$ grm. Mix, and filter after solution. The sputum to be examined should be recently expectorated. Pick out a fragment with a platinum needle, place on slide, and add thereto a minute drop of stain. Mix thoroughly and put on a coverslip and examine. If a little acid green dissolved in glycerine be mixed with the sputum before the acid-fuchsin is used, a brownish hue is imparted to the preparation; and if a double-staining be desired, this may be effected by using in addition to the acid-fuchsin solution the following mixture: Glycerine 40 grm., methylene blue 2 grm., pure carbolic acid 0·5 grm.

While this medium stains all the spirilla infesting the mouth, so that swarms may be observed in the same field, there is no difficulty in differentiating *Treponema pallidum*.

Another method given by the Author is suitable both for fixed and fresh films. This consists of chloroform 40 grm., methylene blue 2 grm., acid-fuchsin 0·25 grm., pure carbolic acid 0·5 grm.

The stained preparations must be thoroughly washed in running water, and if need be in alcohol to remove excess of pigment.

10. A Mild Botanical Stain

The dye annato which is used for normal colouring of cheese and butter stains cuticular tissue well. It should be used as annato extract, dissolved in alcohol, filtered and applied for about 1 hour to sections. It may then be washed in alcohol and be brought down in stages to water or glycerine. Cuticular tissue is stained orange-yellow.

11. To Stain Flagella or Unicellular Animals and Plants

Osmic acid and the iodine-sulphuric acid test for cellulose both make flagella visible but these are hardly to be considered permanent stains. A simple staining method is to mix with a quantity equal in volume to the drop of specimen water, a 10 per cent solution of nigrosine. Allow to dry naturally in a warm place, then mount in balsam. This process causes considerable distortion of the specimens but indicates flagella quickly.

The only staining process for flagella which does not involve drying out the specimens is Nolands. Mix a saturated solution of carbolic acid 80 cc., formalin 20 cc., glycerine 4 cc., gentian violet 0·02 grm. The stain is very powerful and will certainly darken the cells unduly. Nolands' solution is supposed to be unstable but appears to work satisfactorily as a mounting medium if the slide is properly air-tight.

To Stain Hair

The kind of staining necessary in microscopical work is quite different from that usually required in textile work. Microscopical work usually demands that hairs be stained in such a way that all appear different so that they can be recognized and separated, while in textiles, taking for instance a mixture of wool and cotton or a mixture of several wools, all efforts must be concentrated to produce an even colour.

The study of hair is a special problem and the reliable way to do it is to set up a slide library of various hairs taken from various parts of animal pelts and use the collection as a reference.

Hairs vary much in appearance, depending upon the age of the animals, the position on the body and the part of the particular hair taken. In the example of human hair, much depends upon the artificial treatment it has had. A person's hair-dressing history can be read from a single long hair; permanent waving leaves a flattened hair at the place of bending, bleaching or dyeing leaves a colour change in hair where it was growing from the scalp at the time of dyeing; the closeness of the scales indicates the rate of growth of the hair which often relates to the health of the individual. The natural uncut end of new hair when young shows a round tapering end, while an old hair shows a bifurcated tip, as can be seen on many hairs in a paint brush. Chemicals in the body, like arsenic, are deposited in the hair as it grows, therefore the quantity of arsenic in a body and the date of entry can be computed before or after death.

All wool is simply hair, and all hair has scales upon it like tiles on a roof, and in sheep's wool these scales are very apparent even without treatment; the powerful interlocking property of the scales gives wool its fine, strong quality when woven. For this reason, wool may also be felted, that is, simply matted together without weaving. When a wool hair is stretched, the scales tend to stand out and interlock more readily, also when a woollen garment is toused about in a wash-tub, the water and the movement of the fibres cause the scales to mount over each other and so reduce the size of the garment, therefore the less movement a wool garment experiences during washing, the less it will shrink.

The structure of hair is as follows:

(1) The outer skin of a hair is known as its 'cuticle' and it consists of horny material produced from the body skin. Hair is in fact a column of skin and the cuticle is composed of scale of this substance.

(2) The inner parts are called the 'cortext' and contain the pigment cells which in the main give the hair its natural colour, and in humans this is the structure which may be seen to be arti-

ficially dyed. In dyed hair the colour is evenly distributed, while in natural hair the pigment cells are seen separately, but the pigment may disappear with age, although the cells may still be dyed.

(3) In the centre of hair is the 'medulla'. This is a tube of dimensions which vary greatly in different animals, and in the same animal, and often it is not present throughout the full length of a hair. The large cells of a medulla contain mostly air and variations in a medulla do not appear to have any important significance.

The staining methods given are intended to reveal as many features of hair structure as possible, and when a slide collection is to be made for reference purposes, it is important that the method of preparation of the hairs be standardized.

The staining process for hair is as follows:

(1) Cut out a section of hair from the length so that the air in the medulla can get out when the piece is immersed in liquid.

(2) Bleach out the pigment in the following solution:
Hydrogen peroxide one in three solution in distilled water 50 ml., 5 per cent aqueous ferric chloride one drop, ammonia dissolved in water (household kind) to make up volume required. This liquid will do the minimum of damage to the structure of the hair but will ultimately soften it. Light human hairs and fine animal ones need only about fifteen minutes in the solution, but heavy ones like cow hairs need several hours and trial is the only rule.

If the well-known 'blonde' bleach, 1 per cent solution of potassium hydroxide is used, the hair scales will be found to be displaced and the structure of the hair attacked.

(3) Stain in 1 per cent solution of carbol fuchsin used hot. Any of the regularly used stains can be used on hair and the workers must decide which stain gives the best result for their particular investigation, but carbol fuchsin gives excellent general appearance results.

When applying stains to hairs, the basic methods of cleaning the material must be applied as explained under 'Mounting' and a few are repeated here for completeness. All grease must be washed out by immersion in benzol or ether, or a mixture of ether and alcohol. Wash in distilled water before a change of reagent is made (of course this does not apply when dehydrating). Clear the specimen before mounting by soaking it in benzol or other cleaning material described under 'Mounting'.

To Cut Sections of Hairs

The usual methods of paraffin imbedding are quite satisfactory for hairs but simpler means are often useful and two quick methods are given below.

Thread the hairs through the eye of a darning needle, pass the needle through a cork leaving the hairs in the cork, and cut sections of the cork and hairs together with a razor. Examine the hair sections either while still in the cork section, or poke them out with a needle on to a gummed slide.

An even simpler method is to shave with a new razor blade and then shave again in about two hours' time, when excellent sections of beard hairs will be found on the razor.

Notes on the Use of Reference Collections

In many branches of microscopical science a reference collection of material is necessary and this is particularly the case with hair studies when the work may have a background of crime investigation. In order to locate a specimen quickly, it is essential to keep a file of photomicrographs as these can be arranged in a way which makes search easy, and the magnified views presented do not demand the use of a microscope. When the actual comparisons have to be made, microscopical examination under a comparator microscope is essential, but the mass of material to be searched is greatly reduced. For details of storage of slides and equipment, see Chapter 2 which is devoted to these matters.

Negative Staining

The expression 'negative' staining is used as opposed to 'positive' or normal staining to describe the method in which the background of the slide is stained but not the object which it is desired to detect. Naturally the method will not differentiate detail within the object, but often in laboratory work it is necessary only to detect the presence of a known object with the minimum expenditure of time. All normal staining methods need considerable preparation of the material and this time can often be better used.

Before negative staining can be attempted the slide must be entirely grease-free in the chemical sense, and this can now be achieved by washing in standard domestic detergents or teepol. It was formerly necessary to boil the slides in chemicals like chromic acid but the Author has never found this necessary since detergents appeared. It is essential that the background film of stain shall be able to run evenly round the tiny objects on the slide, hence the careful degreasing operation.

The method is best employed for detecting bacteria and their spores, particularly as most spores will not stain well by any process.

One method is as follows:

Mix a drop of the suspension containing the bacteria with a drop of nigrosin stain and spread the mixture on the slide by drawing a film with the edge of another slide. Cover while wet and examine the preparation or allow the film to dry in the ordinary way, after which it can be mounted in balsam.

Other stains can be used as negative stains, for example, strong eosin or Indian ink. Eosin works well with bacterial spores. Only tiny objects can be shown well by negative staining techniques.

Robinow's Method of Showing the Bacterial Nucleus

When bacteria are heat fixed on a slide, considerable shrinkage of the structure takes place, and in addition to this, the outer layers of the cell absorb more stain than the inner ones and so mask the internal structure. The method given below is known as the Acid Giemsa and depends for its success upon hydrolyzing the outer layer of the cell and so preventing the dense staining which masks the normal Giemsa colouring below. To prepare the bacteria use smear preparations and fix in osmic acid vapour for five minutes. Apply hydrochloric acid at 60° C. for 7 minutes, staining with ordinary dilute Giemsa stain at 37° C. for 30 minutes, clear in clove oil and mount in balsam. The process demands chemically clean slides.

To Cut and Stain Sections of Timber

In botanical work it is most important to preserve a microscopical record of specimens, and in the industrial field wood is a usual material which has to be sectioned and mounted, but unless reliable reproducible methods and stains are used, this work is rendered useless due to broken mounts and faded stains after the passage of years, just when their value is becoming great.

In botanical work, microscope slides should be filed in flat metal drawers, which can be kept alongside the dried herbarium specimens; in this way slides are kept in small lots so that any reasonable number can be added to associate with a particular herbarium specimen.

Wood is usually cut into centimetre cubes with a saw, but the blocks have to be softened before cutting with a microtome and this may be done by boiling in water or soaking in hydrofluoric acid. Some woods may be cut directly under a jet of live steam played on the specimen on the microtome. When woods are soaked they do not usually need supporting in the microtome but softer materials need ordinary paraffin wax embedding; but it is rarely necessary to treat botanical matter with special penetrating materials prior to embedding in wax as they are tough enough to retain their characteristics so long as they are mechanically supported.

It has been found that only safranin and haematoxylin are sufficiently long-lasting to be reliable stains, and balsam is the most effective mountant, and if possible it should be used raw as described under 'Basic Mounting Techniques'.

Those engaged in botanical studies must choose for themselves which particular features they wish to record, but in general for woody matter the following features are likely to be most useful and line up with current botanical practice:

(1) The outer layer of bark extending inwards, transverse section.
(2) The vascular structure as seen in a young shoot, transverse section.
(3) A longitudinal section of a young stem showing the fibres and vascular system.
(4) Longitudinal and transverse section, including medullary rays in a large stem.
(5) A transverse section of heart wood.
(6) A piece of the epidermis of a leaf showing any characteristic hairs it may have.
(7) Another piece of epidermis showing stomata.

Most of these sections are straightforward jobs. No. 7 is usually obtainable by lifting a part of the cuticle with a scalpel and tearing it off, when it will usually carry with it the stomata and guard cells together, and can be mounted as a transparency.

Almost any plant can be identified from a section of its stem but it is necessary to prepare a library of comparisons. Some plants have well-defined strong fibres from which ropes are made, examples are hemp, flax, manilla and coconut, also the common plantain provides a good example. The fibres may be soaked out of most plants by allowing them to decay in water up to a certain point obtained by trial, when the fibres are released from the softer matrix by decay. Flax 'retting' is a commercial example of this and is performed in a local pond.

Most of this work which is not purely scientific is of value in detecting adulterations, and an important collection of botanical material is held by Kew Authorities in England and by Yale University in the USA.

The forming of a collection of wood specimens and of hair specimens has something in common, and as mentioned above, it is almost essential that a library of photographs be kept in the herbarium with the actual slides to back it up. It is difficult and time-consuming to look through many slides on a microscope in order to identify an unknown specimen but photomicrographs are easy to display in groups and can easily be filed.

CHAPTER 15

Basic Mounting Techniques

A MICROSCOPICAL OBJECT is mounted on a slide for one or many reasons as follows:

(1) Most objects would be lost unless they were located upon a piece of glass and protected from the atmosphere.

(2) An object is best seen if surrounded by a transparent material in optical contact with it, as there is then a reduced amount of surface reflection and light passes into the specimen, thus causing its interior to be better illuminated.

(3) If an object is filled with a medium of high refractive index it becomes more transparent and can be seen through.

(4) Many delicate objects need the support of a liquid to display their component parts.

(5) Complete penetration of an object by a suitable transparent solid medium preserves it for an indefinite time.

(6) The optically smooth surface presented to the objective by a cover glass optically united with the specimen cuts out optical disturbances due to irregularities on the surface of a liquid or solid.

From these requirements a few general rules may be formulated and modified to suit particular needs.

The best material to use for penetrating an object, supporting it permanently and having a high refractive index, is Canada balsam. However, this material, referred to below as 'balsam', will not mix with water and so the mounting techniques must be such that water is removed. Because of the universal use of this medium, mounting in balsam will be described first, as follows:

(1) The specimen must be cleaned, i.e. unwanted surrounding material must be removed by any process the specimen will tolerate. For example, fatty parts of insects are dissolved in weak potash solution until they are seen to be free from white-looking parts.

(2) The specimen must be dehydrated by passing it through several stages of alcohol strengths up to absolute, but plunging direct

into strong alcohol causes such rapid withdrawal of water that the specimen will distort due to the mechanical forces involved.

(3) Completely penetrate the specimen with a light liquid which will readily mix with balsam. This is known as 'clearing' the specimen and does in fact make it more transparent, in addition to the fact that it allows proper penetration of the balsam. Suitable clearing materials are turpentine, clove oil, xylol, benzol, cedar oil and others, and in general clove oil and turpentine are satisfactory.

(4) When the specimen is well soaked in these liquids (say for a few hours or overnight) it is transferred to a drop of raw balsam and allowed to soak in this in a dust-free atmosphere for an hour or more. In this way the balsam will diffuse through the specimen and bubbles will tend to come up to the surface of the drop. This operation is best performed under a bell-glass with an air pump to draw air out of the specimen, but the pump is not really necessary if clearing is properly done, and if used, the pumping rate must be low or disruption of tissue may result when imprisoned air expands.

(5) When the specimen is seen upon low power microscopical examination to be free of bubbles and dirt, it should be placed upon a covered hot plate at about 100° C. for a few hours to drive off the more volatile constituents of the balsam. Enough balsam must be used to allow a surplus to be present when the cover glass is applied.

(6) When the balsam is hard enough to resist the pressure of a needle point, the cover should be applied, a small weight placed upon it, the temperature increased by about 50° C. and left for a few hours until the cover has settled down either upon the specimen or on to supports. When the balsam is so hard when cold that it can just be dented at the edge with the thumb-nail without chipping, the mount is done.

(7) Surplus balsam can be removed with a cloth soaked in xylol.

Slides should not be ringed with coloured paints and varnishes unless the mount is such that liquid has to be sealed in. These rings of all degrees of elaboration often seen on amateur work prevent the cover from being cleaned easily, foul the front lens of short working distance objectives, and generally present a ridge which catches on things and transmits strains to the mount.

Slides may be marked with dilute water glass (an aqueous solution of sodium or potassium silicate). It should be sufficiently dilute to be handled by a writing pen. The name is written on the slide with liquid and the slide is heated at that place so that the water glass attacks the slide. When rubbed clean, a good impression of the writing will be found etched into the slide.

(8) Label the slide with the name of the specimen, where found, the mounting medium and date.

Balsam goes slightly acid with time and can be expected to cause stains used in biological work to fade after about twenty years. Its refractive index is 1·5 and is so nearly that of glass that it is used to act as glass in optical systems. In immersion systems, glass, balsam and cedar oil are always considered homogeneous.

Some objects, like algae cells and fungi, cannot be got into balsam, usually because they will not withstand dehydration, and other media have been invented for dealing with these, but the method of application remains practically the same. A few important and representative mounting techniques follow.

Euparal

Sandarach, or pounce, a resin derived from *Callitris quadrivalvis*, is a valuable basis for mounting media. The principal menstruum is a mixture of camphor and salol, called for short 'camsal', which forms a colourless liquid having a refractive index of 1·53576. In this menstruum, sandarach is only slightly soluble, the addition of some alcohol or other solvent being necessary. The two alcohols which were found suitable for the purpose were isobutylic and propylic. The mixture of sandarach, camsal and propylic alcohol makes a medium having a refractive index of 1·47892.

Isobutylic alcohol was found to have properties more suitable for microscopical technique; thus it is extremely useful for dehydrating delicate objects, and when used as solvent for camsal and sandarach, forms a medium having a refractive index of 1·47892.

The two foregoing media have the inconvenient defect of dissolving pigments, so that they are practically useless for mounting stained preparations. In a mixture of eucalyptol and paraldehyde is found an efficient substitute for the alcohols, and to the mixture of sandarach and camsal with eucalyptol and paraldehyde is given the name of 'euparal'. The refractive index of euparal is 1·483302. It is commercially available and a suitable solvent is sold.

Invisible Cement for Holding Small Solid Objects on Slides

Gum tragacanth is the best material to use and it should be prepared as follows:

Partially dissolve the finest powdered gum tragacanth in strong alcohol just sufficient to cover the powder. A small crystal of thymol added to the spirit sterilizes and preserves the mucilage from mould, after which sufficient distilled water is added to dilute it, till it becomes a very thin jelly of such a consistency that it will not run from the bottle when tilted. The gum contracts and disappears in drying, so that there is no need to use it too sparingly in mounting. For mounting, say foraminifera in balsam, a little of the same gum may be diluted with distilled water, until it forms a perfectly clear liquid. A drop of this on the glass slip will be sufficiently strong to hold the foraminifera in position; and at the same time will not show, provided the mount is thoroughly dried before the balsam is added.

Mounting Seeds and Germinating Bodies
This work usually requires a quick, simple technique which can be used to make a series of slides to record an experiment rather than an elaborate method of producing a beautiful result.

Germinating seeds and all things of that sort are killed (i.e. fixed) in any of the usual liquids, say mercuric chloride, dilute alcohol, osmic acid, or in the mounting liquid itself. Cavity slides or ring-cell slides are used and marine glue employed for ringing and sealing. The mountant is two parts of methylated spirit (clear) mixed with one part of glycerine and this happens to be a good fixer as well. The seeds, shoots, etc., should be allowed freely to stand in the mountant on the slides for some hours to allow penetration to take place and bubbles to rise. The covers can then be applied. (See 'Mounting in Fluid', page 297, and 'Handley's Method of Mounting Rotifera', page 300.)

Leaves are best mounted for ordinary examination in liquid as above or in glycerine jelly, but if they are required only to show hairs and general cell arrangements for permanent reference, they had better be put into balsam, and for this they need dehydrating through several stages of alcohol. Some delicate matter like petioles will not stand it without distortion but in general glycerine and fluid mounts are insufficiently permanent for works of reference. All glycerine jelly mounts must be ringed with a varnish or marine glue to prevent the glycerine attracting water and softening the mount.

Mounting Fungi: Duddington and Dixon's Method
The workers named above found difficulty when mounting fungi grown upon agar jelly, as fungi hyphea cannot be scraped off the medium without great danger to their structure and the inclusion of agar jelly in a lactophenol mount is very likely to cause air bubbles.

The following erythrosin-glycerine technique allows hyphae structure below the level of the agar jelly to be examined. Blocks of agar 4 mm. square are removed from the growth and fixed for a minimum time of 24 hours in Barnes' solution, as follows: Commercial formalin 10 ml., Glacial acetic acid 5 ml., water 85 ml. It has been found that fungi may be left in this solution for several years without change. The blocks are removed from the fixative and the layer containing the wanted growth is sliced off with a razor blade, washed in water to remove all trace of acetic acid, soaked in a saturated solution of erythrosin for 24 hours, placed in 10 per cent glycerine, dessicated to the consistency of pure glycerine and mounted in glycerine jelly.

Barnes' Turpentine Method
This method has the advantage of more permancy than any employing glycerine jelly.

Prepare the agar block, fix and slice it as above. Stain in erythrosin or Delfield's haematoxylin, place in 10 per cent glycerine and concentrate. When the glycerine has a syrup-like consistency, wash the specimen in 95 per cent alcohol to remove all glycerine, transfer to absolute alcohol to dehydrate.

Make a mixture of 10 per cent Venetian turpentine in absolute alcohol and place in a dish in a dessicator. Transfer, with minimum exposure to the atmosphere, the specimen from the absolute alcohol to the turpentine and close the dessicator immediately. Concentrate the turpentine in the dessicator until it becomes a sticky liquid; it is then not sensitive to moisture and the specimen may be mounted in it in a cavity slide. The turpentine can be hardened a little more on the slide as for a balsam mount.

Armitage's Rapid Method
Prepare blocks of agar as above, slice off a layer about 200 μ thick containing the fungi, fix in absolute alcohol for one minute, stain in saturated solution of chlorazol black E dissolved in absolute alcohol for ten minutes, mount in euparal on a cavity slide, and harden by baking at 35° C.

This turpentine method is violent but has many industrial uses when time is important and recognition more necessary than study. Any fungi can be mounted in this way but one must expect damage to thin walled sporaes. Zoopagaceae which capture amœba can be well shown by this method.

The Canada Balsam Method for High Powers
The blocks of agar are cut and fixed as above and glued fungus side

down with the following cement: gelatine 1 grm., glycerine 15 ml., 2 per cent sodium salicylate 1 ml., distilled water 100 ml.

The gelatine is dissolved in water at 30° C. and the salicylate added; shake and filter through muslin and add glycerine. Place a drop on a cover glass together with a drop of 2 per cent formalin and lower the block on to the drop. Dry out at a temperature below 30° C. The fungi layer is held on the glass by the glue while the agar shrinks, and experience indicates the best time with respect to the shrinkage to cut away the agar with a sharp razor blade, and with practice a section a few microns thick containing the fungi can be left on the slide. Stain with haematoxylin, dehydrate through about fourteen stages in alcohol, allowing two to three hours in each stage, clear by stages in xylol in alcohol, allowing twenty-four hours per stage, finishing with pure xylol, and mount in balsam.

These sections are thin, well-stained, and mounted in a medium of high refractive index are admirably suited to study with the highest powers, and are permanent.

Mounting in Liquid

Specimens which require this are usually fungi and algae, and the medium most used at present is lacto-phenol. The solution is composed as follows: glycerine 78 cc., lactic acid 80 cc., water 100 cc., phenol 50 grm.

A liquid mount clearly must be confined in some way and the usual method employed is to turn rings of cement on to the slide, harden them, place in the specimen in the mounting liquid, and ring again round the edge of the cover so that the fluid is confined by a wall of cement.

Fluid mounts sooner or later give trouble due to leakage, mainly because the expansion and contraction of the liquid fractures the cement. Well slides with concave bottoms should not be used for mounts in materials other than balsam because the concavity effects the correction and apparent centration of the condenser, also the varying depth of the cavity demands constant change of tube length and focus if the specimens are lying on the cavity bottom.

A suitable cement which resists lacto-phenol and adheres well to glass is composed as follows: gum dammar 5 grm., xylol or chloroform 50 cc. Evaporate the mixture down to the required consistency, which should be about that of ordinary varnish at room temperature, and apply to the slide on a turntable with an artist's brush.

Specimens which do not need staining are merely washed in distilled water, dumped into the medium and sealed down. The cover should be drawn down by removing surplus liquid with a filter paper; when dry at the edges the slide should be ringed, and re-ringed after about one month has passed since preparation. It is most desirable that these slides be kept in a constantly heated building, they will then last for many years, but they will not tolerate great temperature changes between day and night as differential expansion ultimately breaks some part and lets air into the mount.

If it is necessary to stain the specimens, suitable dyes can be dissolved in the lacto-phenol solution, the specimen is then subjected to only one reagent and is not allowed to dry and distort between operations.

A suitable mixture used by Sommers is as follows: aniline blue 0·25 grm., lacto-phenol 50 cc., osmic acid 2 cc. of 1 per cent solution. Other dyes like phloxine, cotton blue and aniline green may be substituted in the formula.

Nucleus and chromatophores are readily differentiated by aniline dyes (see 'Staining', page 282).

If the specimens are large enough to be handled, they should be immersed in phials of the dye, but should they be small and delicate, the operations can be carried on on the slide, the liquids being changed by micro pipettes or drawn off by filter paper. If operations are conducted on the slide in a prepared cell, it must be perfectly cleaned up and dry before the cell is finally filled up from a dropper with clear lacto-phenol and sealed. It is impossible to seal a wet-edged cell and the amount of fluid must be carefully judged.

In many cases it is necessary to 'fix' a living specimen so that important changes do not take place in it during the mounting procedure and the fixative is applied immediately after the specimen is 'caught'. Suitable fixatives are as follows:

(1) Formalin (40 per cent) 5 cc., glacial acetic acid 5 cc., alcohol (70 per cent) 90 cc.
(2) Mercuric chloride 0·5 grm., acetic acid 0·5 cc., water 250 cc.
(3) Chromic acid 1 grm., acetic acid 2 cc., water 200 cc.

When sections are to be cut of tiny delicate specimens, fixation and staining must be performed before embedding takes place.

The number of cements and paints which have been recommended for sealing fluid mounts and building up cells is nearly infinite, but the Author has found that ordinary commercial synthetic varnish (made from synthetic resins) is entirely satisfactory, and synthetic enamels are just as good if colours are desired. Commercial marine glue obtainable from the manufacturing opticians is very useful for fluid mounts containing glycerine. Gold size is satisfactory and well tried, but apt to be brittle.

Mounting Delicate Micro-organisms

A good way to mount almost any small insect is as follows: The specimen should be killed in diluted spirit and allowed to remain in it for some hours, it should then be gently washed. Pour off the spirit and rinse with water two or three times, and soak in glycerine much diluted with water, leave it in this for a day or two; the water will evaporate gradually and should be replaced with pure glycerine, the whole process being more effectual if done slowly. Mount in glycerine jelly. The jelly as usually sold is too strong, it should be thinned with a little pure glycerine to avoid shrivelling delicate specimens. Methylated spirit, as sold for burning, does not answer for preparing and preserving specimens, it turns thick and milky in contact with water. Spirits of wine (about 60° over proof, laboratory alcohol) should be got from a chemist, but is rather expensive, and as a substitute common gin or whisky answers for many purposes but is very much weaker.

General Matters Concerning Mounting: Slides and Cover Glasses

It is usual to find specimens in laboratories mounted upon slides of many different thicknesses and this can be a great nuisance because several types of high power condenser will not reach through such slides.

The standard slide for use with dark ground illumination is 1 mm. thick and it appears to the Author that all slides sold should be of this thickness and that the No. 1. cover should always be used. Covers made of modern glass by modern methods are not particularly brittle and there is no difficulty in using the No. 1 kind.

It is usually easier to mount upon the slide than upon the underside of the cover, but there is some advantage in placing specimens upon the cover for high power examination, mainly because the depth of mountant above the specimen is negligible. It is good practice to harden off the mount upside down, resting upon a block under the cover glass so that the specimen tends to rest on the cover rather than on the slide. Covers should always be checked in a strong beam of light for defects in the glass as it is rarely possible to remove a bad cover from a mount without loosening the specimen.

Choice of Refractive Index of Mountants

It is clearly of little use to make a permanent mount in which the specimen and medium are of the same refractive index as the specimen would then be invisible in ordinary light.

Canada balsam makes a number of objects too transparent, therefore an alternative medium of higher or lower refractive index should be used (unless the specimen can be stained). There are many media with lower refractive index, e.g. watery ones like lacto-phenol and dilute glycerine, and some in intermediate positions like the new synthetic resins; these will give greater contrast in a mount but it must be remembered that they limit the possible numerical aperture usable in the system (see 'Homogeneous Immersion Systems', page 37). For the reasons given these cannot be used for high resolution problems, so the demand is often for media of very high refractive index. Several such media exist and are described below.

(1) Styrax is a natural resin derivative, is used like balsam, is clear and permanent and has a refractive index of $1 \cdot 6$. It is a good mountant for delicate vegetable structures, diatoms and insect parts.

(2) Realgar is arsenic monosulphide and can be recognized as a red monoclinic mineral. Its refractive index is about $2 \cdot 0$ but it is most difficult to apply as it must be fused on the slide over a flame and the objects which can be thus treated are limited to diatoms. Even these usually are not properly penetrated by the mountant and false effects result.

(3) Piperine should be used for temporary mounts only, its refractive index is $1 \cdot 64$ but it is a crystalline substance and even when cooled slowly after being held at a temperature of about $180°$ it will crystallize after a year or two.

(4) Hyrax is a synthetic resin with a rather variable refractive index, being about $1 \cdot 63$ when soft to about $1 \cdot 76$ when well hardened. It behaves like balsam and is permanent.

Almost any liquid may be used as a medium in fluid mounts so long as it does not attack the walls of the cells or the specimen and a list of refractive indices of liquids can be found in tables of constants. Paraffin is a good inert mountant.

Hanley's Method of Mounting Rotifera

Mounting delicate animalcules so that a permanent reference is available is a specialist's task. The animals collapse into a shapeless mass when killed suddenly, and so must be gradually narcotized before killing so that their parts remain extended. Mounting rotifera is a typical example and illustrates the methods.

Rotifera must be handled in small pipettes drawn as required from glass tubing. The process of mounting is carried on in a watch glass containing about 8 cc. liquid because if the work is attempted on a slide the small quantity of liquid rises rapidly to room temperature and may kill the organisms prematurely.

The narcotic is benzamine hydrochloride used as a 2 per cent solution in water but an acid water requires more narcotic than an alkaline one. Add about two drops of the narcotic to the dish containing the rotifera and mix with a pipette. After about half an hour, add two drops more and kill the specimens about ten minutes after the second addition with 4 cc. of a 10 per cent solution of formalin. The exact moment for killing has to be estimated and generally this may be determined by studying the specimens with a microscope and observing when they start bumping into each other and other objects without retracting their cilia. They must be killed while the cilia are still active.

A rapid narcotic is made as follows: 2 per cent aqueous solution of benzamine hydrochloride 3 parts, water 6 parts, pure cellosolve (ethylene glycol mono-ethyl ether) 1 part.

This is used quite differently, and the rotifers can be in much less water. For every 1 cc. of water containing the rotifers, from $\frac{1}{8}$ cc. to $\frac{1}{2}$ cc. of the mixture is added and stirred up quickly. The rotifers will show signs of sleepiness immediately, and narcotization is extremely rapid. Synchaeta, for instance, can be narcotized in three minutes.

Rotifera will not tolerate more than a very small quantity of the simple aqueous solution, and in any event there is a maximum beyond which excess will not speed up narcotization. They will tolerate a very great deal of the cellosolve-benzamine combination, however, and whatever barrier exists to more rapid narcotization is apparently overcome.

The results so far obtained with this narcotic have been extremely satisfactory—its speed in particular is helpful, as difficult species that require lengthy narcotization are apt to contract for some normal reason during the process and be unable to expand fully afterwards—but it requires far more trial than it has received before it can be accepted as completely satisfactory.

Washing
After killing, the first step is to get the rotifers into $2\frac{1}{2}$ per cent formaldehyde instead of the weak solution in which they will be resting. They can be collected into the centre of the watch-glass by vibration, and removed with a coarse pipette into $2\frac{1}{2}$ per cent formaldehyde solution in a corked tube. This is turned over several times, to shake up the rotifers, which are then allowed to settle, after which as much of the fluid as possible is removed with a coarse pipette and replaced with fresh $2\frac{1}{2}$ per cent formaldehyde solution. The process is repeated whenever convenient—say six times in all—for unless the narcotic is washed out in some manner, it will form

crystals in the mount later, but there is no need for the tedious transfer from watch-glass to watch-glass by means of fine pipettes.

Mounting

Rotifers can be mounted in cement cells, or in cavity slides. If cement cells are used, the cell is made beforehand and allowed to dry. When mounting, rotifers are picked out with a fine pipette and placed on the floor of the cell. The slip is then placed on the microscope stage and filled very much to excess with $2\frac{1}{2}$ per cent formaldehyde, this excess helping to avoid air bubbles. The mount is then examined under the microscope, and any foreign bodies or small air bubbles removed with pipettes—do not run a needle round the inside of the cement ring to remove bubbles or cement scrapings will appear in the mounts.

Much of the excess fluid is removed with the pipette, being careful not to remove the rotifers also, and a polished cover glass is then placed on the mount with forceps. The cover glass may float centrally on a 'pad' of fluid, in which case the cover must be tapped down smartly with the base of the forceps. If not done smartly enough, the rotifers will be washed out but it is more likely to float half on and half off the cement cell. The surplus fluid is removed with filter paper, changing the point of application as the rotifers move (they will be visible as specks in the light coming through the stage) and when nearly all the surplus has been thus removed the cover can be pushed slowly into place with a bent wire. It is then sucked down by applying filter paper, but before going right down is examined under the microscope and pushed into place with the wire so as to be concentric with the cement ring.

It is important to notice that no wet cement was used on the cement cell. This is quite unnecessary with cement rings as, if the cell is properly made the cover glass when sucked down adheres firmly to the cell. If wet cement is used, the cover cannot be centred once it has been applied, and unless just the right quantity of wet cement is used, the result is unfortunate.

The slide is then placed on a turntable, and a thin ring of Murrayite is run around and over the edge of the cover. When this is dry, the mount can be finished with any of the usual cements.

If mounting in cavity slides, filling to excess and sucking away the surplus will eliminate air bubbles, and one can sometimes dispense with wet cement, but it is safer to run a thin ring of Murrayite around the cavity, as far away from it as possible. If this is allowed to get sticky before the cell is filled, excess fluid can still be used and the cover centred. Do not continue to apply the filter paper to the liquid until the cover is drawn concave as this is liable to result in mechanical failure later.

It is often necessary to preserve materials in bulk before they are mounted or dissected and the following methods used for five different classes of material are representative.

1. Preserving Algae

To preserve without shrinking use Fleming's weaker solution to kill and fix the specimen (10 cc. of one per cent osmic acid, 10 cc. of one per cent acetic acid, 25 cc. of one per cent chromic acid, and 55 cc. of distilled water). Its use for from half an hour to twenty-four hours will not injure delicate tissues. Add 10 per cent of glycerine, allowing each drop to diffuse before adding more. This will prevent the shrinking caused by diffusion currents if glycerine is added too quickly. Add the glycerine till the specimen is well covered, when the fixing solution has evaporated from a watch-glass in which they are exposed for the purpose. Red algae retain their colour almost perfectly, but green algae lose more or less colour, although the chromotaphores retain their shape perfectly and the cells become clearer than in fresh material.

2. Preserving Sea-weeds

The specimen is first laid in a 1 per cent solution of chrome-alum in sea-water and kept there for a period varying from one to twenty-four hours, according to the size and texture of the species. The chrome-alum is then completely washed out, and the specimen placed in a mixture of 5 cc. of 96 per cent alcohol in 100 cc. of water and vigorously stirred. The amount of alcohol is then increased by increments of 5 cc. every quarter of an hour until it amounts to 50 cc. The specimen is then removed and placed in a mixture of 25 per cent alcohol in distilled water, and the quantity of alcohol again increased in the same way, till it amounts to 50 cc. alcohol to 100 cc. water. The same process is again repeated with 50, 60, 70, 80 and 90 per cent solutions of alcohol in distilled water; the specimen being finally preserved in the last.

3. Preserving Diatoms, Mosses, Fungi, and Algae

Use potassium mercuric iodide, and glycerine. Dissolve the salt in concentrated anhydrous glycerine, the refractive index of this medium is $1 \cdot 78$ to $1 \cdot 80$. For preparing mosses for the herbarium use lactophenol gum; a strong solution of gum arabic in water, glucose and lacto-phenol. For desmids use lacto-phenol copper solution; crystallized copper chloride $0 \cdot 2$ parts; crystallized acetate of copper $0 \cdot 2$ parts; distilled water $95 \cdot 0$ parts; lacto-phenol

5·0 parts. This preserves the chlorophyll. For fungi, mosses and algae: carbolic acid 20 parts; lactic acid 20 parts; glycerine 40 parts; distilled water 20 parts.

4. To Study Lichen Structures
Immerse the lichen in a saturated solution of corrosive sublimate in 35 per cent alcohol used hot. This penetrates in a few minutes and prevents shrinkage. If the material is very dry, soak in water for a few days before fixing. Dehydrate in stages of alcohol to remove the sublimate. Stain and section in the usual way.
The method may be applied to any woody or tough growth.

5. Preserving Entomostraca (Scourfield's Method)
Unfortunately, there appears to be no single fluid or combination of fluids that can be regarded as wholly satisfactory, but the following mixture, arrived at after experiments with many alternative combinations, has been found to give good results in many cases and on the whole to be a useful all-round preservative; Absolute alcohol $\frac{1}{8}$; formol (40 per cent) 1/16; glycerine 1/16; glacial acetic acid, a trace; distilled water $\frac{3}{4}$, by volume.

All these materials except the lichen fixative may be also used as fluid mountants, in which application the cells should have solid walls built up with Araldite as the cement.

To Mount Minute Water-living Creatures
Most creatures of this kind, such as free-swimming algae, infusorians and flagellates, are too small to handle and too large to dry upon the slide, like bacteria. In many cases they have to be concentrated while living, in a simple centrifuge. To prepare them, having obtained a sufficient number of specimens, proceed as follows:
Spread a layer of water containing the organisms on the slide or cover glass and kill or 'fix' them by inverting the slide over the vapour of osmic acid, or add a drop of solution composed of potassium iodide 6 grms., iodine 4 grms., and water 100 cc. It is usually necessary to stain the specimens, and a drop of 0·1 per cent methylene blue will colour any mucilage there may be surrounding the cells. A drop of Congo red will reveal most cell wall structure, and if the cell walls have cellulose in them they may be stained with dilute sulphuric acid and iodine. (See 'Method of Detecting Cellulose' in the chapter 'Principles of Staining', page 275.) The specimens may be negative stained in Indian ink or Gurr's negative stain. (See 'Negative Staining', page 288.)
Delicate objects will not in general withstand the kind of dehydra-

tion necessary for mounting in balsam, and they are usually insufficiently dense even when stained to remain visible in this medium, therefore it is best to put them into glycerine jelly as follows.

Obtain the specimens on a slide with the smallest amount of water possible. Add a drop of 20 per cent glycerine and a minute drop of formalin as preservative. Allow the glycerine to mix in and concentrate while the preparation is kept over a warm stove or hot plate for some hours protected from dust. Add some melted glycerine jelly and allow it to diffuse into the existing glycerine for another ten minutes. Apply a warm cover glass and ring with gold size when the mount has fallen to room temperature.

The greatest amount of trouble encountered with the mounting of minute creatures is likely to be to get them in clean suspension. If they are from, say a reservoir, they can usually be concentrated easily in a centrifuge and will be clean enough to proceed with. If they are mixed with mud and detritus, they may often be caused to separate themselves by their tendency to swim towards a light. When they have separated to some extent, they should be drawn off and diluted with distilled water. The process of light attraction or centrifuging should then be repeated and soon they will be found clean enough to mount, but microscopical organisms will not live long in distilled water. (See pages 258 and 300.)

CHAPTER 16

Section Cutting

A MICROTOME FOR cutting sections of materials is an indispensable part of the microscopist's kit. Many types of microtome have been invented and commercially produced, but all follow one of the patterns described below. The general principle of action is always the same, viz. a knife is held in a saddle arrangement which allows it to slide back and forth on rails with its cutting edge forward, or it is given a similar motion by other mechanical arrangements. The specimen is advanced by controlled amounts along an axis at a right angle to the direction of travel of the blade. At each sweep of the knife a thin slice or section is removed from the specimen and the thickness of the slice is determined by the amount the specimen is advanced, the sharpness and form of the knife, the steadiness of the knife slides and the general solidity of the apparatus, particularly of the embedding material supporting the specimen.

Several methods of holding the knife have been devised, those with rails are known as sliding microtomes, while another type—the Cambridge Rocker—carries the knife on the end of an arm which is pivotted at its other end to the frame of the instrument. The action of the knife is very solid, there being only one bearing of the hinge kind. In most designs, the specimen is so held and the blade fixed, thus allowing ribbons to be cut. There is also the rotary instrument in which the specimen in its embedding medium is mounted eccentrically on a spindle, the knife being stationary; the section is thus presented to the knife with a rotary eccentric movement. Such a cutting action is very good as less pushing of the specimen takes place during the cut but the instrument is not so good for cutting ribbons.

For use for occasional section cutting of various specimens, the sliding knife kind of microtome is best, but in many investigations a series of sections through a feature must be taken in order to understand it and these sections must be kept in order. This is possible if the microtome knife or specimen distance of travel, and the specimen advancing mechanism are carefully adjusted on test so that the sections stick together by their edges in the form of a ribbon as they are cut, and can be mounted in order. The Cambridge Rocker microtome is ideally suited to this work and details of the instrument will be found in the catalogues of Messrs Watson. Reichert and other manufacturers also make other kinds of microtomes.

Embedding
Practically the whole of the skill necessary for successful microtomy is centred upon techniques for embedding the specimen in a mass of material to support it adequately while the knife is doing its work. The usual material used is wax of some kind, and the methods of getting the material into it are similar to the methods of mounting described above. All microtomes have different shaped specimen holders so the wax blocks are cast to suit the machine.

Typical Wax Embedding Technique
A wax suitable for an ordinary laboratory is white paraffin wax, but in very warm situations beeswax is better. A penetrating agent which will dissolve in wax is chosen, suitable ones being cedar-wood oil, zylol, turpentine and chloroform. In general, cedar oil and zylol are suitable but both require the specimen to be dehydrated as described in mounting. The method of embedding a specimen in wax for section cutting is given below.

A suitable piece of the specimen is soaked in the penetrating medium for some hours until the specimen can be seen to be quite clear of white patches. The bath is then raised in temperature to about 60° C. which is a few degrees higher than the melting point of the wax. Molten wax is then stirred in to saturation and the bath allowed to evaporate until the solvent has gone and the wax, when allowed to cool, is hard. Endless variations of this method have been published in connection with special embedding jobs but the basic procedure, whatever the wax or the penetrating medium chosen, is always the same.

Notes on Embedding
Other materials than waxes can be used, for example collodion is better for complicated brittle structures and it may be adjusted in hardness by the addition to it of triphenyl phosphate. A rocker microtome usually needs a harder wax than the sliding type.

Wax should not be used at a temperature much above its melting point (45° C.) or the specimen will be broiled, also it must be remembered that the wax melting point rises with the number of times it is heated. After wax has hardened into a solid it continues to harden for an hour or two after, therefore if trials are made with the knife a time is found when the wax is at just the correct degree of hardness for the best cutting of the specimen, but this point must be determined by trial and microscopic examination.

Difficult structures may be impregnated with wax *in vacuo*, but if this is attempted the air must be withdrawn slowly or its expulsion will cause mechanical damage in the tissue. When collodion is used as the embedding medium it need not be removed before staining and mounting the specimen and this is its principal advantage.

Steedman's Embedding Wax

Ester wax is a new ribboning embedding medium. It consists of five substances which can be varied in proportion so that media of different characteristics may be used. A mixture which will meet most requirements is as follows: diethylene glycol distearate 73 g., ethyl cellulose 4 g., stearic acid 5 g., castor oil 8 g., diethylene glycol monostearate 10 g. Its melting point is from 46° to 48° C., section range 2–20 μ at a temperature of 64° F., ribbon range 2–15 μ at room temperature of 64° F., final compression loss 7–6 p.c. at 10 μ. It is soluble in alcohols, ethers, esters, ketones, hydrocarbons, aldehydes, chlorinated hydrocarbons, and in natural oils. Cutting must be done more slowly than with paraffin sections. Flattening must be by floating on tap-water or aqueous staining solutions and completed in a drying oven at 40–45° C., with enough water under the sections to float them. Sections can be stained in the ribbon or as for paraffin wax sections before embedding; xylol, benzene, or ligroin can be used as a premounting fluid, while 'Sira' polystyrene mounting medium is suitable.

An Inexpensive Paraffin Bath

In order that specimens from which sections are to be cut may be infiltrated with paraffin, an embedding bath is necessary. This is usually a very costly item but a satisfactory laboratory apparatus may be rigged as follows: the basis of it is a small potato steamer, obtainable at any ironmonger's shop, which should be altered as follows. The lower portion (*a*), which we will call the heating chamber, must be cut away so as to admit a spirit lamp (7) or gas flame. The perforated bottom of the upper vessel which constitutes the bath (*b*) must be covered with a sheet of tin which must be soldered on to render it watertight. Near the top of the bath a circular plate of tin (*c*) must be fitted having holes (3) drilled in it to admit the test tubes (2) and thermometer (1). This plate should be supported by four small pieces of tin (4) soldered to the inside of the vessel. Having completed these alterations, a layer of cotton-wool or a piece of felt should be placed on the bottom of the bath to protect the test tubes

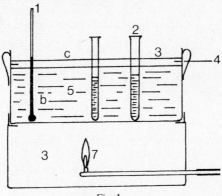

Fig. 1.

from breakage; half fill with water (5) add a chemical thermometer (1) light the lamp, and when the desired temperature is attained (45° C.) place some paraffin wax in the test tubes and put them into the bath; when the paraffin has melted add the specimens. The temperature must be maintained for several hours so that the paraffin may penetrate to the middle of the tissue. Further treatment would be similar to that for ordinary embedding in paraffin wax. After use, the apparatus should be thoroughly dried so that it may not rust, and if this be attended to it will last for many years.

Microtome Knives
A good knife is an ordinary hollow-ground cut-throat razor, as these were invariably of the best steel and can be bought very cheaply as junk in these days of safety-razors and electrical machines. For heavy work on wood and similar medium-hard substances, the greater support afforded to the cutting edge in a smoothing-plane blade is satisfactory, while for really hard objects and metals a knife edged with tungsten carbide is suitable and is commercially available. Whatever knife is used it is essential that it be stiff and solid behind the cutting edge. A blade like that of a safety-razor is useless because the edge is not under control. Extremely thin sections of another kind can be cut with the broken edge of a piece of plate glass (see 'Special Methods of Sectioning, page 313).

Microtome knives do not behave like miniature saws and unless they are used in the hand or on a rotary microtome they always push rather than slice through the specimen, therefore they must be very smooth at the edge. To obtain the necessary kind of edge the blades must first be straightened on an ordinary India hone, then edged on a specially kept razor hone, and finally stropped on leather with rouge and oil, after which process the shape of the blade should be as shown in Fig. 2.

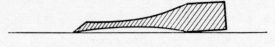

Fig. 2.

A plane blade should be sharpened to the same shape as is usual in a plane. Carbide tipped tools cannot be ground on any hone but can be dressed on a sheet of Bakelite or Perspex charged with micron-size diamond powder in oil. For treating ordinary knives, two large India hones of the 'fine' grade should be obtained and rubbed upon each other to make them substantially flat. They should then be well brushed down and soaked in thin machine oil for twenty-four hours, then they are fit for grinding microtome knives to shape. The next sharpening stage is upon a Water of Ayr stone used with water; finally they are stropped on a soft leather sheet supported upon smooth wood.

Some workers appear to be satisfied with a knife ground upon plate glass with a fine abrasive, but such a method cannot produce the best results because the abrasive leaves the hard glass and sticks in the softer steel. Only stropping can remove the particles of abrasive from the knife they then remain in the strop and do damage there. The usual angle for the cutting faces of the knife to make with each other is 10° (Fig. 124) and a suitable wooden jig must be made to

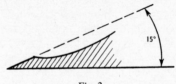

Fig. 3.

hold the knife at this angle on the hones. When stropping a jig must not be used but the blade must be allowed freedom to follow the surface of the leather as it is drawn from heel to toe along the strop. Not any odd strop or stone is good enough for this work. A Turkey hone is satisfactory for straightening and shaping a blade but it is too coarse and irregular in composition to leave a smooth edge. Unless the strop is soft and free from hard particles this too will gouge grooves in the blade. Some workers cannot sharpen a microtome knife in the ordinary way and they usually resort to a modification of the glass plate and rouge method. If the surface of the glass plate is finely ground with about 500 grade carborundum, well washed, charged with rouge and water, and used as a hone for final sharpening, a good result can be obtained. The effect of the hone is mainly one of burnishing away the steel on the hard chipped glass surface with the rouge acting as a pad to moderate the action.

In any grinding operation it is most important that the steel backing piece to the knife be left on when sharpening because this tilts the blade the correct amount for a finishing bevel to be formed on both faces of the blade (Fig. 4).

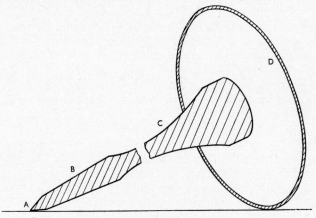

Fig. 4. A. Finishing bevel ¼ mm.
B. Sharpening bevel.
C. Rough grinding.
D. Backing piece.

Steel is not a homogenous material but is made up of irregular concretions of finite size. A good hone will grind all of these components nearly evenly and leave an even edge, but the strop differentiates between the particles, and though it generally smooths off all a little, it certainly creates unevenness if prolonged. If the Water of Ayr stone is carefully used, that is, with a pressure determined by experience with the knife in question, no strop is necessary and a better edge results.

Hard knives should be preserved for soft specimens which offer little resistance, as such knives take a very keen edge from the hone but are liable to break up on the strop and when they encounter hard specimens. A softer knife is necessary for hard specimens (botanical stuff) for, though it has to be sharpened more often, it does not break up, and in general a hard or tough specimen is not so easily deranged by a less keen knife as a very soft one would be. A practical example on a large scale is provided by lumbermen who know that a soft pine tree demands a much sharper axe than a hardwood tree because it offers less resistance.

When knives are sent out by the makers, they have backs to them which are individually fitted and are intended as a guide to the correct cutting angle, and they should be so used for the average sectioning job, but it is the Author's experience that little importance attaches to the cutting angle of the knife so long as the faces make the correct angle with each other (15°) and the lower face of the knife is parallel with the table.

After use, a knife should always be wiped clean on lens paper and put away in a dry box, preferably with the blade wrapped, because atmospheric influences operate readily upon the very thin edge of a knife and destroy it. Water in particular does this and the ordinary shaving razor will be found to last 100 per cent longer time, other things being equal, if it is carefully dried after use and stored in a warm place.

Cutting the Section
When the block with the specimen suitably embedded is mounted in the cup of the microtome, a few strokes with a less delicate knife should be made to level off down to the region sought. Next, the proper knife should be mounted and a few trial cuts taken. The angle of the knife with respect to the table should be set up so that the lower face and the direction of motion are parallel. The back of the knife should then be lifted a further degree or two. The slant of the knife across the direction of motion should be about 10° or square on. The knife, table top and specimen should be lubricated with 85° alcohol or with cedar-wood oil. In general, cutting should be by slow strokes, say a few centimetres per second.

In warm dry atmosphere considerable trouble is often encountered due to the sections becoming electrified and adhering to the apparatus. The best cure for this is to ionize the air and to make it conduct by means of a radioactive isotope similar to that used for discharging looms. Some good can be done by boiling a kettle near by, thus increasing the humidity and causing the charges to leak away.

Mounting Sections
The curly sections can be flattened by floating them upon warm water or collecting them upon warm slides as they leave the micro-

tome. They may be stuck to the slides by means of a thin dried coating of common gum or gelatine. The slides are quickly slipped under the sections, while they are floating, and when all is in place the slides are allowed to dry naturally. The sections will be found to be sufficiently well stuck to resist washing in xylol to remove the wax, staining in non-aqueous media and mounting in balsam.

General Notes on Section Cutting

(1) A frozen section must be cut more slowly than a wax embedded one.
(2) Small changes in room temperature of a few degrees make a big difference to the cutting properties of a wax.
(3) If sections roll up, it is likely that the angle between the faces of the knife or the cutting angle between the faces of the knife and the horizontal are too great,
(4) If the sections wrinkle the wax is too soft or the knife is blunt.
(5) If sections tear, the wax and specimen are not sufficiently well matched in hardness.
(6) Regular furrows across the wax block indicate a damaged knife or one improperly ground.

When embedding in paraffin wax, it is necessary to have the block square with say 10 mm. sides, and inclined to the knife edge at about 45°, that is, matters should be so adjusted that the knife leaves the block at a corner and not at a long side. When cutting ribbons, this advantage must be sacrificed to a considerable extent, but round wax blocks can be used.

When cutting ribbons, the rhythm of the microtome must not be interrupted during the run or uneven sections will result. A thin section is always thicker than the specimen advance mechanism indicates because the section is always a little compressed laterally by the resistance to the knife.

In Fig. 5 (*b*) is the indicated thickness, but (*a*) is the compressed size. The specimen should be set in the block in such a position that the knife has time to 'get cutting' before it reaches the specimen. In general, this means that the specimen is mounted considerably to one side of the block in the direction of exit of the knife.

Special Methods of Sectioning: The Ultra Microtome

A steel knife is at the present time the best device for cutting sections and it must be sharp, smooth (not like a saw) and strong enough to resist being pushed through the material. If merely a sharp cutting edge is required, regardless of strength, the broken edge of a sheet of plate glass is the best available. Such an edge is very smooth because there are no concretions present as in steel, very sharp because the break is a clean cleavage, and is hard, but such an edge is brittle and cannot be used like a knife. If matters can be arranged so that the

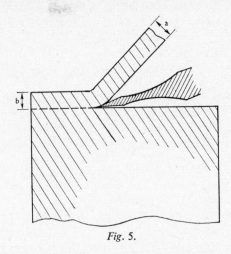

Fig. 5.

glass knife only skims high spots on a specimen, it is highly satisfactory as a sectioning knife, and sections of bacteria have been obtained by employing the following method:

A glass knife is prepared by scratching and breaking a piece of plate glass about 70 mm. square and 10 mm. thick. Several trials are necessary but soon a piece will be obtained which presents an edge of section somewhat as shown and about 25 mm. long. Fig. 6.

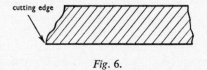

cutting edge

Fig. 6.

The glass knife is mounted vertically in a vice which can be set in position and then securely clamped in place. The job to be sectioned is mounted on the end of a round bar of steel which can be moved in the vertical plane upon a very heavy and well-fitting bearing. Around the rod is a water jacket and thermometer, by means of which the temperature of the rod can be varied. The steel arm and jacket can be moved up and down against a spring control, carrying the job past the knife after the style of the Cambridge Rocker instrument.

If a cover glass or any other object carrying minute particles to be sectioned is stuck upon the end of the steel rod and the knife advanced until it can be seen to be just about to contact the surface, the job may be made to 'stand closer to the razor' by heating the steel bar by means of its water jacket. The coefficient of expansion of mild steel is 12×10^{-6} cms. per degree Centigrade, therefore assuming the rod is 30 mm. long and evenly heated, it expands 0·12 μ per degree Centigrade.

The technique of making the final cut varies with the job, for instance a very thin waste section may be shaved off at room temperature, then without disturbing anything the bar can be heated to advance the specimen the required calculated amount.

The apparatus is best set up on a clean lathe bed where the saddle can be used as a moveable vice to hold the knife and the bearing for the tail of the bar can be mounted in the tail stock. The bearing holding the bar must be adjustable in tightness and opposing cones

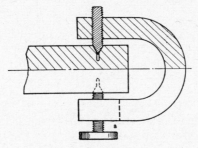

Fig. 7.

work well, Fig. 7. The cones must be short and thick and are best drilled with centre bits so that the cones do not bear on their points; the adjusting screw (*a*) must also be clamped. Stops to control the travel of the bar are mounted on the lathe bed.

This technique is simple but can only be described as 'tricky' to use. The objects must be scattered upon tiny cover glasses or upon the surface of a wax block which has been sheared on the machine. The rod can be lifted in the vertical plane until the cutting face is horizontal, when materials can be scattered on the sheared face of the wax or upon a prepared cover glass mounted on the rod and chemical processes performed. The glass knife will not produce a thin section like that described in ordinary microtomy but operates by shaving off layers from the object, so exposing the internal parts of it. The parts shaved off are lost in this technique and the method must be arranged to keep the object remaining on the mounting block in a condition which allows microscopical study.

An example of an Ultra Microtome is commercially available from Messrs Reickert.

Method of Sectioning Long Brittle Objects
For this work the fine fretsaw blade is the correct tool. Such blades are very hard and sharp and will cut metals up to the hardness of mild steel.

The job to be transverse-sectioned must be supported in a wooden jig and padded out with fine sawdust. Two blocks of close-grained wood are clamped together in a vice, temporarily fixed together, and a hole large enough to accept the job is bored down the interface. A fretsaw cut is now made across the holes as shown in Fig. 8, the blocks are then separated and cleaned up. When a section is required,

Fig. 8.

the object is laid in the groove of the lower block and duly supported with sawdust or other packing. The upper two blocks are placed on top and the saw blade inserted in the slot after the style of a mitre block. The fingers are sufficient to hold the two top blocks in place.

Good results can be obtained with this apparatus without impregnating the brittle materials with gum or balsam. The section should be taken out of the centre parts of the object so that there is support for the waste piece when the first cut is taken. For the second cut, which removes the section proper, there is often no need to use the smaller upper block. If the first cut is sound and unbroken, this side of the section can be stuck to the slide direct and the less good, less well supported second cut can be the one ground away to obtain the required thickness.

Having secured the object to a slide, it may be ground according to the methods described under 'Preparation of Sections of Minerals' but it should be again mentioned here that loose abrasive particles should not be used to reduce soft or brittle substances because they stick in the job and cannot be removed; hones used under a running tap are best.

CHAPTER 17

Methods of Counting Particles

T HE BASIC method of doing this is to obtain a measured volume of the substance in a micropipette, spread it over a slide and count the whole or part directly. Usually a squared slide is used to aid the laborious counting procedure. The method is simple to apply, squares are drawn upon a slide with a diamond or are printed on and the area of each square is accurately measured with a micrometer eyepiece. If liquids are to be used it is usual to have a fence round the squared area so that a definite volume of liquid can be held over a square, under a cover glass. Any dilution can be made up before the material is placed upon the slide but the contents of several squares must be averaged. This is the standard blood cell counting technique. A stout cover glass must be used or concaving will occur as the liquid evaporates thus changing the volume of the sample, also incomplete filling of the cell will cause capillary depression of the cover. Cells should be filled to a convex meniscus before the cover is applied.

If no specially ruled slides are available the area of a particular microscope eyepiece field with a given objective can be evaluated by means of the stage and eyepiece micrometers.

The Lycopodium Method

For this method the spores of *Lycopodium clavatum* are weighed and mixed with weighed quantities of the substance to be examined. When the spores and the material are well mixed together and placed on a microscope counting area, the proportion of spores to any other features can be ascertained. The spores are resistant to many strong chemicals for example dilute acids and alkalis therefore they will pass practically unchanged through a number of chemical processes and retain their identity. The spores, like most of their kind, are of very regular size and weight; the average being 94,000 per milligram.

If a mixture of powders is being counted it is necessary to mix up the powder with the spores using a liquid medium in which to suspend the ingredients, spread the mixture on a counting slide and count several (say ten) squares to obtain the following ratios:

Consider a mixture to be tested of Powders A. and B, the result of the count may be,

    100 spores per 10 areas counted
    40 examples of a feature A,

therefore we have $\dfrac{40 \times 94000}{\text{Dilution of sample}}$ =Number of features A per milligram of spores. As we know the number of spores mixed with the powder A+B, we can find the proportion A to B in the substance under test and the number of features A or B as a function of weight.

All methods of visual counting are most tiring to the observer. Some help can be obtained by arranging a point or a line in the eyepiece and moving the stage with the object across this point in rectilinear motion over the area to be counted, when it will be found that the eye being stationary, experiences less strain than when roving about a fixed slide. Recently an electrical method closely related in principle to television has been set up successfully and is known as the 'Flying Spot Microscope'.

Electronic Counting Methods: The Flying Spot Microscope
Instead of scanning a slide by eye it can now be done by electronic means though to do it the microscope must be used in reverse. The principle of the method is as follows:

The luminous spot on the screen of a cathode ray tube of the type used for television is caused to scan and produce a raster. An image of the raster is formed in miniature by the microscope upon the specimen, the screen of the cathode ray tube being placed about 250 mm. away from the eyepiece; the size of the spot is in this way so reduced that it is comparable in size with the smallest resolvable detail in the microscope. The flying spot scans the object, and any light which passes through the specimen at any instant is modified in amplitude by the part of the specimen at that instant being covered by the spot. If then an ordinary photo-electric cell (photomultiplier) is placed below the specimen it receives pulses of light modulated by the points of density of the specimen, and the electric impulses generated from the light by the cell may be amplified electrically and caused to operate electronic apparatus of all kinds, for example scaling circuits used to count the modulations.

The system can be developed to reproduce the image on another cathode ray tube. To do this it is necessary to synchronize with the first cathode ray tube a pair of time bases on another tube, in this way identical rasters are produced. If the electronic amplifier following the photoelectric cell is now made to drive the grid of the second tube a picture of the specimen is produced after the style of a television picture.

There are several advantages in viewing an object in this way:

(1) Resolution does not deteriorate.

(2) A wide control of contrast is possible by varying the gain of the electronic amplifier.

(3) An electronic counting device placed at the output of the amplifier can be arranged to count only particles of a certain density because increasing density of object shows as increasing output voltage therefore a simple biasing arrangement only is necessary on the counter to cause it to count only pulses above a given size. The screen can be watched while the process is adjusted.

(4) There is no limit to the distance between the microscope and second cathode ray tube as only suitable cabling is required to connect them.

A block schematic diagram of the apparatus appears in Fig. 1. The details are as follows:

Any ordinary cathode ray tube with a short period of brightness persistence is satisfactory for use. The time-bases must be of sawtooth waveform of speeds up to about 10 Kc/s in the X direction and about 500 in the Y direction, the frame being repeated 50 times per sec. The number of particles counted is dependent upon the distance apart of the lines which make up the scanning raster as the size of the spot can be considered constant. If the particles in the field are large greater accuracy is obtained by spreading out the lines by using a fast Y direction sweep but if the lines are close together large particles are counted twice or more times. It can be seen from the above that the apparatus cannot separate particles of different sizes but it can separate those of different densities because a black object causes a greater change of light intensity in the photo-cell than a grey one. The current pulse in the amplifier is also proportional to the signal from the photo-cell and if the amplifier or a piece of subsidiary apparatus is biased in the appropriate way only signals representing certain densities are transmitted. Any selected signals can be counted by means of a commercial electronic scaling unit, details of which are to be found in the catalogues.

If the amplifier is biased to record the presence of particles over a certain density the size of the particles may be estimated from the length of the pulse they produce, the spot being obscured for a longer period by the larger particle. An electronic differentiating circuit will discriminate between long and short pulses and show them as pulses of different amplitude, these in turn operate an Eccles-Jordan trigger biased to operate on signals above a certain level only. Several triggers can be arranged to operate at different pulse amplitudes and so give an idea of the ratio of large and small particles present.

The time taken to scan a frame may be only about 1/50 sec., therefore many counts of different parts of the field can be taken and the fact that the particles may be moving matters very little. An automatic means must be provided to scan the field once only to record a count and this also must be done electronically because of the very short time intervals involved. Several arrangements may be attempted and the Author has found the following one workable and capable of development.

The scanning circuit is allowed to work all the time. The time-base waveform appearing on the Y plates of the tube is amplified, squared and used to turn on an Eccles-Jordan trigger (A). This trigger is normally turned off by an applied voltage under the control of a press button. Release of the button allows the next Y plate wave to turn on the trigger. Trigger (A) operates trigger (B) which is a one-shot multivibrator, and has a time constant equal to the frame period and returns to its original state after one frame. The output of this trigger (B) switches on the counter or scaler to count any signals that may be applied to it from the video amplifier during that frame. The next Y impulse cannot pass through the system to cause another frame or series of frames to be counted until the control button is again operated to reset trigger (A). The actual counting can be performed by neon tube scalers but these will not operate faster than one thousand counts per second.

It is impracticable in this apparatus to select one frame for counting because there is no means of making sure that trigger (A) is turned on at the beginning of a frame, it may be turned on at a point between the start and finish of a frame thus giving a partial count. But if a measured number of frames are counted by holding the switch open for a measured period of time the count recorded can be divided by the number of frames in the time chosen and an accurate result obtained.

If the whole mechanism is slowed down so that one sweep of the scanning spot takes several seconds, ordinary hand switching instead of triggers can be used to separate out one frame and counting may then be performed by Post Office registers if the number of counts is less than five per second, but if the speed is so slow as this d.c. amplifier must be used in the video stage to transmit the slow variations of voltage from the photoelectric cell. Such an amplifier is made by the AVO Company.

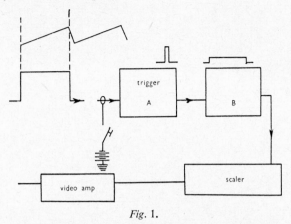

Fig. 1.

Refined Method of Counting Objects such as Blood Cells

The methods described on the previous pages are either costly and complicated or applicable only to special experiments and each of them requires careful examination to determine its inaccuracies in each application. The method recently tried for blood cell counting in which considerable accuracy is required will be used to illustrate several matters requiring care in this work.

An ordinary counting chamber is employed and an image of the part of its floor required for counting is projected by a single combination eyepiece on to the sensitive cathode of a photoelectric tube. A diaphragm on the objective side of this lens defines accurately the area of illumination falling upon the cathode. This area must be of a definite size in relation to the size of the regular particles being counted and in practice is 90 per cent of the average particle diameter in width. The length of the area depends also upon the kind of particles being counted and in blood cell work is about 30 microns (see below). The object being counted is oscillated on the stage of the microscope so that the image of the particles is passed across this aperture. Counting takes place by recording the electrical changes in a photo-cell due to changes of optical density when particles pass the aperture. A commercial scaling unit is used to count the changes and a mechanical oscillator moving the stage in Y direction sweeps with X direction shifts is employed. A speed of about one scanned strip per second is suitable.

In this kind of apparatus either transmitted or dark ground illumination may be employed but several factors influence the decision. With d.g. methods on blood and similar materials a greater signal to noise ratio can be obtained because the background field is dark and the current in the photo-tube is minimum. However, a really dark field is difficult to obtain in practice and very small accidental changes in this darkness due to scatter from stray particles of minute size produces great effects in the results. If transmitted light is used greater steadiness of results can be obtained mainly because the standing current in the photo-tube due to the ordinary field illumination can be used to work a stabilizing system to keep the performance of the photo-tube constant against changes of background brightness. It also provides a useful guide to the sensitivity of the whole system in the same way as does the background noise of a radio receiver. In both systems a d.c. supply to the lamp and a good anti-vibration mounting to the system are essential or unwanted variations at about the speed of counting will be recorded.

The microscopical details of the apparatus are simple and are covered in other parts of this book. The only additions are the projection lens which can be simple and the diaphragm, which in blood counting apparatus is about 4 by 30 microns. This figure quoted above is the apparent size of the image and diaphragm as seen by the photo-tube, i.e. when the image of a blood cell formed by the objective is on this diaphragm it just about fills its width. A typical six millimeter objective magnifies about 40 times, therefore the actual diaphragm slit is still small and should be adjusted on test. The slit mechanism of a Browning microspectroscope is excellent for this purpose.

Ordinary microscope illumination should be used but with a slit in the diaphragm plane of the lamp condenser to provide light to correspond roughly with the slit in the microscope. A side viewing attachment with a light moderator must be employed for visual setting up.

The electrical arrangements of such a system cannot be dealt with in detail in a work on optics but the requirements are as follows. A standard side or end entry photoelectric tube is mounted in a light housing coupled to the microscope draw tube. The photo-tube requires a relatively low resistance chain employing resistors of about 300 kilohms each, followed by a two-valve d.c. amplifier. This assembly and power supplies may be conventional as described in textbooks on electronics or of a special stabilized feedback kind given by Cooke-Yarborough and Whyard in the *British Journal of Applied Physics Supplement* 3, 1954. In either system the output of this group of components is sufficient to operate a standard scaler.

The accuracy of the ordinary visual counting method depends largely upon the experience of the operator and he in turn depends upon certain conventions such as counting all particles with the majority of their area within the counting zone as one and others as none, or, all with certain agreed recognizable features visible as one and those without ignored. When the effects of fatigue are added it will be appreciated readily that accuracies of 10 to 15 per cent are the usual rule. Mechanical and electrical methods also have their inaccuracies because particles may overlap on the slide and may appear only partially in the aperture of the photo-tube. For regular particles like blood corpuscles the effect of particles on the edge of the slit can be calculated but the work is complex. An estimate of edge effect can also be obtained by using slits of different shape though of the same area, and equating the counts on the same specimen. The general idea however is to set the apparatus in action and check it against practical visual counting. In order to do this quickly and accurately an image of the specimen should be projected on to a screen, the area of counting determined by methods described under 'Micrometry', and the image of each particle struck through on the screen as it is counted. It will then be found that accuracy of about 3 per cent is obtainable by the automatic devices. The diaphragms and lights may then be left set up for work on these particles.

For blood counting a suitable dilution is 1 in 2500 by volume and a good dilutant is 0·1 per cent iodine in isotonic potassium iodide with 30 ml. of 0·1 per cent eosin as a staining agent intended to improve the light stopping properties of the corpuscles. A distribution of about 100 corpuscles or particles per linear centimeter will be obtained in this way and will be found to be generally suitable.

Optical Counting
When many counts of particles of approximately constant density and size have to be taken, the amount of light stopped by them compared with the light coming from the clear areas may be measured by means of a photoelectric cell. The method is limited to repetition work with materials composed of simple ingredients.

The field is illuminated evenly by any ordinary means and the whole of the Ramsden disk is allowed to fall upon a photo-cell. No amplifier is necessary because a micro-ammeter is sufficiently sensitive to read output direct and the brightness of the light in the field can be adjusted between the widest limits.

A field of particles is counted accurately by eye and the change of micro-ammeter reading from the reading with a clear field is noted. Several counts with different numbers of particles should be taken and a curve plotted, particle number against current from the cell.

In practice the calibration is found to hold for a wide range of particles of different sizes and colours but a check should be made for each different material counted. When using the method, several parts of each slide should be observed or the average current reading taken of a slide containing moving objects. Colour filters help to resolve some particles by causing their images to be darker than the natural colouring. The illumination source and photo-cell can be considered practically constant when in action together, therefore a set of readings can be reduced to real figures at a later time. Developments of the method for use in manufacturing processes are undoubtedly possible. It is possible here to give only an account of the methods employed, for detailed treatment of the electronic parts, works on that subject must be studied. (See Puckle, 'Time Bases', and Horff, 'Electronic Counters'.)

REFERENCES

PRATT, T. H. (1948). 'The Infra-Red Image Converter Tube', Electron. Engng., 20, 274–8. London: Morgan Brothers Ltd.

ROBERTS, F., and YOUNG, J. Z. (1951). 'The Flying Spot Microscope', Nature, London, 167, No. 4241, 231.

ROBERTS, F., and YOUNG, J. Z. (1952). 'The Flying Spot Microscope', Proc. Instn. Elect. Engrs., Part IIIA, No. 20, 747–57.

SOMMER, A. (1946). 'Photo-Electric Cells', Methuen Monograph. London: Methuen Ltd.

WEIMER, P. K., GOODRICH, R. R., and COPE, A. D. (1951). 'The "Vidicon" Photo-Conductive Camera Tube', RCA Rev., 12, Part 3, 306–13. Princeton, USA: RCA.

ZWORYKIN and FLORY (1952). 'Television in Medicine and Biology', Elect. Engng., NY, January 1952.

CHAPTER 18

Tolansky's Method
of Contour Microscopy

A PROBLEM ARISES in the study of surface marking which is to decide whether or not markings are just colour differences, depressions or raised areas. If the depression or elevation is more than about 2 microns in magnitude and is a sudden change, the microscope fine adjustment can be used either to measure or indicate which condition of surface is being observed; but there are many studies where much smaller changes, as for example etch pits, are under examination. The microscope fine adjustment cannot show differences in position of focus for level changes less than, say, 1–2 microns, nor can it show reliably, steady changes in level which extend across a substantial part of the field of view, the curvature of the high power objective field and the uncertain orientation of the specimen prevent this.

A method for improving results in these studies was invented by S. Tolansky of the Royal Holloway College as a result of his observation of a shadow of a part of a building falling up and down a distant flight of steps. Although the width of the tread of the steps was not visible within the focal sensitivity range of his eyes the zig-zag interruptions of the shadow at each step were clearly apparent. It was clear that here was a method of making visible and measurable tiny undulations within the focal depth of a lens system. A further matter of importance is, that as there is no difficulty in producing a straight shadow edge, a gentle undulation may be made visible and measured against the straight edge. It has not up to the present been possible to do this under the high powers of the microscope and already important work has been done employing the method. Another useful application is to photomicrographs, where, although the microscope fine adjustment may have been used visually for measuring depths and heights, these could not be seen from the photograph however coarse they may have been, because the photograph is merely a flat picture without perspective. If, however, a contour line or shadow edge is included in the picture it is possible to see instantly the lie of the structure and to measure heights within limitations.

In practice a line is used to indicate the contours and this is projected by means of the ordinary vertical illuminator lens system.

A microscope must be chosen which is fitted with an inclined-plane type of vertical illuminator which is adjustable in all directions so that incoming light may be reflected through the objective on to the specimen at any angle. The vertical illuminators made by Messrs Cooke Troughton and Simms, and Messrs Beck, are suitable.

The illuminating system is arranged to produce a dark line across the image of the illuminating condenser as seen in the field of the microscope. The illumination is applied at an angle to the vertical illuminator so that the dark line appears to lie upon the surface of the specimen and because it is projected obliquely, it traces the contours of the surface under examination (see Plate 8).

The Illuminating System for Contour Microscopy

All the special considerations, such as careful blacking of all internal parts of the microscope associated with vertical illuminators must be observed or poor images will result.

A bright source of monochromatic light must be made practically parallel by means of a simple condenser and controlling diaphragm. This beam of light is used to illuminate the vertical illuminator condenser proper. This latter is the one which is imaged by the microscope objective upon the specimen and it must therefore be placed at a distance from the microscope objective, equal to the tube length, that is the distance from the objective shoulder to the eyepiece diaphragm and any extra optical path due to the presence of the vertical illuminator must be estimated when setting up this distance. It is sufficient to set the vertical illuminator condenser the same distance away from the side of the microscope as that measured between the vertical illuminator light entrance hole and the eyepiece flange. Small adjustments can be made on test. As with the ordinary vertical illumination system using an inclined plate of glass, the illuminating condenser must appear in focus upon the surface of the specimen at the same time as the microscope is so focussed and this image must be a clear even disk of light. If it is not, something in the illuminating system is off centre or a lens is insufficiently well corrected for spherical aberration to handle properly the beam being transmitted.

The vertical illuminator condenser has behind and close to it a simple graticule made from 47 S.W.G. wire pulled straight. One of the three wires composing the graticule crosses the aperture diametrically, the others forming chords at 1 mm. separation from the first diametric wire. Only one half of the aperture has wire crossing it. The graticule is assembled upon a ring of brass which fits loosely into the lens mount, the wires being soldered in position after being stretched between external supports over the aperture in the ring. The wires must be straight.

The instrument shown in Plate 3 is illuminated through a spectrometer in order to obtain monochromatic light. The layout is shown in Fig. 1. For detailed treatment of Microscope Illumination see Chapter 7.

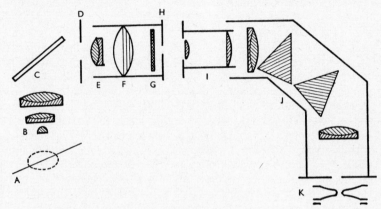

Fig. 1.

A. Line of graticule projected obliquely upon object.
B. Objective.
C. Inclined plate vertical illuminator in microscope.
D. Diaphragm (ordinary).
E. Vert. ill. condenser.
F. Graticule (shown in partial perspective).
G. Ground glass diffuser.
H. Gating slit accepting image of spectrometer slit in monochromatic light.
I. Spectrometer eyepiece.
J. Wide slit spectrometer.
K. Mercury vapour arc.

The contour technique can be applied only when the light falling upon the specimen is oblique. If the line is projected normal to the surface, i.e. through the centre of the objective aperture no displacement of the line in the image can result. The effect may be compared with the absence of shadows when the sun is directly overhead at midday in the tropics. Because, in order to obtain shadows and line displacement with change of surface level, the light must be extremely oblique, a monochromatic line source is essential for useful results and even apochromatic objectives need this assistance to their corrections. Graticule wires are spaced over half the aperture only, because when maximum obliquity of the illuminating light is arranged only half of the vertical illuminator condenser is useful. (See 'Setting up for Contour Microscopy', page 328.)

The vertical illuminator condenser which, together with the wire is imaged in the microscope field, should be a well corrected assembly in order to produce a clear image of the line. This lens can be an ordinary plano-convex one of about 70 mm. focal length but in the apparatus here described is a 100 mm. achromatic microscope objective of 10 mm. diameter which is a suitable size for the field of view of the microscope.

Matters which affect the Practical Use of the Contour System
It has been shown above that the image of the dark line must be projected upon the specimen obliquely and if there are several lines forming the graticule the obliquity of the pencils of light forming the images of the several lines must vary. The displacement of the image of a line as it passes over irregularities on the specimen varies with obliquity, therefore the displacement of the image varies with different lines on the same graticule. When the images of three lines of the graticule appear parallel upon the specimen and are substantially the same distance apart it must not be assumed that contour heights on the specimen as measured by image displacement in each of the lines are the same. Each line arrives on the specimen at a different angle, therefore the contour heights mapped out by each line and indicated by image displacement in that line must be separately evaluated. The only reason for employing several lines in the graticule is that when the microscope is set up, one line may be in a better position in the field than the others. For instance, a line at the edge of the field at maximum obliquity may be visible in a low power eyepiece but not in a high one. A choice must be made before measurements or photography are started and once made must be adhered to during the run.

Setting up for Contour Microscopy
The system is set up exactly as for normal vertical illumination. (See Section 'Study of Surfaces', page 221.) When so set up an image of the vertical illuminator condenser will be seen upon the specimen when the microscope is focussed and across this image will be seen the wires of the graticule. The vertical illuminator condenser together with its graticule must now be adjusted so that the image falling upon the inclined plate in the microscope is oblique. The microscope field will then be found to be unevenly illuminated towards one side of the the field of view, so the image of the illum-

inant must be brought back to the centre by means of the adjustments on the inclined plate of the vertical illuminator in the microscope. The limit of obliquity is reached when the illuminating beam passes off the inclined glass plate and further adjustment of the lamp inclination for particular jobs may be necessary for the best effects. The object of the adjustment is to project a reasonably clear image of a selected line in the graticule on to the specimen so that the contour effect is visible to the eye in the microscope. A slide containing specimens of crystals or an uneven surface of some kind should be used to experiment upon or a piece of metal etched with acid makes a good experimental specimen. When the system has been set up it will work in the same way for any other specimen and so long as the optical bench is not disturbed, a calibration once arranged will remain correct.

The image of the graticule line will not be so clear as the detail in the specimen because the line is a projected image upon the specimen, but maximum sharpness of this line image can be obtained after setting up by making a small movement along the optical bench of the microscope relative to the illuminating system.

When the graticule lines are vertical in the illuminator they will fall nearly horizontally upon the specimen owing to the action of the inclined glass plate of the vertical illuminator and unless this plate or the graticule is accurately adjusted in rotation about the optic axis a double image of the line will appear in the field, this being due to reflections from the two sides of the glass plate. When the apparatus is adjusted as above so that the reflections from the sides of the glass plate combine, a clear dark image of the graticule line results. Experiments have been made with half aluminized vertical illuminator plates but no useful improvement is to be obtained.

It was mentioned above that some unevenness of illumination in the field of view must be expected owing to the obliquity of the light and if a low power (X6) eyepiece is in use, an even field cannot be obtained. Two courses of action are open. Either a higher eyepiece may be used in order to present to the eye or photographic plate a complete picture of the best part of the field, or the photomicrograph may be masked when the enlargement is made. It is usually better to use the high power eyepiece (X20) but this will not work well unless the objective has high aperture ($\cdot 9$ NA). A low aperture system cannot produce an illuminating beam of sufficient obliquity nor an image which will stand eyepiecing more than X15. High aperture objectives usually have short focal length, therefore high eyepiecing sometimes produces unduly high magnification with consequent loss of definition and depth of focus. A compromise of medium (X15) eyepiecing and plate masking produces the best results from most specimens.

Experiments have shown that the Contour System is capable of indicating differences of level which cannot be seen with the eye and so cannot be measured by means of the fine adjustment. A calibration for contour work can be made by photographing a specimen which shows contours, sectioning the specimen approximately along the line of the contour, photographing the sectioned face in the normal way and examining the photographs together at the same magnification. Great magnification may be applied to the sectioned face after the exact position of the section is noted on the contour photograph. The larger steps can then be measured and other steps may be obtained by interpolation. The apparatus must not be disturbed after calibration.

This method of calibration is sound but is difficult to apply to an irregular specimen because it is difficult and often impracticable to advance the sectioning exactly as far as the photographed contour line. It is easier to make a section across a specimen which shows contours, set it up in the ordinary way for contour microscopy and set the line as close to the edge of the specimen as possible. The image must be photographed for record after which the specimen is turned through 90° in the vertical plane and examined as an ordinary section with the vertical illuminator. This is also photographed at the same magnification for record and any differences in level now showing as steps, are measured with a micrometer at much higher magnification if necessary. The heights of the steps may now be transferred from the ordinary section view to the contour view and a calibration obtained.

If much of this work is intended it pays to make up a calibrated slide consisting of a piece of waxed polished brass upon which parallel needle scratches very close together have been made in the wax, and very lightly etched. One or more of the scratches will show contour effects on the microscope and the lines being straight and even can easily be sectioned across to enable the actual depth of the etch to be measured near the contour line. It is possible to measure a linear distance in a microscope of about $\frac{1}{2}$ a micron and it will be found that a step of this height causes a very clearly defined kink in a well-set-up contour line. If, when the contour system is in use the kink is measured with an eyepiece micrometer, it will be found that a twentieth of this distance can easily be seen, thus from a comparatively coarse calibration step a very sensitive measurement can be made.

When the microscope is focussed accurately upon a flat part of the specimen the graticule line will also be in focus, but if the surface has different levels, it cannot all be in focus at one time unless the level changes are small. When the microscope is focussed accurately upon a feature the graticule line will not be in focus upon other features at other levels and the system cannot provide useful results unless the graticule line can be seen passing over the various levels. It follows that a micrograph must be taken at an average focus so that the line and features may be seen together, but a micrograph which shows the contours may not be the best micrograph for showing fine surface detail. The Contour system works best upon a specimen having level changes of less than 5 μ, and if the level changes are greater, lower aperture objectives and less obliquity of the incident beam are necessary. In practical work it is often necessary to take photographs of the specimen at different focusses and use the contour photograph for measurements only.

It may be observed that the microscope fine adjustment can be used to measure coarse level changes of the order of 5 μ. This is true, but the fine adjustment cannot be used to examine a photograph and any device which makes contours visible upon a photograph in a measurable way is very useful even if there were nothing else to gain and the resolutions were low.

Composite Photographs

It is necessary for most microscopical work upon surfaces to take many exposures at high power across a specimen. Low power photographs are seldom useful, and a method has to be devised to produce a high power picture including the whole of the specimen. A track across the specimen is decided upon and a series of photographs is taken with an overlap of exposures of about half the field of view. The negatives must be printed with or without enlargement when they may be fitted together so that the features on the specimen match. When they match, the contour line position is related to the completed composite photograph. A calibration may be obtained by any of the methods described above.

To Determine the Obliquity of the Contour Line

It is not practicable to calculate the angle of obliquity of the part of the incident light which contains the contour line because the adjustments are all critical and angles in the vertical illuminator and illuminating system are practically impossible to determine with sufficient accuracy for useful calculations to be made. A test specimen should be used which shows sudden changes of level of about 10 μ; an etched metal plate is suitable.

The step must be placed in the centre of the microscope field of view and its height measured in microns by means of the fine adjustment. The specimen is then placed beneath the chosen contour line and the displacement of the line measured in microns directly with an eyepiece micrometer. The angle of obliquity is then obtained by geometrical construction and a calibration obtained from these measurements. Fig. 2.

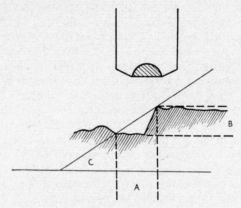

Fig. 2. A. Displacement of contour line (in microns).
       B. Height of step (in microns).
       C. Angle of obliquity.

CHAPTER 19

Microscopical Apparatus for the Study of Transparent Objects

IF AN object is opaque like a piece of metal or lump of stone its surface can be examined only by using the methods given in the chapter 'The Study of Surfaces'. Only a surface, and no internal structure, can be seen in this way. If a transparent object is studied with light injected from the top, a small amount can be seen but only if the structure is very dense compared with the surrounding mass, the illuminating rays make the upper parts of the specimen bright and so mask any light which is returning from the deeper parts. This effect is very noticeable when a motor-car head lamp is used during a foggy night. Most specimens have a sufficient amount of small internal detail to scatter light like the particles of a mist.

More of the inside of a semi-transparent structure, though it may be more difficult to interpret, can be seen by examining the specimen by light which has passed through the substance.

In these conditions any haze of light due to scattering by the parts first and strongly illuminated occurs at the bottom of the specimen, not between the objective and the part being studied. Image forming light from lower regions has to pass through upper regions on its way to the objective, but as these upper regions are not highly illuminated, the image is not fogged though it may be affected in clarity ('Microscope Tube Length', page 61). In some regular structures like diatoms false images do occur.

Light is helped on its passage through a structure if the structure can be packed with a material of higher refractive index than air. If a structure like an insect's leg is examined by 'holding it up to the light', i.e. as a transparent object, little light can pass through because the internal parts reflect light from their surfaces and little arrives at the other side. If the structure is filled with a liquid or solid of refractive index, say $1 \cdot 3$ or $1 \cdot 5$, most of the internal reflections are absorbed, the light arriving at the viewing side is modified mainly by the optical density and colour of the internal parts. This is the main reason for mounting microscopical preparations. In the chapter General Microscopy, page 37, the matter of mounting for maximum resolution is amplified to cover the necessity of selecting the mounting medium to have at least as high a refractive index as the immersion medium in which the lens works. (See also chapter 'Basic Mounting Techniques', page 293.) In order to study the methods of examining transparent objects, a typical structure consisting of a block of transparent material, say, resin, with parts of an insect embedded in it, will be taken. Such an object is not a regular structure, transmits a good quantity of light, has internal detail, no scattered opaque parts, and is in fact much a typical specimen.

The Effect upon Light of Typical Microscopical Structures

When light falls upon a surface other than a mirror, it is scattered in all directions in one hemisphere practically in equal amounts. If then a lens is used to collect this scattered light in order to form an image, the lens may receive light from as great an angle as it is designed to accept, the only limitation being the aperture of the lens itself.

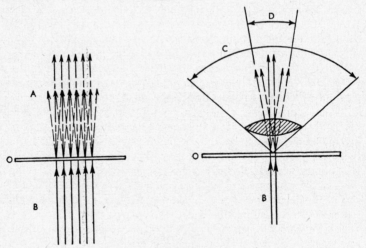

Fig. 1. O. Semi-transparent object.
A. Image-forming light.
B. Illumination.
C. Lens angle.
D. Used angle.

Such an opaque object needs no aid from accessory apparatus to cause it to supply any objective with a cone of light of sufficient angle with which to resolve all details possible. The only problem is to focus a sufficient *quantity* of light upon the specimen.

The matter is very different when transparent objects are studied

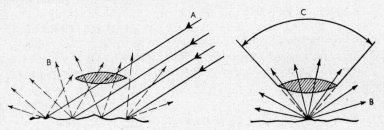

Fig. 2. A. Incident light.
B. Reflected light.
C. Lens angle.

because in this case most of the light passes through the object and only a small amount is deflected. Any light reflected from the back of the specimen is lost.

It will be seen from Fig. 1 that with transparent specimens only a small part of the possible angle of the lens is being used and it is shown in the chapter 'General Microscopy' that when a properly corrected lens is not used at its fullest possible angle, that lens cannot show all the detail theoretically possible for its aperture.

Some objects are relatively dense and regular, and do in fact scatter a large amount of light at large angles with the normal. Such objects can be well seen without accessory illuminating apparatus, but in general, the light emerging from transparent structures needs to be made to have greater angle with the normal in order to develop the full resolving power of the objective. This is done by illuminating the specimen with light at an angle to the normal, and the usual apparatus for doing this is a lens placed below the specimen known as a substage condenser.

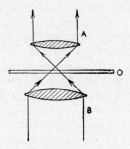

Fig. 3. O. Object (transparent).
A. Objective.
B. Condenser.

The angle of light entering the objective is now primarily determined by the aperture and focus of the condenser and the function of all microscope substage condensers is to apply light to the underside of a specimen at a controllable angle. A condenser of as high an aperture as the highest aperture objective in use is an essential part of a microscope.

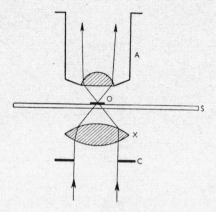

Fig. 4.   O.   Object.
          S.   Slide.
          A.   Objective.
          X.   Condenser.
          C.   Diaphragm.

This apparatus may consist of a simple lens of focal length about 40 mm. and diameter 30 mm. Such a lens placed below the specimen and carried in a sliding tube or on a focussing rack or spiral will concentrate a beam of light from the illuminant on to the specimen and will focus there an uncorrected image of the source of light. The cone of light from a simple lens of the geometry mentioned has an angle of about 90° and therefore will illuminate moderately well an objective of about this angular aperture, the power of such an objective being about 12 mm. focal length. It is necessary to fit a diaphragm of some kind below the simple condenser in order to control the angle of the cone of light reaching objectives of less than 90° angle so that they are not swamped with light they are not designed to handle. A condenser should not normally be used in a position other than focussed upon the specimen, for if it is defocussed (in order to alter its working angle), disturbances in the cone of light passing through specimen and objective result in patchy distribution of light in the zones of the objective which in turn lead to inaccurate images.

The image formed by a simple lens contains many aberrations, the chief of which are spherical and chromatic. Because of the spherical curvature of a simple lens the images formed by different parts of it cannot be in the same position. Fig. 5. If an approximate focus is

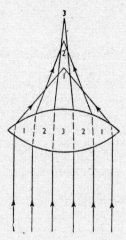

Fig. 5. 1′, 2′, 3′, Images of point source at infinite distance.
1, 2, 3, Zones of lens.

taken, say, that of zone 2 image 2, and the eye or objective placed there, only a ring of light with a spot in the centre can be seen coming from the lens. This is due to the fact that image (3) is behind the focus and image (1) is in front. Fig. 6. If a focus nearer (3) is taken

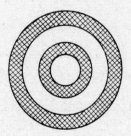

Fig. 6.

matters are different because the spherical curvature of the lens causes less spherical aberration to rays nearer the axis, therefore the centre spot is bigger and more uniform. Fig. 7.

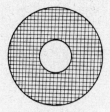

Fig. 7.

The focus of zone 3 is now so far away that its image is only a diffuse area of light. The maximum useful angle which can be provided by a lens is represented by the diameter of the 'solid' centre cone when used with any particular illuminant. This angle usually measured in terms of Numerical Aperture is known as the Maximum Solid Cone of a condenser and is nothing to do with its angle as measured for total Numerical Aperture tests which ignore aberrations.

A low aperture well corrected lens may have all its total numerical aperture available as a solid cone while a lens with very high normal measurement of NA may have only one third of this total aperture available as a solid cone. A further complication arises due to the fact that a large source like an opal electric bulb will fill a large number of poorly corrected zones, while a small source will fill very few and so make the lens look much worse on this test. It is for this reason that a solid cone test should be made with a small source like a bunched filament lamp, or, with the actual illuminant to be used so long as this illuminant is always used for comparisons.

The chromatic aberration of the simple lens causes coloured fringes to appear around all images from whichever zone they are taken, the fringes being wider in the outer zones.

The effects of these aberrations in a condenser, upon the microscope image, are as follows:

(1) The useful angle of the condenser is its maximum solid cone and all light from the outer less corrected parts of it (zones 2 to 3) cause glare and fogging in the image because they illuminate structural parts of the optical system, and project light into the objective at angles which it may not be intended to handle. These high angle beams are, however, of use in dark ground illumination.

(2) Colours are superimposed upon the specimen.

(3) A good focus of the illuminant cannot be obtained. In later sections it will be seen that actual focus of the illuminant on

the specimen is not necessary but its limiting field diaphragm must be accurately focussed in order to limit the area of the specimen illuminated.

The ways of improving the performance of a condenser with these limitations are as follows:

(1) The larger the source the more the zones' images run into each other, therefore the central solid cone is increased in angle. Large sources like opal electric bulbs should be used when the condenser is poor.

(2) Colours may be controlled by the use of colour filters and if large sources are used the fringes may fall outside the objective field of view.

(3) Use of the condenser diaphragm to limit the cone of light to make the focus sharper.

The practical tests for correctness of the set-up with simple condensers are as follows:

(1) The substage condenser diaphragm should be reduced until a clear image of the source, a mark upon the source, or its diaphragm, can be obtained when microscope and condenser are focussed upon a small clear specimen or mark upon a slide.

(2) The diaphragm should be opened until, with the source in use, the even disk of light in the centre of the objective just begins to break up into rings (Fig. 8, A and B). This effect is observed by removing the eyepiece and looking down the tube at the objective back lens after it has been focussed upon an object in the normal way.

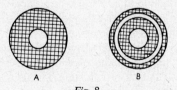

A   B

Fig. 8.

(3) A small change of focus of the condenser is permissible so long as the even disk in the objective back lens is not destroyed. When the even disk is at its maximum size, this is the most that can be got from the condenser in transmitted light. The maxi-

mum aperture of objective which can be illuminated reasonably well by this simple condenser is that which can be filled with an even disk of light up to about half its back lens diameter when tested by method (2) above.

Simple Condensers with Non-spherical Surfaces

If a lens can be made with a parabolic curvature of surface instead of a spherical one there is no spherical aberration when the lens is used to focus parallel light, but chromatic aberration remains. It will be seen on page 496, 'Lens Making', that only a spherical curvature can be generated accurately by natural means. It has not been possible up to the last few years to make a parabolic optical surface of high curvature, but it is now commercially practicable to use a combination of moulding and polishing to produce a parabolic lens of focal length 25 mm. and diameter 50 mm., good enough to be used as a magnifier and such a lens will give an image quite good enough for use as a condenser. The parabolic component may be designed to be used alone or in combination with a plano convex spherical lens of about four times its focal length. The combination forms an excellent lamp condenser as well as a substage condenser.

A condenser with large lenses, 50 mm. in the above example, cannot be brought up through the ordinary mechanical stage, therefore it is important that such condensers be used on microscopes with plain stages or be of focal length at least 25 mm. so that there is room for the stage structure to clear the mount. It is not at present possible to make small parabolic lenses with sufficiently accurate curves for use in conventional condensers and objectives which are given high apertures by employing small diameter lenses with short focal lengths.

A parabolic lens such as that described above gives a solid cone of light from an ordinary bunched filament bulb of more than 90° angle, i.e. about 90 per cent of its measured aperture. This cone will illuminate satisfactorily objectives up to a power of 8 mm. with a very large amount of light useful for projection purposes.

Parabolic lenses cannot be made in the variety of sizes usual in spherical examples owing to the profiling method of manufacture; changes in the focal length of a system can be arranged by placing convex or concave spherical lenses below the parabolic one. This is admissible in a condenser and little damage is done to the quality of the image formed of the illuminant, but the treatment cannot be extended to produce changes of focal length greater than 25 per cent but often this amount is adequate to allow a system to reach through a thick mount or clear a mechanical obstruction.

Uncorrected Substage Condensers

The best known substage condenser is the Abbe type which may consist either of two or three lenses of spherical form, uncorrected for chromatic aberrations and only partially corrected by selection of curvature, for spherical aberration. Figs 9, B and C.

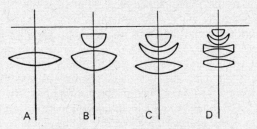

Fig. 9. Refracting Substage Condensers.
A. Simple.   B. Common Abbe.
C. Aplanatic Abbe.   D. Achromatic.

The measured mechanical working distance of the Abbe condenser is about 5 mm. but a piece of dead glass is left upon the top lens (*a*) Fig. 10 in order to reduce the working distance to about the above figure so that the condenser can be used in oil contact with a slide of normal thickness. As this system is an uncorrected

Fig. 10. a. Dead glass.

one there is little point in oiling it to the slide except where an attempt at dark ground illumination is being made. The dead glass may be ground off to give about an extra 2 mm. of working distance to the condenser, thus allowing it to focus through a thick slide or trough without in any way affecting its optical properties. No great accuracy is called for in this grinding and the ground surface may be made clear by cementing a cover glass over it with Canada balsam.

The numerical aperture of the Abbe condenser is about $1 \cdot 2$ when oiled on, but the solid cone when used with an ordinary opal electric lamp is about $0 \cdot 5$ NA.

The Abbe condenser is the maid of all work of the routine laboratories and the medical world for it is robust, insensitive to centring, may be taken to pieces easily and without danger, and may be used as a low power condenser when its top lens is removed. When it is oiled on to a slide and an immersion objective is in use it makes a good annular illuminator when its central region is stopped out, the illumination is then in the hollow cone about NA 1 to NA 1·2. (See also page 348.)

The Abbe condenser has many limitations: it delivers an uncorrected beam to the object, hence the field of view may be extensively coloured, especially when oblique light is employed. The solid cone of illumination is small and will not work properly an objective of higher power than 6 mm. and NA 0·5. A clear focus of the lamp field diaphragm cannot be obtained. Owing to its uncorrected output, centring and squaring to the rest of the optical system are unimportant. However, allowing for all these limitations, the Abbe condenser designed ninety years ago fulfils its functions admirably, and with a large light source like a paraffin lamp or opal electric bulb will do all the histological and most medical students require from a microscope, even with a 2 mm. immersion objective.

Corrected Substage Condensers
It was shown diagrammatically under the heading 'Simple Substage Condenser' that in order to develop the full resolving power of an objective, light from the specimen being examined must fill the whole of the objective aperture. In order to achieve this when transparent specimens are being examined, a condenser must be employed which both provides illumination at the same angle as the objective and provides a solid cone of light within this angle. In other words, the system must not have spherical aberration beyond a certain tolerance. If corrections of spherical errors are attempted, chromatic errors are dealt with at the same time. A condenser never gives as good an image as an objective because it would be a waste of effort to make it do so, but a dry condenser of NA 1 should, with a light source of about 25 mm. diameter, give a solid cone of light of 0·9 NA. A check may be made in the following way:

Set up an objective of 0·95 NA or near this, at the correct tube length with a small marked specimen like a scratch or bacterium in focus. Focus the condenser so that an image of the light source to be used is correctly in focus on the specimen. Remove the eyepiece and look at the objective back lens. As the substage condenser iris diaphragm is opened to its maximum extent the patch of light will increase outwards from the centre zone of the objective towards the edge of the back lens. At about three-quarters of the back lens diameter, with the illumination source an even disk about 25 mm. diameter, the patch of light will fade at its periphery. The aperture at which fading commences represents the maximum solid cone

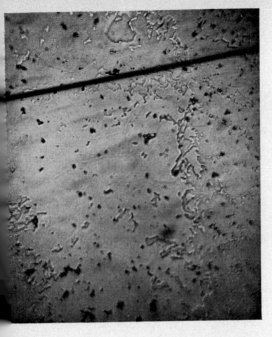

A. Very light etch.

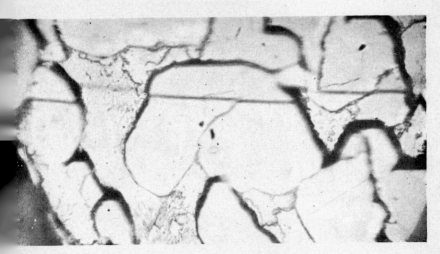

B. Strong etch showing pits several microns deep.

PLATE 8, A and B
Contour Lines Projected on to the Etched Surface of Germanium.
Photograph magnification 3,000 diameters, 2 mm. apo. objective and ×15 eyepiece. Contour line displacement, $\frac{1}{2}$ micron height difference = 1 mm. horizontal
displacement.

C. Contour line on etched germanium. (Conditions as D.)

D. Line on etched metal specimen $\times 2,000$ showing shift in line representing a change in height of $\frac{1}{4}$ micron. 2 mm. apo., 1 mm. $= 1\mu$ in horizontal magnification.

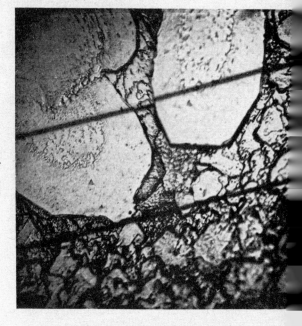

PLATE 8, C and D Contour Microscopy.

obtainable from that system. A modern dry achromatic condenser should illuminate fully to its edges on objective of 0·85 NA. The angle of the solid cone available from a condenser is obtained by setting it up as described above and comparing the illuminating power of the condenser with that of a known objective, using a light source of pre-arranged size. Once the maximum solid cone angle is known, it can be assumed that all NA readings below this will represent solid cones of illumination and the condenser iris diaphragm can be calibrated for a particular condenser if required.

The Practical Advantages of Achromatic Condensers

One advantage has already been mentioned, namely that a solid cone of the same aperture as the objective is necessary for full resolution to be attained, usually a clearer image of the lamp diaphragm is provided, therefore a better focus of the source may be obtained. This is an advantage, not so much because of improved resolution but because it is a considerable aid to work when changing objectives to be able to leave the image of the illuminant or its diaphragm focussed upon the specimen or upon some particular part of it. A different objective may then be brought into focus easily, using the illuminated area as a guide to position—with lessened danger of missing the focal point on the object (should this be transparent) and hitting the slide. The centration of the system may be checked at the same time.

The fact that the microscope field of illumination can be clearly limited by means of a good image of the lamp field diaphragm makes it possible to limit scattered light, and to prevent light from parts of the specimen outside the objective field of view from throwing light into the system at angles which the objective was not designed to handle. The effect of illuminating more of the specimen than is covered by the microscope field of view is to cause glare from internal reflections in the lens mounts and consequent loss of contrast in the image. Some objectives are much more susceptible to this trouble than others, depending upon their internal arrangements.

Light can be applied to the specimen obliquely by placing stops behind the achromatic condenser. The image of the source will remain in the same position whatever the obliquity of the image-forming rays and will not show much colour. The possibility of being able to use critical oblique light justifies the purchase of an achromatic condenser even if no other reasons are acceptable. Nearly-transparent specimens in liquids, powders on surfaces and living bacteria can be well seen in transmitted light only when the light is oblique and colour-free.

Construction of Achromatic Condensers

Many optical and mechanical forms of the achromatic condenser have appeared in the past and some will be mentioned in later pages. The present day construction is all that it should be and consists of a short mount about 40 mm. long containing three lens groups. The top lens is solid like the front of an objective and the remaining two groups are pairs or triplets, the components of each group being cemented together and mounted in metal sleeves. The sleeves fit together by means of screw threads to build up the complete condenser

Most achromatic systems are useful for low power work with the top lens removed but of course their corrections cannot be good in this condition but are good enough for most low power work. Condensers may be unscrewed for experimental purposes or for cleaning without any fear of errors in reassembly, the centring of components is important but the mechanical accuracy of the large mountings is good enough to return the lenses to nearly their original position. An objective must not be demounted in this way because it will not go back except in the hands of a specialist.

Some condensers are made with back lenses about 25 mm. in diameter while others are little larger than objectives—about 8 mm. The angle of light provided by the two condensers may be the same but the amount of light passing through the system is greater the larger the lenses, also the large lens arrangement allows greater working distance. Fig. 11. A condenser may have any focal length and aperture

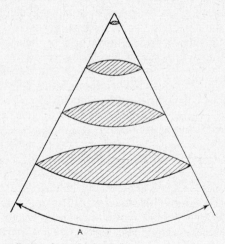

Fig. 11. Different sizes of lens giving same angle of illumination.
A, angle of system.

exactly as in an objective and the shorter the focal length the smaller the image of the illuminant. The actual quantity of light passing through a microscope is only of importance in microprojection, the electric lamps usually used in visual work are often too bright and produce too much light. The large mechanical parts of the mounting of a condenser with large lenses often prevents the mechanical stage from experiencing its full travel as the top lens mount may be about 25 mm. diameter at the top, while the fully mechanical stage has an aperture only 50 by 40 mm. It has a thin top plate but unless the condenser is working through a very thin slide its working distance is insufficient to allow this plate to pass over the condenser mount in order to carry a point on the slide about 25 mm. from either end, into the microscope optic axis. For this reason the type of condenser which has a mechanical top diameter of about 15 mm. is the more useful accessory. For complete treatment of this matter see 'Microscope Stages', page 144.

Position of the Substage Diaphragm
It is most important that the diaphragm be placed within about 8 mm. of the condenser back lens. Usually condensers are mounted upon diaphragm-containing mounts but often substages are so arranged that the diaphragm is about 25 mm. below the condenser. If the diaphragm is far below the condenser it cannot completely control the pencil of light impinging upon the lens; it then becomes useless for its purpose. Secondly, the condenser cannot project an image of the diaphragm in such a position in space that it can be magnified by the microscope and used as a centring aid. The first matter is the more important but there is no reason why the diaphragm should not be put in the correct place, i.e. within 8 mm. of the condenser black lens. (See 'Dark Ground Illumination', page 348.)

Practical Use of High Power Condensers
Condensers of good quality are designed to work through a particular thickness of slide but when using high power dry examples, slide thickness is rarely considered important. The condenser does not form an image of the specimen in the way the objective does, therefore a little more spherical aberration is not of any importance and an immersion condenser will work quite satisfactorily without oil at NA 1·0 and solid core of about 0·95. When it is in oil contact with the slide, difficulty is often experienced in maintaining contact when the slide is thin because of the large quantity of oil present. The thickness of oil may be reduced by oiling pieces of glass between slide and condenser, but these are very apt to slip off and should in

some way be made captive to the condenser. One successful method used in a laboratory where condensers were often oiled on, was to cement glass disks of different thicknesses to brass tubes designed to slide loosely over the condenser. Both sides of the glass disk must be oiled to the condenser and slide respectively. The disks were found to be difficult to clean after use and commercial Golden Syrup was later used as the immersion medium as this will not run off and is easily wiped away with water.

It must never be assumed that the diaphragm and condenser on a substage assembly are on the same optical axis. When a new condenser or new mount is brought into use an ink spot should be placed in the centre of the condenser top lens with the aid of a turntable or lathe, the condenser mounted and the spot focussed with ascending powers of objective. When the spot has been centred by means of the substage screws or sliding changer adjustments, the diaphragm should be reduced to a pin hole and its image focussed with the same objective. The image should not be more than 10 mm. out of centre of the field of view when viewed with a 4 mm. objective and ×6 eyepiece. If it is out of centre by say half a microscope field diameter the mountings should be examined and possibly rejected. (See 'To Line up a Microscope for Critical Work', page 369.)

When the top lens of a condenser is removed for low power working it is not unusual to find that the centring of the modified system is quite different. If this is noticed the condenser should be checked for levelling as its axis may not be parallel with that of the microscope. Such a fault is common in old swing out substages and loose Akehurst changers.

Historical Methods of Illumination by Means of Apparatus below the Stage: The Swinging Substage

This apparatus came into use about 1870 and was probably designed by Dr Piggot of the USA for use when large angle highly corrected achromatic condensers were not available. It was known at this time that advantages were to be had in resolution by using as the illuminant, light applied at a great angle to the axis of the microscope and if this could not be obtained by means of a high angle lens, the lens assembly itself was inclined. The mechanical arrangements were as follows:

The microscope was built with a very thin stage which could be tilted with respect to the microscope axis with the object at the centre of tilt. This stage was usually insufficiently rigid. The substage and mirror were mounted on an arm pivoted on the limb at the same point as the tilting stage and having an arc of movement with radius say 130 mm. and centre that of an object on a typical slide on the stage. The condenser was usually an objective of about 12 mm. power with a small conical front lens capable of being focussed on the under side of the specimen in the usual way. The whole contraption could be swung from side to side about the object, with or without employing the tilt of the stage, through an angle of about 160° without appreciable change of focus of the condenser. Oblique illumination up to 170° in air could thus be obtained. It was not possible to use oil immersion condensers with a swinging substage.

About the year 1880 there was a luxurious growth of swinging devices upon the microscope. Stages tilted, substages swung, stands rotated upon their bases, the specimen was made the centre of a radial stand inclination slide and some swinging substages could be brought above the stage for top light. None of these devices contributed anything to the stability and usefulness of the microscope as a research tool and all have passed out of use now that properly corrected achromatic condensers are usual.

A swinging mirror fitted to a naturalist's microscope should not be neglected. If a concave mirror is mounted on a jointed arm capable of being extended to, say, 150 mm., pivoted just below the stage, it may be used to concentrate light on a variety of specimens like mosses, lichens and entomological specimens, from above or below the stage. Dark ground effects can be obtained upon objects in troughs of water and in Petri dishes.

It is much easier to illuminate an object from its underside than it is to apply sufficient light to its top surface. The achromatic immersion condenser and diaphragm will do all possible jobs with mounted transparent specimens, but the matter is different with opaque objects. Here many experiments may have to be made before the structure sought is made visible and there is no particular accessory which will do it all. (See 'The Study of Surfaces', page 202.)

Dark Ground Illumination
When an object is rendered bright by a beam of light which does not itself enter the eye, the object is seen bright against whatever background may be present. If the background be made dark, very tiny objects may be made visible if the light is sufficiently intense, and this part of the subject is treated in greater detail under the heading 'Ultra Microscopy'. In many ways the system of dark ground illumination is similar to that of top light but in microscopy it means 'Illumination by a hollow cone of light obtainable by placing a stop in the substage condenser to cut out the central rays up to and, slightly exceeding, the angle of the objective'. A condenser of greater

angle than the objective is essential but its solid cone need not be large because only the outer zones are in use.

Some objects like animalcules in pond water are made beautifully visible by this means; they appear as silvery silky structures against a black background, but there are several advantages and disadvantages in the method and these will be treated in detail.

Methods of Obtaining Dark Ground Illumination

When low powers up to 16 mm. focal length and 0·4 NA or thereabouts are in use, an ordinary condenser either simple, non-achromatic, or corrected may be used with an opaque stop placed as in

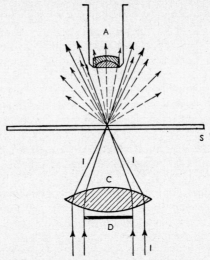

Fig. 12. I. Illuminating rays not entering objective.
   A. Light scattered by object, entering objective.
   D. Light stop cutting out rays up to objective aperture.
   C. Condenser here shown as a spot lens.
   S. Slide.

Fig. 12. The device shown is known as a 'spot lens' and is quite satisfactory for powers up to 25 mm., but for higher powers a condenser must be used. In order to obtain better control of scattered light (which may spoil the dark field) stops are best placed between the components of condensers. The Abbe condenser is particularly good for DG illumination mainly because it has a high angle and large lenses; the stop may be made of black paper and be placed immediately below the condenser top lens. It may be kept in place by a tiny speck of glue and its size found by trial.

If the stop is to be placed below the condenser, the required size may be measured for use with a particular objective and condenser by the method described under 'Rheinberg Differential Colour Illumination' for obtaining the size of the central coloured disk.

When a well corrected lamp condenser and small source are in use, giving a solid parallel illuminating beam the stops can be placed immediately below the condenser in a carrier. This is the better arrangement because stop sizes can be easily changed; the carrier usually takes the form of a swing-out ring about 30 mm. diameter usually provided on the lower part of the substage condenser mount and is known as a 'stop carrier'. A frame may be made as in Fig. 13 to fit this carrier, when various sizes of stops can be made from any opaque material, with holes in their centres to fit over the pin A.

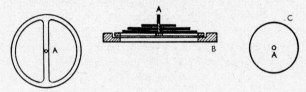

Fig. 13. A. Pin and holes.
B. Stop carrier.
C. Stop disk.

Disks of glass with opaque disks painted upon them may be used as stops. On old microscopes, a revolving wheel of stops was provided, associated with an achromatic condenser. A device called a Travis Expanding Stop (after Mr Travis of the Quekett Microscopical Club) was manufactured and consisted of a reversed action iris diaphragm, i.e. the leaves expanded from a central hub instead of closing in from a periphery. There were no disadvantages with this device and it is a great pity that it is not now made. The expanding stop and iris diaphragm could be used together to obtain controlled annular illumination, and the wheel of stops for oblique light.

To set up Dark Ground Illumination using Stops in Condensers

A specimen should be selected consisting of an isolated dense object upon an ordinary glass slide; a test diatom is a good example and the specimen and slide must be very clean. First set up the microscope, using the objective required and the condenser without a stop, to study the specimen as a transparency, with an image of the light source focussed upon the specimen. Open the substage iris diaphragm wide, remove any coloured filters there may be and without changing the condenser focus, introduce stops until by trial one is found of such a diameter that the direct light is completely shut off leaving the specimen brightly illuminated on a dark background. If the condenser is an uncorrected one, small changes of focus of the condenser may improve results. In general, the dark ground condenser should produce an image of the light source upon the specimen just as accurately by means of its outer zones, as by its central zone which is used for examination of transparent objects. This method of setting up is usually quicker and better than that which must be employed to obtain the correct diameter for Rheinberg disks.

It is impracticable owing to light scatter to illuminate satisfactorily objective powers greater than 8 mm. and aperture 0·7 NA by means of condensers and stops. It is also of little use trying to obtain an even dark ground illumination from a point or small source of light. Any zone of the condenser will form an image and if the centre zones are blacked out the size of the image of the light source is just the same as when the whole condenser aperture is in use. It follows that an even dark ground field can only be obtained from a source and condenser of such a physical size that their image fills the field of the objective when set up for transmitted light. A large diffuse source is not usually bright enough for this work and the lamp and lamp condenser system is more suitable. Unless an even field is obtained stray fans of light and bright artifacts result.

Reflection Type Dark Ground Condensers: The Wenham Paraboloid (Fig. 14)

This instrument is not now manufactured but is the prototype from which modern devices have been developed. It consists of a block of glass of the shape shown, about 40 mm. diameter at the base, about 25 mm. high, made from a solid cylinder of clear glass of these dimensions with a hole drilled along its axis for the stop spindle. A hemispherical cavity about 6 mm. radius is ground in the top on the axis of the hole by means of a spherical tool charged with carborundum. The cylinder is then mounted in a lathe and centred on a mandrel passing through the hole and locating in the spherical depression. It is turned and worked by hand, with a file-shaped brass tool and a drip feed of carborundum-powder and water, to a paraboloidal curve set up on a template laid out by means of straightforward geometrical construction to suit the size of paraboloid being made. The focus always comes about 2 mm. above the upper surface of the block to allow the instrument to work through a thick slide or trough. All surfaces are polished, none needs silvering. A good paraboloid can be made from Perspex but the surface is relatively soft, though it can be polished with metal polish.

When the stop is carefully used to admit only the most oblique rays and the apparatus accurately centred, good dark ground effects with 4 mm. objectives aperture 0·7 NA can be obtained. It has no

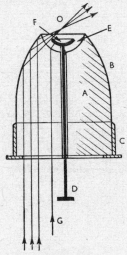

Fig. 14. A. Glass block.
        B. Paraboloidal surface of revolution.
        C. Brass tube mount.
        D. Ajustable stop moving along axis.
        E. Hemispherical polished cavity.
        F. Light stop.
        G. Light of low N.A. cut out by stop F.

equal for dark ground illumination of low and medium powers and if the upper hemispherical cavity be optically united with the slide with water, even better results can be obtained with medium powers.

The common paraboloid does not form a good image of the light source, therefore the lamp and its condenser should be set up correctly with an ordinary condenser on the microscope. The paraboloid is then substituted and its focus and stop arranged on test to give the best results. For highest angle illumination the stop must be fully up but for the best results (with objectives of medium aperture only) the stop should be nearly fully down.

The Immersion Paraboloid
This instrument is intended for use with high power dry lenses up to 0·9 NA. It will not work with immersion objectives unless their numerical aperture is reduced to below 1·0.

It consists of a solid block of glass approximately 18 mm. diameter at the base and 8 mm. diameter at the top and about 8 mm. high with sides of paraboloidal curvature. The instrument differs from the preceding paraboloid in that its top is flat and light-stops are built into its structure. Fig. 15, A and B.

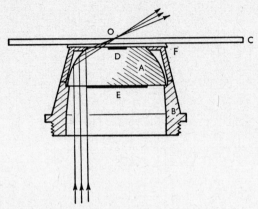

Fig. 15 (*a*).

A. Solid paraboloid.
B. Brass substage mount.
C. Slide of definite thickness, usually 1 mm.
D. Blacked out depression often containing centring spot.
E. Metal foil fixed light stop.
F. Oil seal between paraboloid and mount cap. (Surface of A must be in air except top where it is oiled evenly to slide lower surface.)

Owing to the fact that extremely oblique rays must pass into the specimen in order to produce dark ground effects with high aperture objectives, the paraboloid is oiled on to a slide of particular thickness, usually 1mm. The focus is well defined and the image of a source 25 mm. diameter, is only large enough to fill evenly a 4 mm. objective. The apparatus must be set up as follows:

A specimen made up of a suspension of powder or of bacteria should be used as a test object. A very thin layer should be spread upon a well-cleaned slide of the thickness stated on the condenser and a well-cleaned cover glass placed upon it and pressed down hard. Only capilliary thickness layers are suitable owing to the sharply defined focus of these paraboloids. The paraboloid should now be oiled on to the slide and the light adjusted to fall upon the lower surface of it. If the lamp has a lamp condenser it is desirable to first set up the microscope for an ordinary transparent object with an ordinary condenser, in order to make sure that lamp, lamp condenser, mirror and substage condenser are properly in line. The lamp condenser must show as an even disk of light when the microscope condenser is set up to illuminate an ordinary transparent specimen. If the specimen is now examined with the 25 mm. objective a very bright spot of light (the paraboloid focus upon the powder or bacteria) should be seen near the centre of the field. It is possible to alter the paraboloid focus slightly within the oil drop; this should be done

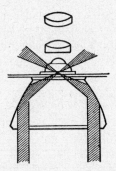

Fig. 15 (*b*). Paraboloid dark ground illumination.

and the spot of light should focus through a definite minimum size, The spot shows as an image of the light source on the objects when at its minimum size but it is rarely recognizable unless the specimen is too thick with particles for the apparatus to work properly as a dark ground device; a slide of chalk dust in water will demonstrate the focus of the illuminant and the result of using too thick a suspension. The paraboloid presents a surface to the slide about 8 mm. diameter and when oiled on, transmits, if moved in a direction normal to the slide surface, an amount of force easily capable of bending the slide. When adjusting the paraboloid for exact focus the movement of the substage must be slow in order that the oil may give, and allow the slide to take up its natural position.

It is unlikely that the paraboloid focus will be in the centre of the field of the objective used unless the microscope has been lined up before, so when satisfied that the apparatus is correctly set up as above, the substage centring screws or Akehurst Slide adjustments should be used to centre the paraboloid image to the highest power objective in use. (See 'To Line up a Microscope for Critical Work' page 369.)

The images of specimens which are obtained when high power paraboloids are in use are never so clear and distinct as when low powers or transmitted light are employed as all specks and impurities are made luminous above, below and to the side of the field of view which scatter light into the field and fog the image. Only apochromatic objectives are successful; others have not good enough corrections to stand working at full aperture upon tiny bright specimens. Tube length correction is critical but even with these disadvantages good images of living bacteria and blood cells can be obtained and a number of biological discoveries have been made by employing this method. Very bright lights such as arcs and ribbon or bunched filament lamps must be used as the smaller the difference in opacity or refractive index between the specimens and the liquid in which they are immersed, the brighter the light which must be applied to make them reflect enough light out of the medium with which to be seen. It is brightness and not quantity of light which must be sought.

The Concentric Dark Ground Illuminator (Fig. 16 (a) and (b))

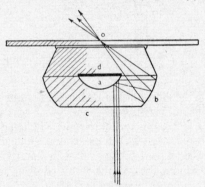

Fig. 16 (*a*). Concentric illuminator.

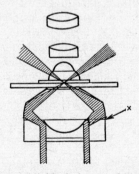

Fig. 16 (*b*). Cardioide dark ground illuminator.

This also is a device designed to concentrate rays into a very small intense illuminating area suitable for use with high power dry (or stopped down immersion) objectives. Like the paraboloid, it focusses light by reflection, thus avoiding any errors due to chromatic aberration. In general the concentric illuminator forms a smaller spot of light than the paraboloid but a paraboloid could be designed to give similar results if greater physical size were acceptable. In Fig. 16(a) the concentric illuminator can be seen to be constructed from two disks of glass. The lower disk has an optically flat lower surface (c) through which light enters, and a polished concave depression (a) which forms a reflector. The periphery has a reflecting surface (b) worked upon it as shown. The upper disk of glass only connects optically the lower disk to the slide and carries a stop (d). Owing to the fact that reflections take place from two surfaces, these may be arranged geometrically to give negligible spherical aberration, though the curves are both spherical, the manufacture of this device is therefore simpler than that of a paraboloid.

The same procedure for setting up obtains as for a paraboloid, and it must be emphasized that accurate centring is essential. Separate centring mounts are sometimes supplied with dark ground condensers for use on microscopes without proper centring substages or Akehurst slides. The position of the image of the light source cannot be altered by altering the position of the mirror, as it can when an ordinary large lens condenser is used with transmitted light. In the transmitted light example a recognizable image in the microscope, though incorrect, can be obtained with the condenser 3 mm. off the axis of the microscope but nothing recognizable will be seen when a dark ground condenser is 0·25 mm. off axis.

Reflecting condensers are designed geometrically, simply according to the laws of reflection, and do not present any difficulty.

The Focussing Dark Ground Illuminator

The concentric type of condenser can be made with the upper and lower disks of glass mounted separately so that, with small changes in the shape of the surfaces a focussing dark ground condenser is formed. If thin slides are in use a small amount of focussing range can be obtained from both the paraboloid and ordinary concentric types but unless the optical path between the paraboloid and the specimen is exactly the correct length and the object is really mounted upon the slide surface, the results will not be the most satisfactory. The built-in focussing arrangement introduces slight light scatter, but this is outweighed by the advantage of being able to focus the illuminator on the wanted plane in, say, a histological

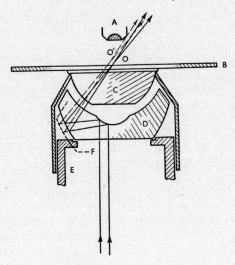

Fig. 17 (a). A. Objective.
B. Slide.
C. Stationary part.
D. Moving part.
E. Sliding mount, usually lever actuated.
F. Position of D for focus 0^1.

or tissue culture preparation or any made-up slide. Fig. 17 (a) and (b).

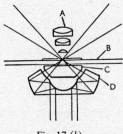

Fig. 17 (b).

Messrs R. & J. Beck have pointed out that this condenser will, when used with a slide of minimum thickness (about $\frac{1}{2}$ mm.), produce rays of sufficient obliquity to illuminate, upon a dark ground, objects studied with immersion lenses working up to $1 \cdot 2$ NA. With the top component removed a good, dry, dark ground illuminator is left which will work up to objective NA of $0 \cdot 7$.

Dark Ground Illuminators for Use with Oil Immersion Objectives of NA 1·0 and above

All the above-mentioned dark ground condensers are intended to illuminate objectives up to NA 0·95, though some of the condensers may be made to give slightly better performances in special circumstances. The dry, high quality apochromatic objectives with correction collars should be used for this work. Immersion objectives must be stopped down but they do not lose all their advantages because of this as an immersion objective usually gives a better image for the same aperture than a dry one, and disposes of the need for tube length adjustment. (It must be noted that a correction collar on an oil immersion objective is present so that an odd length microscope such as one with a binocular body can be accommodated. Once set up, no further change of the collar is necessary while it is used on that particularly microscope.) See 'General Microscopy: Tube length adjustment', page 61.

An illuminator was designed by E. M. Nelson to provide light of very high angle, so that specimens mounted in media of higher refractive index than 1·5 can be illuminated at angles corresponding to NA 1·4 and above to the maximum possible of about NA 1·5. If an object is mounted in water it cannot accept light at an angle greater than the equivalent of NA 1·3, but for objects mounted in media of the refractive index of balsam, cedar oil, styrax and realgar an objective of less than 1·4 NA, say in practice 1·35 NA, can be illuminated by dark ground. If the specimens are in water the objective aperture must be reduced to about 1·2 NA because the water medium can accept from the condenser only rays below NA 1·3. A small amount of aperture must be left between the NA applied by the condenser and the NA accepted by the objective for if this is not done the field of view cannot be dark because illuminating rays enter the objective. The smaller this difference of numerical apertures, the lower the level of illumination. In practice, it is best to mount an adjustable diaphragm above the objective and set up the system on test. In this way a balance can be obtained between the advantage of high objective NA and the disadvantage of a small amount of light. Bacteria usually show best with more light and lower objective NA whereas diatom resolution demands high NA in both condenser and objective. Bacteria have nearly the same refractive index and density as water, while a diatom is dense and can be seen in much less light.

The Nelson Cassegrain Condenser

The name 'Cassegrain' has been used to distinguish this condenser from others because its design is in many ways similar to that of the Cassegrain telescope. Fig. 18.

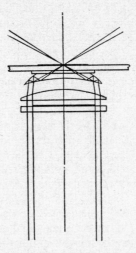

Fig. 18. Diagram of rays through Cassegrain condenser.

Owing to the great obliquity of light produced, a larger source than usual is desirable. The instrument does not produce an image of the source in the ordinary way but the concentration of light is very good. The best source of light is a bright diffuse one such as that provided by a 100 watt pearl electric bulb or any similar bright source like those used by photographers for flash purposes.

As with other dark ground immersion condensers, slide thickness and centration are critical and the set up procedure is the same as that given for the immersion paraboloid.

The Cylinder Dark Ground Illuminator

This device is intended mainly to illuminate evenly a large field for, say, a 80 mm. objective when large objects are studied. It consists as shown in Fig. 19 of a hollow cylinder of brass or other bright metal about 40 mm. diameter and about 50 mm. long with its inside surface well polished. A small bunched filament lamp is mounted on a sliding block as shown and reflections from the inside of the tube join and produce a bright region above the end of the tube. The position of the spot can be adjusted between small limits by sliding the lamp block up and down.

A stop must be provided to prevent light proceeding direct up the

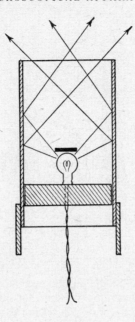

Fig. 19.

axis of the cylinder. No special finish is required inside the cylinder but a high polish is desirable.

A high power form of the cylinder illuminator[1] has been made in which the lower end of the cylinder has inserted, point towards the stage, a polished Perspex cone. Internal reflection in this cone causes light to impinge upon the inside of the cylinder at a well defined angle Fig. 20 and illumination of NA 0·86 is possible.

Perspex optical parts of sufficient accuracy for illuminators may be turned with ordinary lathe tools, surfaced with emery paper and polished with metal polish. Perspex may be welded with chloroform but an optically clear joint will not be obtained.

The Practical Use of Dark Ground Illumination
In the sections describing the construction of dark ground condensers the type of image they produce has been touched upon, but it may well be asked what the advantages are in the technical field.

When an object is dense compared with its surrounding medium it can be seen easily in that medium whatever the type of illumination. No one has difficulty in seeing a book on a table in air, but they may

[1] W. G. Hartley *J. R. micr. Soc.* Series III, Vol. LXX p. 282.

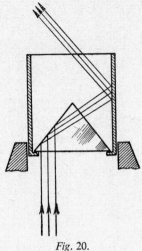

Fig. 20.

have difficulty in seeing a glass plate in a washing-up bowl in water. The plate in water illustrates the fact that transparency and density are not related. If the microscopist wishes to study the plate in water he must either stain it so that it becomes visible in ordinary light or he must apply light to it obliquely in such a way that any tiny differences in refractive index between it and the water, cause the maximum disturbance of the light, and this is best done by making the light so oblique that none enters the objective or eye as the case may be, as shown in the introduction to this section. The surface of the objects and a few optically dense spots below the surface will reflect a small proportion of the incident light out of the medium into the microscope and any other small or large optically dense particles in the preparation will also reflect light in proportion to their size and optical densities. The resultant microscope image is for this reason usually much obscured by stray light, and mainly the surfaces of the objects are visible.

The dark ground condenser goes some way towards turning the ordinary microscope into an ultra one. In the section 'Instruments and Use of the Eyes' it was shown that a particle must be about half the wavelength of light in diameter before it can disturb light sufficiently for the diffraction patterns resulting to form a true image of the shape of the particle (see 'Ultra Microscopy', page 479). But a particle, however small, can be made to reflect some light even though an image cannot be obtained: an example is provided by dust in a sunbeam, much of which is in size below optical resolution.

All the small bright sparks of light from objects in the mount rendered detectable by the ultra microscope action of the dark ground illumination tend to illuminate the dark field though the shapes of the causal particles cannot be resolved. As the particles turn in the liquid by Brownian movement and present different shaped surfaces to the light, the sparks of light twinkle and may cause confusing effects. The safe rule for sorting out these appearances is to know the resolving power of the microscope and so understand what size in the field the minimum resolvable particle must be. The shapes of all smaller ones can then be disregarded as true images. (See 'General Microscopy', page 31.)

Although little can be learned about gross structures by dark ground methods (owing to surface illumination), much can be learned about the presence of small external detail and of tiny objects. Under low powers the movement of flagella and particles in liquids cannot be better shown than by dark ground methods, while in the high power field objects like bacteria blood cells and diatoms can be well shown in the living state. The form of bacteria can be seen but it is unsafe to try to recognize species, though the immersion dark ground condensers lead to the discovery of spirocheats which are extremely small bacteria which could not at the time be stained, a particular example being that of syphilis. It is always wise to examine unknown fluids or extracts by high power dark ground methods in order to make sure as far as is possible that nothing is being missed by the staining techniques.

Cleaning Glassware

When high power dark ground systems of any kind are in use every blemish and particle of dirt on the slide, cover glass and in the oil is rendered visible. Old slides and covers are rarely clean enough to use and it is recommended that a stock of new slides be obtained. In general the slides provided by the makers, wrapped in paper have surfaces free from scratches and imperfections, but the same cannot be said of cover glasses. It is necessary to make several test preparations with covers taken at random from a new box in order to be reasonably sure that the batch is good. Inexperienced workers are recommended to try making a preparation of plain water with a cover glass which has been touched with the finger, for once this kind of dirt has been seen, errors are much less likely in future work.

Slides are best cleaned in a 10 per cent solution of hydrochloric acid in which they should be left to soak in a rack to keep them apart from each other for an hour or two. They should then be washed in tap water and finally in distilled water. The best drying material is an often washed cotton handkerchief as leathers become greasy, wool is naturally so and silk is non-absorbent. After polishing, slides and covers should be held in the focussed beam from a high power microscope lamp when dirt and blemishes will become clearly apparent. It is almost certain that a static charge of electricity will be generated on the glass as a result of the drying and polishing process, but when a cotton cloth is used, this is reduced to a minimum. After the first vigorous cleaning the slide may be left for a few minutes to allow the charge to leak away, after which a finishing light wipe will usually clean it without generating another high charge. The static charge upon any material may be removed entirely in a fraction of a minute by exposing the material to radioactivity from, say, a 50 millicurie source of beta several millimetres away.[1] This method is used to remove the charge from paper-making machinery and textile looms (which collect dust) but is unlikely to be necessary during ordinary microscopical work.

Cover glasses, owing to their thinness are much more difficult to clean than slides and much laboratory work is done with oil immersion objectives so that the need for cover glasses is removed. Much, however, must be done with them, and though experienced workers can clean covers between thumb and forefinger, the method is bad and leads to many breakages. The following old fashioned way is still best. A flat piece of wood, about 150 by 50 mm., is covered with a piece of chamois leather, pinned down with drawing pins around the sides of the block, and a similar piece of wood has a pad made of a cotton handkerchief backed by a piece of dry felt (not carpet underfelt which is oily). Selected covers are laid upon the leather where they will stick while being rubbed with the cotton. They are then turned over and the other side polished. It is often useful to charge the cloths with optician's rouge to improve the polish of the covers, but this should never be necessary with commercial products of the present day. After this main cleaning process they may be wiped with a clean cotton cloth between the fingers and to do this the cover should be held between thumb and forefinger by its edges; a cotton cloth, which must be as flexible as a handkerchief, is held between the thumb and forefinger of the other hand, allowing a bight of cloth of, say, 50 mm. The cover is then cleaned by moving the cloth with moderate pressure diametrically across the cover, finger and thumb moving together, the cover being rotated occasionally. On no account should an attempt be made to clean a cover by moving a finger and thumb in opposite directions with the cover between them. This attempt will either break the cover or

[1] Specialist advice about the handling of such a source must be obtained.

leave marks adherent to the surface where the cloth was stationary on one face. The hands must be washed before attempting to clean lenses and slides for serious work, or grease from the pores will pass through the cloth. The modern commercial detergents are quite satisfactory as cleaners for slides.

A slide may be tested for cleanliness sufficient for microscopical work by dropping one drop of distilled water upon it when, if clean, the drop should spread evenly up to a diameter of about 25 mm. It has been said that when the slide is perfectly clean, the breath will not condense upon it but the Author has not been able to clean a slide in such a way that this effect can be shown where no static charge is present. When a slide has been well rubbed in a dry atmosphere breath often will not condense, though the slide be visibly dirty and cool, but it will do so when the slide is discharged. The water drop test is quite satisfactory and is recommended.

All lenses and prisms in an optical system must be clean or sooner or later an obscure, poor result will occur. When one is cleaning laboratory glassware violent chemical methods are used and it is perhaps unnecessary to say that lenses and their mountings must not be cleaned with chemicals. Most lens components are still cemented together and fixed into their mounts with Canada balsam and this is readily attacked by alcohol and zylol. In general, experience has shown that saliva used with a perfectly clean, often washed, cotton handkerchief, makes the best lens cleaner after the dust has been swept off. Lenses become greasy due to condensation upon their surfaces, they also become coated with dust due to ordinary exposure and both these contaminants must be removed with care and not just rubbed off in housewife fashion. The hands must first be washed, then a really freshly washed handkerchief is taken in the hand or is used to cover a finger, depending upon the size of the lens, is moistened with saliva and used to sweep across the surface of the lens carrying before it any dust which was upon the surface. The cloth over the finger is now changed for a fresh part and the lens polished in the ordinary way. A lens must not be rubbed with any cloth until the abrasive dust upon its surface has been swept away as above, for should this dust be trapped under the cleaning cloth, the surface of the lens will be scratched, and though there may be no scratching visible to the naked eye, minute marks will be visible when light passes through the lens and an accumulation of these causes fogging of the image. If a lens is really dirty, as opposed to being in need of polishing, and if this lens may be removed from its mount, it is quite safe to wash it in soap and water or with a detergent.

Lenses in some positions, an example of which is an objective back lens, cannot be properly cleaned without dismantling the component and this should not be done except by an expert person. Such a lens may be improved by sweeping its surface with an artist's camel hair brush, but it must not be cleaned by screwing a cloth down the back of the mount because this will only cause circular markings which are much worse in their effect on an image than undisturbed grease or dust. When a lens has been left with its back exposed for many days or weeks it must be dismantled for cleaning. (See 'Objectives', page 54.)

Cotton cloth has been found suitable for a cleaner because it can be washed easily, is very flexible, does not shed fibres and will absorb grease but it will not clean a lens if it has been much handled before use.

To check a lens for clarity it should be held in the parallel beam of light from the high power microscope lamp when all grease and specks which mar a dark ground or vertical illumination system will then be clearly seen by the light such contaminants scatter. In ordinary optical work more harm is done to an image by grease upon the lens than is done by dust, the reason being that grease is in optical contact with the refracting face while dust normally only stops or scatters some light, but in dark ground and vertically illuminated microscopical systems equal trouble is caused by dust because the scattering of light destroys contrast in the image and so obscures very fine detail.

No lens or plate of glass can be entirely free from bright spots when tested in this way but a test should be made with, say, the field lens of a good modern eyepiece. The result will then be a standard of reference against which other pieces of apparatus can be tested. The results are always surprising because so much dirt and so many imperfections become visible, especially in ordinary laboratory apparatus.

Immersion Liquids

The medium connecting the cover glass with the objective must always be either cedar wood oil or one of the compositions made up to have the same refractive index and dispersion, but the liquid used to connect the slide to the illuminator is not so important. A dark ground condenser usually has a top about 12 mm. in diameter and when this top is connected to a slide by a film of oil of capillary thickness, the resistance to movement of the slide is great, also no focal adjustment of the condenser by means of the substage is possible, but if a slide is chosen of such a thickness that half a milimeter or thereabouts of working distance is left, a large amount, several drops, of oil are required to connect it with the slide. This quantity has a tendency to run over the edge of the condenser and so disconnect a region of the illuminator. When the microscope is inclined, a similar thing happens more quickly. These undesirable happenings can be prevented by the use of golden syrup as the condenser immersion medium, as described above, but it must be fresh and contain no sugar crystals to scatter light. There are other advantages in golden syrup too: oil often gets transferred by the underside of the slide to the surface of the stage (see 'Stages', page 144) and to the substage condenser-changer slides. Golden syrup can be removed very easily with water and does no harm if it dries hard.

The ordinary oil bottle with a dip stick is good enough as a container for occasional immersion work but one is liable to transfer bubbles to the optical path. Immersion oils (and golden syrup) are best kept in an ordinary cycle oil can of the press-side, plastic construction with its spout fitted with a flexible synthetic tube such as a cycle valve tube or one of the many synthetic tubes of diameters ranging from 1 to 25 mm. available under the name 'Systoflex', in the electrical industry. The transparent types are inert to almost all materials and are of great use in laboratory apparatus. Immersion oil must be wiped off lenses after use without fail as there is no really safe liquid which can be used in quantity to dissolve hardened oil off a lens without danger of penetrating the mounting. Cedar oil will dry hard in about two weeks and should this have happened, the safest way to remove it is to dissolve it in a larger quantity of new, fluid cedar oil. When the whole of the oil has been wiped from a lens a handkerchief moistened with saliva will remove the remaining traces and so prevent the surface collecting dust.

Miscellaneous Apparatus Associated with Dark Ground Illumination
It is possible, by inserting coloured filters in place of the opaque stops in dark ground systems, to arrange the field to be one colour and the object illuminated by oblique rays, another. This arrangement is known as Rheinberg Differential Illumination after Julius Rheinberg who invented it. If a mixture of dark and light ground illumination is used, the diffraction patterns produced by the objective, due to each type of illumination, will mix and spoil the image, but in Rheinberg's system different colours are used and mixing is prevented. Fig. 21. No special advantages over other systems have been claimed for that of Rheinberg, but the results are beautiful in appearance. The sizes of the disks are arranged by trial for the condenser and objective in use and are usually made of coloured cellophane, as sold for Christmas decorations.

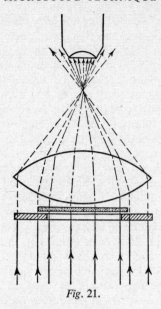

Fig. 21.

To set up the system, proceed as follows:

(1) Use a test specimen, say a diatom, and set up the microscope in the usual way for transmitted light.

(2) Remove the eyepiece and adjust the substage diaphragm so that it just cuts the edge of the objective back lens when viewed down the tube. The condenser is then working at the same NA as the objective.

(3) Select the darker colours for the central patch, say, blue, and cut a disk slightly larger than the substage diaphragm aperture.

(4) Mount this disk in the centre of a disk of clear glass which will fit the substage filter or stop carrier.

(5) Complete the coverage of the clear disk with an annulus of the other colour, say red, and place the composite disk in the substage stop carrier.

(6) Open the substage diaphragm wide, and light from the outer zones of the condenser, passing from the red area of the disk, floods the object without entering the objective, while blue light from the central area forms the blue background, as for ordinary transmitted light with a blue filter.

Variable Intensity Dark Ground Illumination
A similar system to that described above may be set up with polaroid

sheet. The centre zone is made of polaroid with its axis in one direction, and the outer zone is arranged to have its axis at 90° to it. A plain polaroid analyser is used in or above the eyepiece. When the analyser is not present ordinary light field conditions exist. When it is present it may be rotated to select light from either the central or marginal zones of the illuminator. If it is crossed with the central zone, dark ground conditions exist; if crossed with the outer zone, transmitted light conditions are obtained. All intermediate positions can be used and interference between the diffraction patterns at the back of the objective is avoided because the light from the two zones is oppositely polarized, but if the object is birefringent, the arrangement cannot work and allow straightforward interpretation of the images. Allowance must be made (i) for extra light due to birefringency of the specimen (some light from the central zone will pass into the analyser though this be crossed, when the specimen is birefringent), (ii) for directional effects due to birefringency in the object, (iii) for some polarization from high angle lens surfaces. These matters must be examined on test but it is possible to make suitable objects more clearly visible by this method and it has been useful when searching a mass of material for certain particles as optimum illumination for the wanted objects may be set up at the start and the search proceeded with.

It should be noted that no claim has been made that the Rheinberg and Polaroid systems improve resolution, but they do, under certain circumstances, improve contrast. (See 'Miscellaneous Apparatus' where the polaroid system is described in more detail under 'Mullinger's Illumination', page 508.)

The Trough Illuminator (*Fig.* 22)
This consists of a Perspex or glass trough built about 50 mm. long by 25 mm. high and 6 mm. thick. The usual structure being a plate of 6 mm. glass with a U-shaped piece ground out and microscope slides cemented on so as to cover the sides of the U.

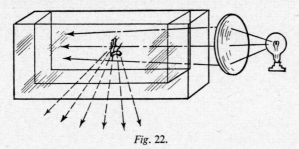

Fig. 22.

Light is sent in at the polished ends of the trough and objects, like animals in water, are well illuminated for low power study. With care, only one stratum of the tank can be illuminated at a time, thus removing a great amount of the scattered light encountered when conventional dark ground illuminators are used with thick tanks. The trough may be made of Perspex welded with chloroform and will then withstand sea water for many weeks.

Glass tanks are best built up from semicircular holes about 50 mm. diameter in 6 mm. glass plates with slides cemented on with Araldite. The holes are trepanned out of a sheet, about 80 by 80 mm., and then separated into two equal parts. Araldite made by Aero Research Ltd, Cambridge, is the best adhesive and will resist sea water and any other liquid, except hot acetic acid, indefinitely. (See 'Miscellaneous Apparatus: Henderson's Trough', page 513.)

For use on a microscope with a horizontal stage, tanks may be flat and open at the top. This kind of tank is to be preferred because more air can get to its occupants and needles can be inserted easily; small Petri dishes are very good. (See page 165.)

CHAPTER 20

To Line up a Microscope
For Critical Work

REGARDLESS OF what a research microscope and its equipment looks like, it must have the following facilities:

(1) A coarse and fine adjustment to the body tube.

(2) A draw tube, preferably operated by rack and pinion.

(3) Objective changers of the centring type.

(4) A rotating mechanical stage.

(5) A centring focussing substage with iris diaphragm, constructed with sufficient travel on the rack and pinion for the changer slides to clear the stage when changing condensers (6).

(7) A plane mirror.

(8) A Köhler illuminating system as described under 'Microscope Lamps'.

(9) A camera as described under 'Photomicrography using a Microscope'.

The reasons for these demands are as follows, only some points require comment:

(2) A draw tube is required to adjust the tube length of the microscope to suit objective-specimen conditions (see pages 61 and 48).

(3) It is necessary to mount all objectives in a battery so that when they are interchanged on the microscope the optical axis of each falls upon the optical axis of the microscope. Also all objectives must be on the same axis as the eye-piece and they must be on the axis of rotation of the stage.

(5) The iris diaphragm which controls the condenser aperture must be mounted solidly in the substage and be in the optical axis of the microscope. The substage need have only pre-set centring screws.

(6) All condensers must be mounted on changer slides so that each may be centred to the axis of the microscope, i.e. to the substage diaphragm, which is also in the axis of rotation of the stage. The iris diaphragm must be so placed in the substage

that it comes immediately below the condenser back lens when the condenser is in place. (See 'Microscopical Apparatus for the Study of Transparent Objects: Position of Substage Diaphragm', page 345.)

(7) For illuminating low powers, a low power condenser and plane mirror should be used. The concave mirror produces a poor image and has no place in accurate microscopy.

(8) and (9) The matters are fully considered in the respective sections.

To set up the above conditions, proceed as follows:

(1) Arrange the lamp and microscope in the positions desired. It is usually best to work with the lamp on the left of the microscope as it can then be reached easily and the hands, when operating the microscope, do not get in the way of the illumination. It is best to use the mirror rather than the lamp direct because this provides useful flexibility when adjusting the light.

(2) Clear all lens systems off the microscope, insert a piece of dark glass somewhere between the lamp and eye, and adjust the lamp, lamp condenser and microscope mirror so that a clear round disk of light (the lamp condenser) is seen through the empty tube of the microscope.

If the lamp has not been previously adjusted, its condenser must be arranged to focus an image of the source at the same optical distance away as the microscope condenser. It should then be inspected through dark glass for centration (see 'Microscope Lamps', page 115). The axis of the illuminating system and the mechanical axis of the microscope now coincide approximately and form a starting point.

(3) Place a mounted object, small and easily seen, upon the stage, and a 12 mm. objective on the microscope *nosepiece*, not on any changers, and focus the specimen. It does not matter at this stage what the conditions of illumination are, it is necessary only to see the object through the microscope.

(4) It may now be assumed that objective and eyepiece are on the axis, because mechanical errors in making the threads on a microscope tube are rare. Rotate the stage and it will probably be found that it does not rotate accurately upon the axis determined by (3) above. This is not the fault of the manufacturer but is due to the fact that no two objectives have exactly the same optic axis position with respect to their mechanical parts. The difference is not sufficiently great to

introduce errors in the eyepiece-objective axis relationship, but the error causes objects to fail to appear in the field of view when a low power objective is changed for a high one.

Having chosen an axis determined by a particular medium power (4 mm.) objective and eyepiece, adjust the rotating stage to rotate accurately in this axis as nearly as can be determined with the objective in use. If possible, clamp the stage centring screws in this position. Clamp the object also for future reference. Eyepiece objective and stage axes are now nearly coincident.

(5) Adjust the centring screws of the substage so that the iris diaphragm appears to be in the axis. It is sometimes possible to focus its smallest aperture with the objective directly if the body tube can be lowered sufficiently. If this can be done, centring the substage is easy and direct, but if it cannot be reached, centre it as nearly as possible by removing eyepiece and objective, looking through the tube towards the substage and adjusting the iris position until it appears concentric with the nosepiece aperture. The iris needs to be opened to about half its maximum aperture for this experiment. Owing to its construction an iris diaphragm can be trusted to remain concentric while its aperture is varied.

Select a condenser change slide and set its centring screws so that the condenser mount appears to be centrally placed with respect to the diaphragm. Select an ordinary achromatic condenser and place an ink spot in the centre of its top lens. This may be done in a lathe or on a mounter's turntable. Mount the condenser on the changer, rack it up towards the objective and attempt to focus the ink spot. If it cannot be found, move the condenser by the centring adjustment of the changer until it is found.

This procedure ensures that objective, eyepiece, stage and condenser are on a common axis.

(6) Close the substage diaphragm to a pin-hole, lower the condenser and look through the microscope for the aerial image formed by the condenser of the pin-hole. It will probably not be found immediately because the substage iris may not be exactly in the axis of the condenser and the rest of the system. To find the image of the pin-hole, operate the substage centring screws until it is found.

(7) It will now be discovered that the ink spot on the condenser is no longer in the centre of the field. Operate both the changer centring screws and the substage centring screws

until both the ink spot on the condenser and the image of the iris diaphragm are on the axis of the microscope. If possible, clamp the substage centring screws in this position.

(8) The exact axial positioning of the mirror is unimportant because it has a plane surface.

(9) All the parts of the microscope are now on the same axis and other components one at a time may be centred upon their respective changer slides to match the set-up on the particular microscope in question.

(10) Mount the remaining objectives on their changers and substitute them one at a time for the objective used in the lining-up procedure. Set up each objective by means of its changer centring adjustments so that the image of the object appears in the centre of the field of view of the microscope and also on the axis of rotation of the stage. If the mechanical parts of the instrument are as good as they should be, a 2 mm. objective so set up should return to place sufficiently accurately for the image of the object to appear constantly within about one quarter of a field diameter when the objective is removed and replaced. Of course, lower powers will easily satisfy this requirement.

A rotating stage should be sufficiently good to ensure that an object does not leave the field of view of a 2 mm. objective when it is rotated through 360°.

(11) A similar procedure should be adopted with the condensers. It is best to place a spot of ink on the top lens of each condenser and adjust the changer centring screws until this is central in the objective field. Dark ground condensers must be done this way and are often marked by the manufacturer.

Having set the iris diaphragm on to the optic axis of the instrument (5) and (6), its aerial image formed by any substage condenser (except dark ground ones) may be used for centring purposes.

Place the condenser to be centred on its changer slide, focus with the microscope the image of the iris diaphragm closed to a pin-hole, and adjust the centring screws of the changer until the image of the pin-hole is in the centre of the field.

When this work has been done on the equipment of a particular microscope, it need never be altered again unless someone adjusts the substage or stage centring screws instead of the changer screws when fitting a new condenser or objective.

When a new objective is about to be fitted to its changer, care must be taken to check that the centre of rotation of the stage still coincides with the axis of the existing objectives. If it does not, the whole line-up of the microscope should be checked in case adjustments have been altered. Centration of the stage does not effect in any way the optical performance of the microscope, but if not correct is a great nuisance when polariscopic work is being done.

Quick Check of Centration of Components

Set up the microscope to study a transparent object (see chapter 'The Study of Transparent Objects', page 333). Remove the eyepiece and look at the back lens of the objective. When the substage iris opening is altered in size its image will be seen in the objective back lens and this image should be exactly concentric with the objective back lens mount. If it is not, either:

(1) the condenser is not centred to the microscope axis;
(2) the objective is off the axis, probably due to its changer centring screws having been altered;
(3) the substage iris is off axis due to the substage centring screws having been altered, or a concave well slide used.

Condition (2) may be checked by observing whether or not the rotating stage is on axis. If it is, then the objective is correct and the stubstage is in error. If it is incorrect it is best to perform the whole lining-up procedure.

If the substage is in error, sections 5, 6 and 7 of the lining-up procedure must be repeated. It is quite likely that adjustment of the centring screws to the substage only require readjustment, as it is unusual for the centring screws on the condenser changer to be altered, but someone may have substituted a different condenser, in which case alteration of the substage will throw out of adjustment all other components.

There is much to be said for not having milled head controls to the substage centring arrangements. Pre-set, locking adjustments are better because they do not become deranged accidentally.

Lining-up the Lamp with the Microscope

Most of this has been covered in the section 'Microscope Lamps' but is repeated here in the interests of completeness.

Having lined-up the microscope, a dark filter should be placed in front of the lamp to protect the eyes, and the lamp beam directed at the microscope mirror.

With the microscope set up with a substage condenser, 25 mm. objective and a simple object like a diatom on the stage, adjust the mirror until a full beam of light passes into the microscope. Focus the light by means of the substage condenser on to the object so that an image of the lamp condenser is seen superimposed upon the object.

Adjust the position of the lamp and the focus of its condenser until a perfectly even disk of light is seen as the image of the lamp condenser; it is against this disk of light that the microscope image is seen. The disk is varied in size by means of the lamp diaphragm and the power of the substage condenser. A low power substage condenser produces the largest image of the lamp. It is unwise in the interests of accurate microscopy to alter the apparent size of the illuminant by altering the focus of the substage condenser.

The microscope is now ready to use. A substage condenser is chosen which will illuminate with the image of the lamp condenser, a substantial part of the field of view of the objective in use; a typical eyepiece is a X8. A larger amount than about three-quarters of the field should not be illuminated except in projection work, and the area should be regulated by means of the lamp condenser diaphragm.

To adjust the working aperture of the system, focus the microscope, remove the eyepiece, observe the image of the substage diaphragm in the objective back lens and adjust the substage diaphragm so that about three-quarters of the back lens diameter is filled with a solid pencil of light. Adjust the tube length of the microscope as described in the section 'To Set Up a Microscope to the Correct Tube Length'. Adjust the brightness and colour of the light according to the methods described under 'Microscope Lamps' and 'Colour Filters'.

When a microscope is to be set up for use with dark ground illumination or vertical illumination, the lining-up procedure described above is exactly the same in the first instance. When the various condensers and illuminators are substituted for the transmitted light apparatus, reference should be made to the sections 'Dark Ground Illumination', 'The Study of Surfaces' and 'To set up a Simple Vertical Illuminator'. Pages 61, 115, 131, 202, 220.

If the microscope is a binocular instrument of the high power type (see 'Binocular Microscopes'), line up exactly as above but use one tube only. If the binocular is of the breeches tube kind, remove the prism and line up on the direct image tube, and it should be noted that the correct working of the Wenham binocular is entirely dependent upon correct line up of the illumination, the image of the light source must be correctly focussed upon the specimen or much stray light will occur in the tubes, also the objective must be handling a three-quarter pencil of light or uneven illumination of the field and lack of stereoscopic effects will result. In the Powell and Lealand

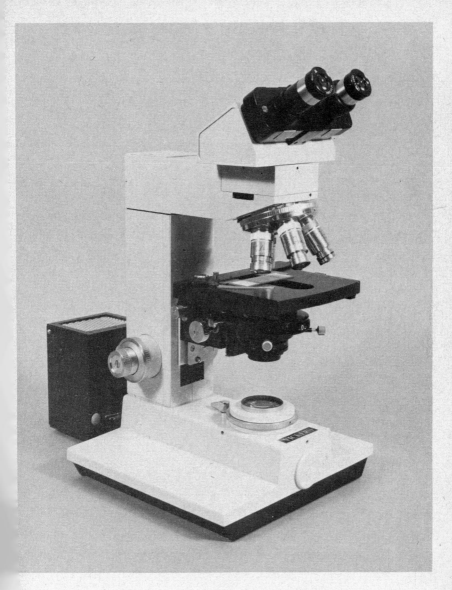

PLATE 9
A Commercial High Power Binocular Microscope
by Messrs. R. and J. Beck Ltd.

instrument, the insertion of the prisms displaces the illumination to one side and this amount must be corrected by altering the position of the whole substage until the previously lined-up substage iris diaphragm again becomes concentric with the objective. This effect is a nuisance with this kind of binocular and it demands milled heads instead of pre-set controls to the substage as a whole.

There is no particular lining-up procedure with the low power Greenough forms of instrument because these have no substage.

The Stephenson instrument works best without a substage, and if required for transparent specimens at low power, a sheet of opal glass, illuminated from below, will give the best results. However, a condenser may be used with a diaphragm below it perforated with two holes about half an inch diameter side by side, so placed to favour a beam of light entering each tube along its axis. The arrangement is not satisfactory and the condenser working aperture cannot be controlled.

CHAPTER 21

Binocular Microscopes

I N THE chapter 'Instruments and Use of the Eyes', matters governing the choice between parallel and diverging tube binocular microscopes were discussed. In the following sections, particular types of instrument will be examined, together with their uses.

In Fig. 1 is shown a group of sketches gathered by Mr F. C. Wise[1] of all the manufactured binocular systems which have been applied to the microscope. Only a few of these are still used because the process for depositing a thin semi-transparent reflecting film of metal like rhodium or aluminium upon the face of a prism to make it divide light evenly between two paths has made a great difference to binocular design and has made many old useful systems mere curiosities.

The Principle of the Binocular Microscope
When a pencil of light leaves the objective of a microscope it normally passes through the eyepiece into the observer's eye. If the pencil is divided by an optical device, half of the pencil can be made to enter each eye; but the division may be made in two quite different ways. The diameter of the objective pencil may be used as the dividing line so literally halving the aperture and sending half into each eye, or the pencil may be divided by reflecting off a portion of the light from the whole aperture so that each eye receives half the total amount of light but receives it from the whole of the objective aperture.

From the section 'Numerical Aperture' it will be seen that resolution will be reduced if the objective aperture is reduced; it follows

[1] Journal of Quekett Micr. Club. Series 4, Vol. VII, No. 6, p. 305.

Fig. 1. (1) Wenham high power.
      (2) Powell and Lealand.
      (3) Stero attachment Reichert.
      (4) Leitz.
      (5) Zeiss.
      (6) Zeiss. Prisms (A) rotate for interocular adjustment.
      (7) Stephenson.
      (8) (*a*) Porro prism. (*b*) Later Porro. (*c*) Roof prism erector.
      (9) Inclined eyepiece prisms.
   (10) Reichert high power.

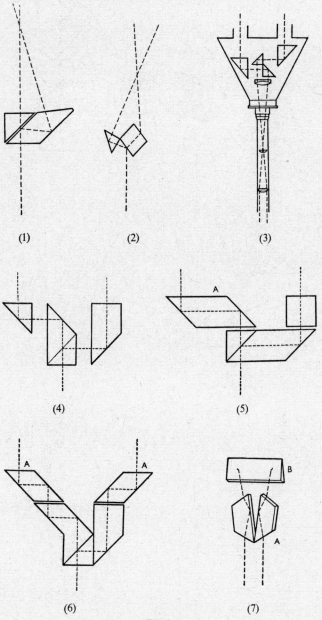

(1) (2) (3)

(4) (5)

(6) (7)

Fig. 1.

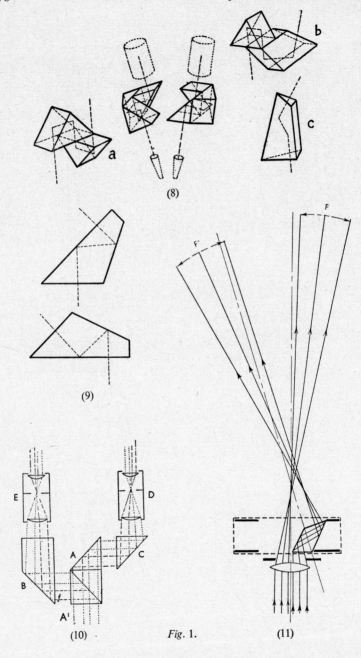

(8)

(9)

(10)

Fig. 1.

(11)

therefore, that binoculars for use with high powers must be of the type where light from the whole of the objective aperture reaches both eyes. This type is easy to recognize by setting up the microscope, removing one eyepiece and looking at the objective back lens. The 'whole aperture' system will reveal the whole of the back lens.

There are other important optical differences. If a diametrically divided pencil has its right half sent to the left eye and vice versa, the observer sees a stereoscopic image and the simplest and best way of doing this is by means of the Wenham prism, diagram 11 in Fig. 1. The drawing is self-explanatory, and it should be noted that the angle of divergence in this instrument is determined by the ground angles of the prism and no amount of tilting will alter it. The prism is always made for a particular microscope (or mass produced set) and they cannot be interchanged. The Wenham binocular passed out of use when long body microscopes were discontinued, as a tube length of about 300 mm. and divergence angle of less than 14° is necessary for comfort. Shorter tubes meant greater angles of divergence of the tubes, and this angle is the same as that of the convergence of the eyes when using the instrument, so, clearly, there is a limit to convergence. Some Wenham binoculars were made with shorter tubes during the transition period long to short tubes, which demanded large convergence of the eyes, but they were not satisfactory and passed out of use.

The Wenham binocular belongs to the days of the late nineteenth century, when microscopes were much larger than those of today, of 250 mm. tube length fitted with objectives which had large diameter lenses. If used on these stands with large (sometimes 30 mm. diameter) low power eyepieces there is produced a naturalist's microscope which has never been surpassed. The Wenham instrument is not made today and its place has been taken by the small twin-lens types described below. The Wenham form does not work well with objective powers greater than 12 mm. and numerical aperture about 0·6. The prism is best placed within 6 mm. of the top of the objective mount when all apparatus is on the microscope, and if the lenses are in short mounts so that the back lens is nearer the prism so much the better, as higher powers can then be used without distortion of perspective. The prism is always made to slide out of the axis when high power monocular vision is required.

Some space has been taken to describe this instrument mainly because it is the only well-known example of the diametrically divided pencil stereoscopic type. There are others, one of which is the Stevenson type (diagram 7, Fig. 1). Recently a single lens stereoscopic form has been revived very successfully by Zeiss and is described under 'Low Power Binoculars'.

The High Power Binocular

This name has come to be applied to the 'whole aperture' type of instrument because in this alone can high power objectives be used at full aperture. The high power form is not stereoscopic because both eyes see the same image from the objective, and high power forms of the instrument cannot be built up from twin tube microscopes because the short working distance of high power objectives prevents their coming close enough together to focus upon the specimen.

A basic form of high power binocular is shown in Fig. 1 (10). A number of variations on this pattern are possible but the basic design is the same.

The dividing prism A+A' is composed of two cemented parts A and A', the shape of which are clear from the diagram. The hypothenuse of A' is coated with a metal like aluminium by evaporating this on to the glass from a wire filament of the metal, heated electrically. The operation is performed in a high vacuum and an extremely high degree of cleanliness is demanded. The amount of metal deposited is such that the light, when attempting to pass through the assembled prism, divides equally between the eyepieces. When the metal is deposited before the prisms are cemented, a transmission of about 40 per cent of the incident light produces about the correct result when assembled. The metals deposited are chosen to give an image free from background colour. Prisms B and C require no treatment as they are used in the totally reflecting position.

Adjustment for interocular distance is performed mechanically by sliding the eyepiece tubes together with their respective prisms B or C to and from the objective axis of the instrument. The rays of light from the objectives are travelling horizontally between prisms AA' and B or C, therefore a change in length in this path does not alter the collimation of the instrument.

Interocular adjustment may also be made by inclining the eyepiece tubes, together with their respective prisms B and C, but this requires the prism to be turned through half the angle that the eyepiece tube moves. Both arrangements are used satisfactorily, but the parallel motion system is clearly the simpler. When prisms are used to transmit and divide image-forming pencils of light, the thickness of glass in each leg of the system must be the same. Glass has a refractive index of, say 1·5, therefore a piece of glass 25 mm. thick is optically equivalent to air 40 mm. thick. If, then, there is more glass in one leg than in the other, the images will focus at different levels above the objective. This is inconvenient for the eyepieces and correcting thicknesses of glass are always fitted in the tubes or are associated with the prism structure.

In the section below describing the Powel and Lealand instrument, a compensator is described which is arranged to 'raise' the image in one tube by half an inch to off-set the effect of glass in the other tube.

In the Wenham instrument, the path length difference is so small that a slight alteration of level of one eyepiece is adequate to correct the system, alternatively the field lens of the eyepiece in the direct tube may be lowered on its threads by about 2 mm. but no more or the magnification over the fields in the eyepieces will become different. An eyepiece form of binocular attachment is available to fit standard microscopes which varies little from the layout shown for the high power system. So far as its optical operation is concerned, the whole dividing prism assembly may be placed in any part of the optical path and the only inconveniences are mechanical ones. It will be noticed that the high power binocular does not erect the image.

To set Up a High Power Binocular Microscope

The microscope should be set up in exactly the same way as described in the section 'How to set up a Microscope'. The presence of the dividing prism does not alter the microscopical requirements at all but one tube only should be used while doing this.

The extra matters to be attended to are:

(1) The interocular distance must be correct for the observer. To check this, de-focus the objective so that the field is bright but contains no image; apply both eyes and adjust the interocular milled head until the eyes easily fuse the fields of the eyepieces together into one circle. (See 'Collimation Errors', page 390.) When the objective is again focussed, the eyes will be comfortable.

(2) The separate focal adjustments for each eye must be correct. To obtain this adjustment, the eyepiece which has the focus adjustment collar on it should be blacked out with a piece of card so that the eye above it may be kept open without seeing the image. The microscope should then be focussed accurately with the other eye comfortably in use. Transfer the card to this eyepiece and set the adjustable eyepiece to suit the other eye, i.e. to focus a clear picture without readjusting the micro-

scope focus. When both eyes are now applied to the microscope it will be found to be in adjustment. Having in this way decided the correct settings of the graduated eyepiece adjustment collar, the setting may be noted and it remains unchanged for any change of objective. For the greatest accuracy, the setting-up should be performed with a medium power objective because a high power system does not provide such a sharply defined image in the eyepieces as does a medium one.

(3) Some observers have difficulty in fusing the images in binoculars while many cannot use a converging type and others cannot use the parallel kind. But as all observers with normal eyes can see to a distance and all can read at 250 mm. distance, it follows that all observers can use both kinds of instruments if they remember that the parallel tube type requires the eyes to relax, i.e. to look at infinity, while the converging type requires the eyes slightly to look 'into' the instrument in the position of reading. If these points are remembered, no difficulty will be experienced. (See 'Instruments and Use of the Eyes', page 9.)

When a binocular instrument of any kind has been operated for several minutes, it should be possible to withdraw the eyes and use them immediately to look at ordinary things in the room. If any difficulty at all is experienced in seeing stereoscopically or in focussing objects, there is something wrong with the binocular even though it appears satisfactory in use.

There are several other high power binocular systems in existence which are shown in Fig. 1 in sketch form, and the working of all these should be clear from the sketches. No special points arise concerning any of them, except that of Powell and Lealand. Examples of this instrument may often be found in the hands of amateur naturalists, and though made at the earliest in 1860 and at the latest about 1910, the stands are still capable of performing every task demanded by modern microscopy. The binocular arrangement is one of those made obsolete by the metal evaporation technique but they may still be used quite successfully if a neutral tinted glass be placed in the through tube to equalize the light with that in the side tube: an alternative arrangement is to evaporate a 10 per cent absorption covering of rhodium on to the reflecting face of the dividing prism as such an amount increases the reflection of light from this face sufficiently to equalize the light in the tubes. A rhodium coating can be cleaned with a soft cloth.

Examination of the light-dividing apparatus will show that the optical path in the side tube is about 12 mm. longer than in the through tube and this means that equal eyepieces cannot be used successfully unless they have very long eye-points which allow the side tube one to be mounted about 12 mm. lower than the other without unduly inconveniencing the eyes. The difference in image size and focus is less noticeable when high powers are employed but the system still has this basic fault. It may be compensated by fitting a block of clear glass 40 mm. thick into the draw-tube of the side tube of the body; 40 mm. of glass is nearly equal in optical path length of 25 mm. of air, and compensation is thus obtained by apparently lifting the image in this tube by about 12 mm. without any change in magnification. Normal matched eyepieces can then be used, the only disadvantage remaining being the undesirable weight at the upper end of the microscope. (See page 174.)

For further details of a general nature concerning setting up microscopes, see 'To Set up a Microscope for Critical Work', page 369.

Twin Objective Binoculars

In recent years this form has become common for almost every laboratory and industrial purpose requiring magnification. It has the advantage of very good stereoscopic qualities, robust construction and easy mounting without the need for special substage apparatus. The maximum power is about 200 diameters, the numerical aperture is low in order to obtain great depth of focus, and an erecting system is built in and combined with the interocular adjustment. The form is typified by the Greenough binocular and this model is here described.

Fig. 2 shows the internal optical arrangement which is similar to the ordinary binocular telescope. Often the erecting prisms are placed at opposite ends of short tubes, but in most microscopes they are mounted upon each side of a stiff perforated metal plate in order to keep the microscope short. The objectives are specially constructed with very thin narrow mounts in order to place the lenses close together, allowing them to focus the same spot on the object, and this requirement limits the power of the instrument. The divergence angle of the beams from the objectives is about 15°, and a pair of objectives is mounted together on a changer slide at this angle and they are not again separated. The eyepieces are paired but are in others ways standard.

It will be seen from examination of the figure that the prism assembly may be rotated about the objective axis without displacing the image. Because the position of the prisms off-sets the eyepiece from the axis, the rotation of the assembly provides an easy means of interocular adjustment.

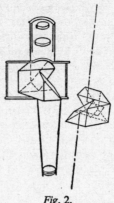

Fig. 2.

Operation of the Greenough Binocular

It is unnecessary to go through a procedure like that demanded by true microscopy in order to set up these instruments. They are usually illuminated by top light but may be used with transmitted light, in which case they are best illuminated by placing the whole instrument on a sheet of ground glass and placing an electric bulb beneath. Operations can then be carried out using the ground glass (ground side down) as a stage. Should this become too hot, light which has been reflected from a distant source by means of a common mirror should be applied to the ground glass. Daylight from a large window, with no condensing apparatus on the microscope, is by far the best illuminant for a low power stereoscopic system.

The Greenough bodies may be mounted upon any kind of extending arm provided that it is sufficiently stiff. For an extension of about 300 mm., the supporting arm must be about 25 mm. diameter steel to give sufficient rigidity. Some commercial forms of the mounting have inadequate hinge fittings for inclination at the end of the arm, thus wasting the rigidity of the arm. The most useful kind of mounting for industrial work is a plain round rod about 12 mm. diameter, about 80 mm. long, fitted to the block of the coarse adjustment slide. By this rod the main bodies and their coarse adjustment are secured to the arm with a standard laboratory clamp. The whole is thus fully adjustable, more rigid than hinges, and may be clamped to anything.

Much laboratory work which is done under a Greenough type binocular is performed on the bench with window light or top light, and some kind of steady stage is demanded for this work. For most purposes the microscope stand should be placed upon the same board or slab upon which the stage rests, for if this is not observed a displacement of the microscope stand with respect to the object takes place whenever it or the object is touched. It must be remembered that in these instruments the stage, which is really the bench, is not connected with the microscope body. A suitable slab is a thick drawing-board of about 300 by 600 mm. dimensions. It is easy to adjust the height of this to suit the worker by placing blocks under it. Dishes and similar large containers will slide satisfactorily upon the wood surface and a suitable sliding mount for small parts is made from a piece of plate glass about 12 mm. thick and 100 by 100 mm. in size upon which a piece of plasticine is placed. The job is stuck at any required angle into the plasticine and is moved about in a smooth and firm manner by sliding the glass block upon the wood. Guide rails or a T-square can be placed on the table as a guide for the glass block when searching, or for use when returning an object to the same place for study.

Disadvantages of the Greenough Form

Owing to the slanting direction at which the objectives point at the object, the field of view of a flat object is out of focus at the left- and right-hand sides; the focus is too high on the right and too low on the left for the left-hand tube, and vice versa for the right-hand tube (Fig. 3), the errors increasing with distance from the axis. This is no great disadvantage when irregular objects are studied, but workers should not be asked to operate these instruments for many hours at a time, particularly when flat surfaces are being examined, or the result will be bad headaches.

It is also bad practice to use wide field eyepieces on the Greenough form because these encourage the eyes to use the whole field of the instrument and exaggerate the focal errors described, in positions away from the field centre. The best eyepieces are Huyghenian X6 or X8, which give the best possible definition and limit the field to reasonable proportions.

While the Greenough is a most useful hack instrument, there are advantages in comfort to be had from the single objective type described below. The Greenough also has the advantage of comparatively low price.

Single Objective, Low Power Binoculars

The first important example of this type was the Stephenson, designed about the year 1870. The instrument was not often made, probably because the long tube microscope with the Wenham fittings was then much in use. During the last thirty years the Greenough type has eclipsed it, but recently Messrs Zeiss have reintroduced successfully

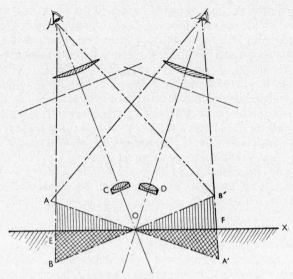

Fig. 3. A, A′.   Image plane of body D.
B, B¹.   Image plane of body C.
O.   Position at which both objectives are in focus on flat specimen (X).
X.   Flat specimen surface.
AOE.   Specimen below focus of D.
B′OF.   Specimen below focus of C.
BOE.   Specimen above focus of C.
A′OF.   Specimen above focus of D.

a model described below, in many ways similar to the Stephenson.

The Stephenson instrument is shown in Fig. 1 (7). It produces an erect image, is stereoscopic and does not suffer from tilted fields of view inseparable from twin objective types. It is a successful and comfortable instrument and deserves recognition.

The dividing prisms A are mounted so that they can be adjusted to suit exactly the angle of divergence of the tubes of the instrument; the angle of inclination is determined by the prism B which completes the inversion of the image and provides a small amount of inclination adjustment. Usually the tubes are fixed in inclination and divergence, and the two adjustments mentioned above are for the purpose of collimating the instrument. The stage is fixed horizontally and usually has attached the legs upon which the instrument stands. Any low and medium power objectives can be used but perspective becomes exaggerated with powers higher than half an inch. As with all low power binoculars, top light gives the best results.

This short description has been included mainly because of its historical value, as the Stephenson instrument is not now made but may sometimes be seen in dealers' shops.

The Zeiss Low Power Binocular

This is by far the best low power binocular instrument which the Author has used. It employs a single objective about 25 mm. in diameter of special design, being both aspherical in surface and of a shape which is not a solid of revolution. It is in effect two equal

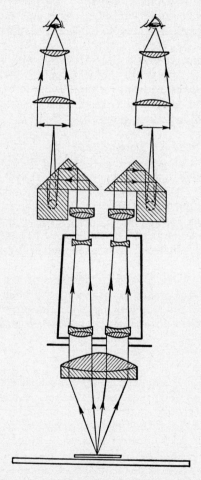

Fig. 4. Zeiss stereoscopic binocular, single objective type.

segments of aspherical lenses joined together for the purpose of obtaining the best possible correction for each of two pencils of light passing through at an angle of divergence from the axis. Diaphragms are used to isolate the pencils from each other. Only this objective is used for all powers but each half of the objective is followed by one of several small telescopes which presents an image to an ordinary eyepiece. The telescopes, one for each half of the objective, are ganged in a changer compartment and to change power the telescope mounts are rotated into position above the objective by internal mechanical arrangements. Inclination and erection are built-in and the stage is horizontal. In use, slight colour is to be seen at the outer edge of the field of view in each tube but this is not apparent when both eyes are used.

The instrument presents a clear flat image with fine definition up to the field edges, no tilt in the field, and no greatly exaggerated perspective. It is a real advance in this class of instrument and is one of the fruits of the development of aspherical surfaces. Unfortunately, its price is high, including import duty.

Collimation of Binocular Microscopes
When the details on page 381 'Setting up a Binocular Microscope' have been attended to, it often happens that the eyes are not comfortable and it may be that stereoscopic vision cannot be obtained. These faults are due to lack of collimation of the tubes of the instrument, and this is brought about by mechanical derangement of the lens and prism mountings. The faults must be put right, as even if present to a very small degree great fatigue will result from a single observation.

The usual error is one where the images do not lie on the same lines as those which pass from the objective through the centre of the eye pupils of the observer and in practice this error requires one eye to point upwards or downwards in order to accommodate the microscope. The fault can easily be seen as follows:

(1) Set up the microscope as described above and if the eyes cannot fuse the images, set up one tube at a time.
(2) Deliberately alter the focus of the eyes, i.e. if the binocular is of the parallel tube kind look at an imaginary closer object; in fact, look 'into' the microscope instead of 'through' it, or, should the instrument be a converging tube type, rest the eyes at infinity, i.e. look 'through' the instrument. When this eye focus change is made the images in the tubes will separate and be seen by the observer side by side and an error in level will

then be clearly apparent. If the images could not be fused in the first place they would be seen separately anyhow.

An error much less easy to find and recognize is that due to the angle of divergence of the light from the light-dividing prism not coinciding with the mechanical axis of the tubes (which determine the optical axis of the eyepieces). This error can be accommodated by the eyes fairly easily due to the fact that the eyes are meant to converge in natural use but it makes for poor microscopy because field tilt results. The fault may be detected as follows:

(3) Set up the microscope as described above, page 381.
(4) Use as an object a large flat section, say a plant stem or petrological section, and focus it accurately.
(5) Observe whether or not the eyepiece diaphragms combine to make a complete circle, but while doing this care must be taken not to alter the focus of the eyes. A little practice will teach the observer to see the diaphragm as a 'frame' of the microscope picture in the same way that an artist is conscious of the frame of painting. If the microscope picture is seen with both eyes in use in a diaphragm frame which is oval or figure-of-eight shaped, a collimation error exists somewhere in the system.

A general test of collimation of an instrument may be made as follows:

(6) Secure the services of an observer who does not need to wear spectacles.
(7) Rest the eyes outdoors by looking into the distance casually for about five minutes.
(8) Use the microscope for about five minutes.
(9) Look away from the microscope at objects upon the table; the eyes should see the objects without any sign of misalignment, focus difficulties or errors in stereoscopic perception. An observer's eyes often learn to tolerate instruments which are considerably misadjusted but the fact is revealed when the eyes are used again normally.

It is known to the Author that some observers have become so used to misadjusted binocular instruments of all kinds that they have learned to ignore the image in one eye and the tests above will not show up such a matter. The collimation of a microscope does not require alteration for different observers, therefore in laboratories the microscopes should be checked occasionally by an experienced observer with normal eyes.

Causes of Collimation Errors

In modern instruments these are always mechanical, but in old Wenham and rarely in Powell and Lealand forms the error is optical due to the prism faces being not square with the prism axes. Of mechanical errors the following have all been experienced by the Author:

(1) Prisms lifted from or moved in their seatings by a knock received upon the microscope body.

(2) The tubes bent or moved under their set screws due to observers struggling with stiff eyepieces.

(3) The interocular adjustment (of any kind) becoming slack.

(4) Detachable bodies failing to seat properly upon the nosepiece portion.

(5) Interchanging of pairs of objectives between Greenough-type stands.

(6) Eyepieces very loose in their tubes. (In binocular instruments they should be lightly spring-loaded by means of a slotted draw-tube.)

(7) Objectives tilted in their mountings or threads crossed. This applies mainly to binocular telescopes.

(8) Eyepiece lens cells improperly screwed into the tubes. Their fine threads are easily crossed especially when painted.

(9) On detachable-prism instruments, the prism belonging to a different example of the instrument being inserted.

To Collimate the Greenough Type of Instrument

The objective pairs are not normally interchangeable between instruments but in laboratories changes are often made and errors in collimation result. To correct these, proceed as follows:

(1) Check that the eyepieces do not wobble in their tubes and that the interocular adjustment is firm. If these matters require attention the work must be done first. If eyepieces are loose in their tubes it is usually due to the sprung draw-tube having become distorted. If it is not sprung and the usual eyepieces are in use, the draw-tube should be slit along its axis and bent inwards sufficiently to grip the eyepiece lightly. The effect of wobble in eyepieces can be observed by setting up the microscope and gently moving them while making an observation, and if the collimation is affected, they must be tightened.

If the interocular adjustment appears sloppy, it can be tested in the same way. As this adjustment varies with design of instrument, the nature of the repair is dictated by the mechanical design of the instrument. Some are an obvious dovetail slide design which may often be made serviceable by coating with thick grease pending a proper repair. Those of the Poro prism design (where the prism boxes rotate upon the bodies) are apt to become loose on their bearings and the clamping rings will be found inside the prism boxes.

(2) If the eyepiece diaphragm test (5) page 389 shows the collimation error is small, say small enough to be accommodated with discomfort, it can be safely assumed that any remaining error is in the objective pair. Small set screws will be found on the mount and these should be adjusted a little at a time until the collimation tests described above are satisfactory. The adjusting screws are preset and may be found covered with a filling of some kind which should be replaced as a deterrent to interference by the unskilled.

(3) If the instrument is more than say half a field diameter out of collimation, it is likely that a prism has become displaced; in this event the prism boxes should be opened and a visual inspection made. There is usually a spot of sealing compound placed upon the clamps and if this is broken it is likely that the prism has moved. In all modern instruments the prism seatings are machined with stops to locate the prisms, so a displacement is thereby rendered unlikely and is easily rectified when discovered.

(4) If an instrument has experienced a complete disturbance it must be reset entirely. In the case of the double-bodied instrument, each must be set up separately as an independent microscope, and the only part of the procedure which requires special attention is the correct positioning of the erecting prisms. To do this, obtain a straight-filament bulb like a projector lamp, set it up with the filament exactly vertical by projecting an image of the filament on to the wall; project the image through the prism system of the microscope and check that the image is still vertical having passed through the prisms. Repeat for the horizontal direction. The prisms must be set by trial until the conditions above are satisfied. Next, look through the prism systems against a diffuse light, one at a time, to check that there are no obstructions in the path. Place the objective on the instrument, focus the straight filament on to the object position and observe the path of the light through the tubes and prisms on to a wall or screen beyond. The path taken by the light must correspond with the mechanical axis of the tubes, and if it is in error, say 25 mm., at a distance of 1,200 mm., and the objectives appear to have their

axes in the same plane, it is probable that a prism is not flat on its seating, but if the prisms appear correct, then the objectives are badly out and the divergence of their pair must be checked with the lamp in the way described above and must be made the same as the mechanical divergence of the tubes, also the axes must be in the same plane.

When all these matters are satisfactory, the instrument must be finally collimated on test as described on page 388.

Not all details above apply to all instruments but the basic procedure is always the same and the details applicable to each case should be selected.

When setting up laboratory apparatus it is common practice to use paper as a shimming material and this is quite sound because paper is made under great pressure and will not compress. Thin typing paper is 75 microns thick and ordinary writing paper is about 150 microns.

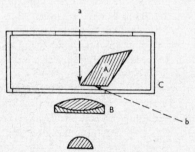

Fig. 5. A. Wenham prism.
　　　　 B. Objective.
　　　　 C. Prism box.

To Collimate the Wenham and Other Instruments with Breeches Tubes
As mentioned above, the Wenham prism must be made to fit the angle at which the tubes are placed to each other, as no amount of tilting the prism will make any difference to the angle of divergency of the pencils. The usual error in collimation is due to the tubes not being squared-on to the prism. Most Wenham bodies screw on to the arm of the microscope and sometimes the stop has become displaced, sometimes the prism is loose in its box. To set up, proceed as follows:

(1) Make sure that the prism is the correct way up in its box, it should be as shown in Fig, 5.

(2) Place the lower flat face (*b*) horizontally and with its edge (*a*) half-way across the objective aperture when the carrier is pushed into the 'binocular' position.

(3) Set up the Microscope with a substage condenser in use, use as an illuminant a straight filament bulb, focus the image of the bulb filament on to the object position, remove the eyepieces, and observe with a paper screen the images of the filament formed above the two tubes. Check that they are on the axes of the tubes for all positions of the viewing screen. If they are not, turn the tube or prism box until they are. If it is the divergence angle of the pencils that is wrong, the wrong prism is in use or the wrong tubes have been fitted.

(4) It sometimes happens that the prism box is in the wrong place as shown in Fig. 6. In this case the filament images above the tubes are always in positions parallel with the tube axes but displaced therefrom.

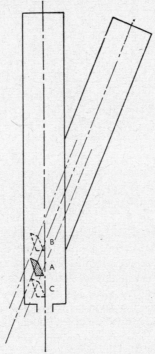

Fig. 6. A.   Correct position of prism.
          B. & C. Incorrect.

In all cases of wrong prism position, surgery is required to effect a cure. The divergence angle of the prism should be measured and drawn by the method given in (3) above. This geometry should then be applied graphically to the mechanical structure of the microscope and it will be seen whether the tubes require to be raised or lowered with respect to the prism.

If the angle of divergence of the tubes does not match that of the prism, there is no solution other than obtaining another prism which might, by good luck, fit the microscope.

The Powell and Lealand form is tackled in the same way, but as the high power form of prism is adjustable for tube divergence angle, matters are easier. (Fig. 7.) Prism (A) may be set in its mount at any

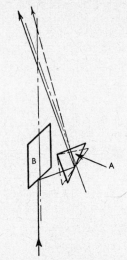

Fig. 7. A. Adjustable prism.
B. Fixed plate.

angle so as to direct light along the axis of the side tube, but it is still necessary for the prism box to be the correct distance from the breeches point of the tubes and for it to be the right way up.

The dividing prisms of the Stephenson instrument are separately adjustable for divergency angle and the inclination prism or mirror is likewise adjustable by various obvious mechanisms. To line it up, proceed as described above for other binocular instruments.

CHAPTER 22

Microprojection

UNDER THIS heading is described apparatus for projecting an image upon a screen so that an audience can be entertained or instructed. Film projection is not covered in this chapter.

Microprojection apparatus is little different from a microscope with an exceptionally bright light, the same principles are involved with only a few modifications of detail, and the apparatus is shown in outline in Fig. 1.

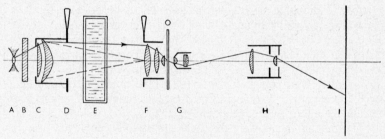

Fig. 1. Microprojector system.

The main difference between the microscope and the microprojector is that the microprojector must be able to give a field of view fully illuminated over its whole area after the style of the cinema screen, and in order to do this the lamp condenser must be about 30 to 40 mm. diameter and optically good enough to appear as an even disk of light when illuminated with the source employed. (See 'Microscope Lamps', page 125.) This condenser must also be of short focus so that its light grasp is great, but when a lens is placed about 40 mm. from a carbon arc, a piece of clear quartz B should be placed as shown to protect the lens from hot particles. Heat absorbing glass is not suitable because it absorbs so much heat that it bursts. The heat absorbing glass, if used at all, should be in the position of the water trough.

The lamp condenser and its diaphragm must be in centring mounts so that the arc or other source can be accurately lined up. It is the even disk of light in the lamp condenser which is imaged by the microscope and provides the field of view on the screen, therefore this part of the apparatus must be correct.

The quantity of light which passes through a system is dependent upon the physical aperture of the lenses, so it is best to choose a substage condenser which has large lenses, and the Holoscopic made by Messrs Watsons is one example with a back lens about 20 mm. diameter. The same remark applies to the objective, and some of the older ones by Ross and Zeiss will be found to work well.

It is possible to project an image with only the objective on the microscope and no eyepiece, but it is not recommended because there is then no tube length correction. The advantage is that the field of view is small, therefore a long throw of, say, 10 metres is possible without great enlargement of the screen field and consequent loss of light. An eyepiece should be used for this work but it should be specially made up with clean simple lenses of 80 to 200 mm. focal length arranged in a Ramsden form; the diaphragm between them should be movable for reasons given below. Such an eyepiece does not magnify more than $1\frac{1}{2}$ times but it projects a small well-illuminated field such that a disk of light about 2 metres diameter is available on a screen about 6 metres away. This is a convenient size for most class-rooms, and if the diaphragm is movable it can be arranged to be in focus on the screen all the time, thus providing a well-trimmed edge to the projected picture which does so much to create the impression of fine definition.

The magnifying power which can be usefully employed upon a particular apparatus depends upon the brightness of the light source available. Assuming optical matters are correct as outlined in this chapter, a carbon arc source will illuminate satisfactorily with a throw of about 6 metres a 2 mm. immersion objective, and a source composed of a motor-car headlamp bulb slightly over-run will illuminate a 6 mm. objective. The ribbon filament lamp is about the same in brightness as an over-run wire filament lamp but is a much better shape and more robust. (See 'Microscope Lamps', page 115.)

Any other form of source at present available is unsuitable for microprojection, but a watch should be kept on the market for a development of the high pressure mercury arc bulb which may possibly be as bright as a carbon arc, but to date the high pressure mercury lamp is not bright enough for projection.

Having got this apparatus together, the remaining work consists of practical application.

The lamp may be simply mounted but it must be well screened to prevent stray light in the room spoiling the contrast in the picture. Filament lamps should be burned holder downwards so that the heat is circulated in the way intended by the makers, also the glass of the bulbs is clearer at the sides than at the end. If the bulb is used holder upwards there is considerable chance of melting the lead contacts of the bulb, or at least softening them so that sooner or later an intermittent contact appears. A lamp housing after the fashion of Fig. 2

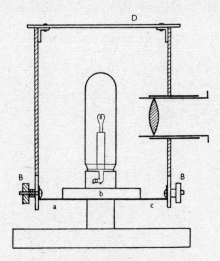

Fig. 2. B. Thumb screws in slots in can C.
       D. Tin case with push-in top.
       a. Bent metal base.
       c. Base (a) telescoping into can D.
       b. Batten holder.

is suitable. The batten holder (b) is mounted on a base (a) which slides in a roughly telescopic fitting (c) in the canister. The bulb is centred to the condenser by moving the telescopic lid about in any required direction and it can be made stiff enough to stay put in any position. More elaborate lamps and condensers have been described under 'Microscope Lamps', page 115.

The canister is best mounted on an optical bench saddle. The microscope can be of any conventional pattern but preference should be given to a large stand with plenty of room round all its parts. The classical instrument of about 1900 is usually suitable, mainly because in addition to its size, apparatus can be easily fitted to it and its adjustments have long travel. If the instrument is of the Ross bar pattern, so much the better, because the tube can be removed leaving all adjustments in place and any experimental or wide tube apparatus can be fitted.

When delivering a talk and using microprojection as an aid, it is nearly essential to obtain continuity in the performance; different slides require different power and the same slide often requires to be shown under different conditions; one condenser will not alone illuminate all objectives and give an even-sized screen illumination with each, it is therefore essential to have devices on the microscope which allow changes to be made very quickly and without fuss. The ordinary sliding Akehurst condenser changers are satisfactory but a rotating plate carrying condensers like objectives is the real answer to the problem of quick changes and there is no difficulty in making such a plate so long as it is for a large microscope which has plenty of clearance below the stage. Accurate centring of each condenser is not vital so long as the high power one is lined up with its objective. Fig. 3.

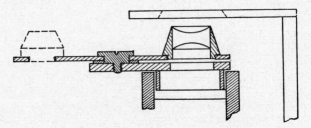

Fig. 3. Condenser changer for large apparatus.

A revolving nosepiece is the most suitable changer for objectives, for although it has certain faults explained under 'Microscope Stands' it is the quickest. When the illumination is set up using these devices objective and condenser can be changed together in a second of time, there being no need to operate the lamp condenser field diaphragm at all when condenser and objective are properly matched.

If the microscope is one with a thin stage, it is possible to rotate the condensers without much change of focus of the substage being necessary. If screw-cutting their mounting holes is a difficulty, the condensers can be clamped in place on the plate or pushed through plain holes with a strip of gasket cork as packing.

It is necessary for success to reduce slide changing time also to a minimum and Mr H. S. Henderson demonstrated to the Quekett Club a rotating slide changer which consisted of a plate of 300 mm. in diameter, again like an objective changer but mounted on an out-rigger on the stage. Slides can be clamped on this by ordinary stage clips to the number of twenty or thereabouts and previously centred to the microscope optic axis under their own holding-down clips previous to the lecture.

Troughs to hold liquids are not easy to handle unless the projector is vertical with a horizontal stage, so in the ordinary vertical stage design troughs are best set up on long racks which slide against a guide rail across the stage. While one rack is being loaded in order by an assistant, the other is being shown, and should fixed points only be demonstrated the troughs too can be preset in position.

If the objectives and condensers are chosen to match each other there is no need to change tube size or eyepiece as the eyepiece should be chosen to give the size of picture required for the particular room and should always be a low power one if not the projection type.

The Author constructed a vertical microprojector so that specimens in petri dishes and on hanging drop slides could be projected. There is no fundamental difference between this and the horizontal type, but it is easier to place objects loose on the stage and this is the main reason for its use. A first-surface mirror is used above the eyepiece to change the direction of the beam, Fig. 4. The instrument is stood on a low three-legged stool and is operated from a sitting position, but there is trouble sooner or later due to deposits from the filament upon the glass of the lamp causing light loss and overheating. There is also danger of the whole apparatus becoming too hot when used for a long period due to there being no proper cooling draught to carry the rising hot air away from the specimen and stage. Commercial designs can be of the vertical kind but the lamp should be in the normal position, its light being deflected into the microscope by means of a mirror. In the instrument shown in Fig. 4 variation of NA is obtained by altering the focus of the substage condenser. The eyepiece diaphragm trims the field of view. All designs of projector require a plate of opaque material about 200 mm. diameter to be placed somewhere under the stage to shield the screen and audience from stray light coming through the stage aperture, the plate can with advantage be mounted on the nosepiece end of the body tube.

The ordinary cloth home-cinema screen is satisfactory for microprojection but so is a sheet of clean white drawing paper. Much more light can be obtained by projecting the image from the back of the screen, in this case a large translucent screen can be bought or may be made from new clean tissue paper, but this screen is not good for anything other than entertainment work because it is too coarse for fine detail. A sheet of glass about 4 by 4 metres ground evenly on one side by means of a block of glass, water and 300 carborundum, is the very best for back projection, it is not fragile when framed and takes about two hours to make. If back projection is used, the throw

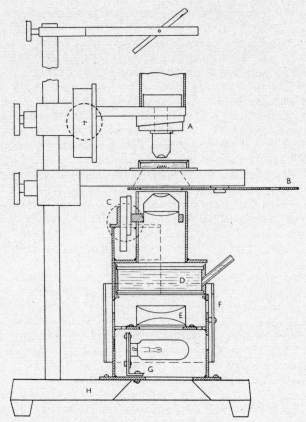

Fig. 4. Partial sectional view of vertical projector.

A. Objective sliding changer, tube and eyepiece added when wanted.
B. Anti-scatter diaphragm.
C. Condenser enclosed in screening tube, rack actuated from outside.
D. Large water trough for cooling.
E. Lamp condenser.
F. Screen for ventilation holes.
G. Lamp on simple bracket, fixing holes all slotted for preset centring
and focus, screws accessible through vent holes with D and F
removed.
H. Heavy retort stand 1 metre high.

of the projector can be much less and the large screen field given by
normal eyepieces can be used with advantage.

To Set Up a Microprojector

The principles of microscopy as set out in the appropriate chapters hold in microprojection. Place the objective and a specimen on the microscope, focus an image of the lamp condenser diaphragm on the specimen by means of the substage condenser, focus the specimen properly and observe the nature of the disk of light on the screen. At some stage in these proceedings select an eyepiece which is suitable for the screen. The substage condenser should be of such a focal length that the image it produces of the lamp condenser through the microscope on to the screen just about fills the field of view. Should the image of the lamp condenser be much larger than the microscope field of view (as shown by the necessity to reduce the lamp condenser diaphragm size before it can be seen in the field), light is being wasted and a shorter focus substage condenser should be chosen for that objective. If the image of the lamp condenser is too small to fill the microscope field no light is lost but the picture is limited, and a longer focus substage condenser giving a larger image is required.

If the disk of light on the screen is uneven bilaterally, the lamp is not centred to its condenser. (It is assumed that the microscope is properly set up before this.)

If the disk has an uneven distribution of light radially, the lamp condenser is not in focus with the lamp or is not sufficiently well corrected to do the job. (See 'Microscope Lamps', page 125, for an appropriate test.)

To Operate a Microprojector

When demonstrating to an audience of any kind, it is quite essential that slides and apparatus be changed in the minimum time, and if some trouble is taken in the first place to use a limited number of objectives with condensers to match them, nothing except these two components needs changing when slides and magnification are changed. As described above, the normal limit of the screen field of view is by means of the adjusted eyepiece diaphragm and this is not again disturbed.

The screen disk of light is the image of the lamp condenser, and the image of its diaphragm fully open should be just larger than the defined field of view. Closure of this diaphragm should reduce the screen field of view and the diaphragm should be in focus coaxially with the field. The substage condenser diaphragm should not be visible in the field of view and will operate as usual by reducing the numerical aperture of the microscope; it is thus a control of contrast. There are then only the revolving nosepiece, the condenser changer mechanism and the substage iris to operate.

If it is necessary to use a pointer to indicate features in the image, the most convenient is a stout needle operated against the lamp condenser surface, but if this surface is not accessible, a long ordinary pointer had better be used because the eyepiece end of a micro-projector is too sensitive mechanically to allow an eyepiece pointer to be used.

Notes on Making Large Ground Glass Screens

If an image is formed on ground glass and viewed from the back as in a reflex camera, a much larger amount of light is transmitted than is reflected from the same image on a white screen. Not so much light is seen from viewpoints off the optic axis but viewing is satisfactory over a cone of about 60° about the axis. When light is limited as in microprojection, the advantages of back illumination outweigh the disadvantages of limited viewing position.

To make a ground glass screen, obtain a sheet of plate glass not less than 1 metre square, as smaller sizes appear woefully small even in private houses. Obtain the use of a flat table top or sheet of concrete sufficiently large to support the sheet without undue over-hang; the supporting surface need not be specially flat because no great weight is to be applied to the glass. Obtain a block of glass similar to that used in schools to demonstrate refraction of light (100 mm. square and 10 to 30 mm. thick) to use as a scrubber. Obtain some fine carborundum powder, No. 300, which should appear like grey flour, wet the sheet, sprinkle freely with carborundum and scrub with the block in long sweeping strokes. There is sufficient powder and water on the surface when the glass block does not drag on the sheet, and the grinding action can be felt easily. Recharging is necessary at frequent intervals and the whole is a dirty job.

After about an hour of this, wipe and wash down the sheet and examine it when dry for even grinding. When it appears to be satis-factory it must always be kept free from grease and dirt and it will then retain its even bluish-white appearance, but should it become dirty it should be washed with commercial detergent.

The screen should be framed as a safety precaution.

CHAPTER 23

Phase Microscopy

THE MICROSCOPIC image is formed in the instrument by interference between a beam of light which passes through the object and the beams of light which have been diffracted by the object and by the objective. Both these diffractions must be considered when the final effect seen by the eye is calculated.

When thinking about image formation it is necessary to remember that diffracted beams are coherent with the incident beams causing them, and therefore, interference is possible. This occurs at all points in the field under consideration. Images therefore depend for their visibility upon the diffraction-causing properties of the object under observation. Diffraction is caused when there is a difference in refractive index between an object and its surrounding medium, or a difference in opacity between them.

If all diffracted beams enter into combination and form an image of an object, that image must be in all respects exactly like the object, and if the object is transparent differing only in refractive index from its surrounding medium, the image of this object at exact focus will likewise be transparent and invisible.

Such an image can be made visible by stopping out some of the diffracted rays (often performed by the objective itself) or by altering the focus of the instrument. It will be seen that the black and white dot effects observed on diatoms are really out-of-focus images of this sort.

It appears from the above that a microscope is an unsatisfactory device for examining objects of a semi-transparent nature unless the differences of refractive index can be used to form an image in black and white.

A phase microscope is a device which causes difference of refractive index between an object and its surrounding medium to be made visible in the form of an ordinary black and white image. The term 'phase contrast' has come to be used because the differences in phase of some rays in the light bundle passing through the system are used to give the necessary 'contrast' between the object and the background.

When examining the system of phase microscopy it is not necessary to do away with the older treatment by geometric optics, but one must understand its limitations. Light passes through a lens system in a 'geometrical' way, but treatment will be found in most text-books of physics which shows that the sinusoidal vibrations which excite the sensations of light cannot form a point of brightness any-where, but the distribution of light in an image must be of a harmonic nature and this sets the limit to resolution in any optical instrument.

Some explanation therefore is necessary of the terms 'frequency', 'harmonic motion' and 'phase' before an understanding can be obtained of phase differences in light. In a talk delivered by the Author to the Quekett Club in 1950, an electrical analogy was used to help those not familiar with the physical aspects of light. It has been recommended that the same treatment of the subject, though elementary, be used here, particularly because the demonstration can be set up, and many microscopists are not physicists.

When a piece of wire is moved to and fro in a magnetic field (near a magnet) electric current is produced in the wire. The direction of flow of this current depends upon the direction of movement of the wire. When the wire is stationary the current stops. When the direction of movement of the wire is reversed, the flow of current through it is reversed. Therefore if the wire is mounted on a spool and revolved, the electricity will start in one direction, stop and commence in the other direction as the wire moves in and out of the magnetic fields (Fig. 1).

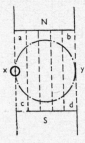

Fig. 1. A rotating wire crossing lines of force in a magnetic field. The portion of rotation from a to b produces a current in one direction through the wire, whilst the portion of rotation of the wire from d to c produces a current in the other direction. When at either position x or y no current is produced because the wire is not crossing lines of force.

The 'radiation' of electricity from a device of this kind (called an alternator) follows the curve shown in Fig. 2.

Light is radiated from an atom in a light source in a similar way, but this comparison must not be pressed further. This to and fro motion of any body, or variation of intensity as shown by the sine curve is known as harmonic motion.

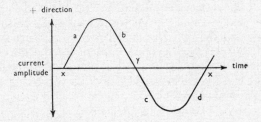

Fig. 2. Graphical representation of the current generated in a wire rotating in a magnetic field for one revolution.

The 'frequency' of a radiation is the number of complete cycles (*x* to *x*, Fig. 2) which take place in 1 sec. In the case of mains electricity it is 50 c.p.s. and for light about 10^{19} c.p.s.

If instead of one wire we consider two independent ones fixed on the same spool and therefore revolving at a fixed distance from each other, another wave will be produced by this second wire exactly related to the first in spacing. These waves are 'coherent' because they are related to one another, but they are 'out of phase' by an amount depending upon their separation which may be stated in degrees of rotation. If the wires were not on the same spool but revolved at random, they each would produce a wave, but these two waves would not be related to each other: i.e. they would not be coherent. Normal light emission is of this type. Atoms radiate independently, and only coherent light, i.e. a dioptric or central beam and its diffraction waves, can 'interfere' with each other and form an image. Non-coherent light cannot cause any interference effects.

Light is a sensation caused by sinusoidal vibrations of a certain frequency from a mass of excited atoms and can usually be considered as a single wave for purposes of optical proof, and this will be done in the following explanation of the phase-microscopy system.

This system requires that light of diffracted beams shall undergo a shift of phase angle compared with the dioptric beam producing them. All phase systems depend for their action upon this. We must now consider practical requirements and arrangements.

Diffracted beams of light from an object must fall outside the central beam and so are generally considered by microscopists to occupy the outer zones of an objective. If therefore we arrange that, say, the outer three-quarters of the objective aperture is covered by a phase-shifting device, some diffracted beams from some objects will experience a phase change, but it will be seen that this depends upon the closeness of the structure observed which decides the spread of the diffracted beams. The transparency of the object will also have an effect upon the ultimate images. Devices of this sort are open to the objection that they interfere with the light passing through the back lens of the objective and therefore must cause some unwanted effects such as reduction of aperture and distortion, due to irregular transmission and possible stopping of diffracted beams. Some of these effects are inseparable from the phase-contrast system, but before examining practical arrangements it is essential that the qualitative part of the theory is understood.

In order to explain properly the operation of the system we must resort to vector diagrams. Consider again Fig. 1. The position of the wire relative to the spindle upon which we have considered it to be mounted may be shown as a point travelling in a circular path around another point (the spindle axis). The vertical distance between these two points may be taken as a measure of the amplitude of any wave generated or 'radiated', but here it is the position of the wire that we are concerned with. The diagram may now be drawn as shown in Fig. 3. Likewise another wave developed by a second wire can be represented by an additional vector as shown in Fig. 4.

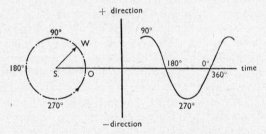

Fig. 3. The generation of a wave shown vectorially together with its projection on a time base. s the point about which rotation takes place, w the wire moving along dotted path.

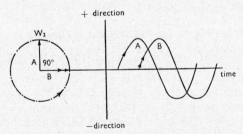

Fig. 4. Two rotating wires separated by 90° are represented by vectors A and B, and are projected on a time-base. The fixed separation of two such vectors may be any number of degrees.

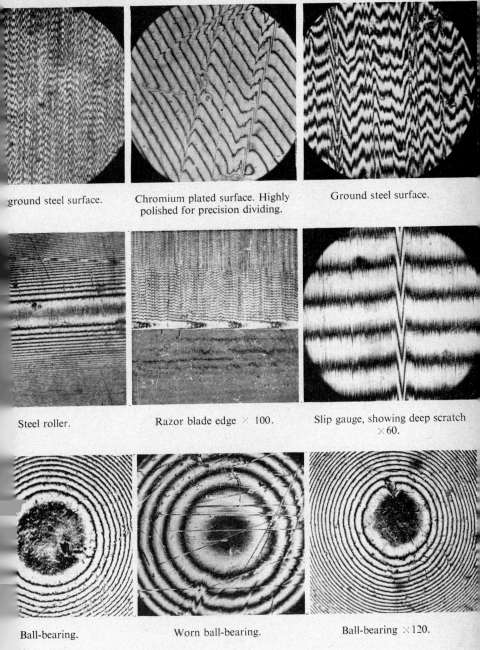

ground steel surface.
Chromium plated surface. Highly polished for precision dividing.
Ground steel surface.

Steel roller.
Razor blade edge × 100.
Slip gauge, showing deep scratch × 60.

Ball-bearing.
Worn ball-bearing.
Ball-bearing ×120.

PLATE 10
Examples of Appearances of Surfaces under the Hilger
Interference Microscope.

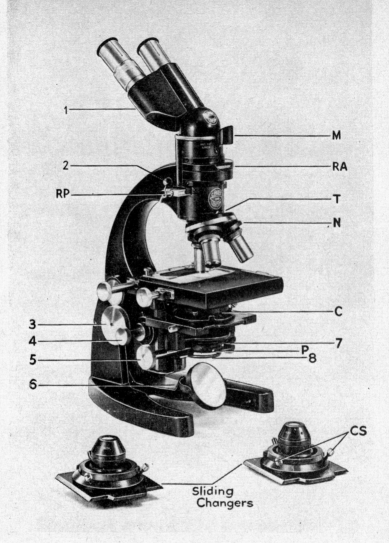

Sliding
Changers

PLATE 11

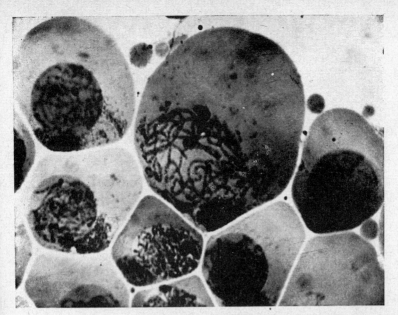

Upper: Analyser of interference microscope set for positive contrast.

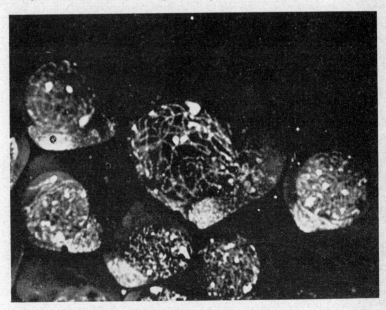

Lower: Analyser set for negative contrast.

(i) Cell at *lower* left centre. Resting stage showing fine chromosome threads. Note difference in contrast of nuclear sap (low refractive index).

(ii) Cells at *lower* right centre and extreme right. Leptotene stage with fully-developed single chromosome threads.

(iii) *Upper* centre. Pachytene stage. Homologous chromosomes have paired at corresponding points and the transverse bands along the length of the thread can be seen. Massive X-chromosome at top of nucleus.

Courtesy of Dr E. J. Ambrose, Chester Beatty Research Institute, London.

PLATE 12

Chromosomes in Living Cells of *Locusta*. Taken with ×40 shearing objective.

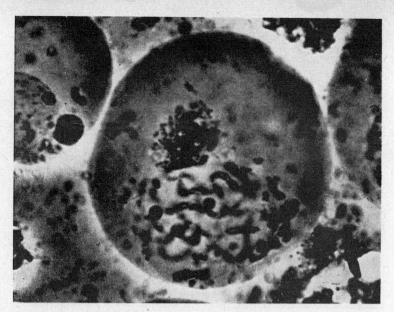

Upper: Positive contrast.

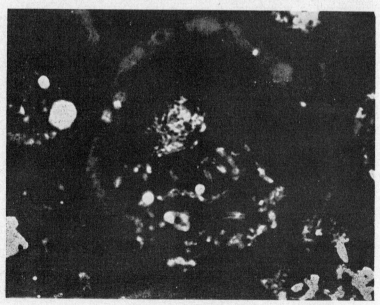

Lower: Negative contrast.

Central cell is at pachytene stage. Nucleus in lower part of cell. The two homologous strands of the paired chromosomes are clearly resolved. They can be seen to be associated at corresponding chromomeres. In one pair, which runs horizontally across the nucleus, the separate strands appear to be relationally coiled.

Courtesy of Dr E. J. Ambrose, Chester Beatty Research Institute, London.

PLATE 13
Chromosomes in Living Cells of *Locusta*. Taken with × 100 double focus objective.

In this diagram the two vectors are shown 90° apart. Reference to Fig. 1 will show that a complete cycle or, when referring to radiation, a complete wave, is between the points x and x. Half a wave is between x and y, therefore it will be seen that in Fig. 4 one wave is a quarter of a wave behind the other. It is better to speak of 'phase differences' in degrees rather than of fractions of a wave-length. Remembering that light is a sinusoidal vibration we will now consider what happens when light passes through a phase microscope.

It has been said earlier that to form an image light beams must interfere. This interference is brought about by the combination of the central beam, and the beams caused by diffraction in the object and the objective. Only coherent light can interfere, and if the random phase relationships of light vary with respect to each other many times each microsecond, so do their respective diffracted rays, therefore all remains in order and a steady image is formed.

Consider as an object a grating, but instead of opaque bands, let the bands be of material having the same transparency as the mounting medium but of different refractive index. This grating will normally be invisible, but will cause diffraction effects due to differences of refractive index only. No differences of brightness will occur between the bars as a result of absorption of light as with an ordinary grating.

Let rays of light coming through this object be represented by vectors. First, in Fig. 5 let a represent the phase of a wave of light coming through the mounting medium only. Let its brightness be oa.

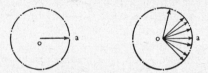

Fig. 5. The vectorial representation of the phase relationship of light beams (see text).
Fig. 6. The vectorial representation of the phases of light beams. oa the reference point for the object and the mountant of the same refractive index.

Secondly, any waves of light which have passed through those parts of the object which have a different refractive index from the mounting medium will arrive either a little earlier or a little later. If the refractive index is higher, they will be later and if lower, earlier. These times of arrival may be shown on the vector diagram as in Fig. 6. This diagram represents the state of affairs when an object is used which has parts of higher and lower refractive index than its mounting medium. The eye cannot possibly see phase differences of light (note frequency 10^{19} cyc./sec. and speed in space of 300,000,000 m./sec.), but the brightness seen at any point in an image is the brightness due to the 'average' phase of light at that point, i.e. the results of vector addition of all phase angles arriving there.

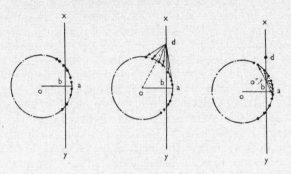

Fig. 7. The vectors (see Fig. 6) shown as points to avoid confusion.
Fig. 8. The angle of phase difference which gives the greatest contract
= <o'oa.
Fig. 9. Amplitude of the phase shifted central beam reduced.
o' now appears at o".

Let Fig. 6 be redrawn as Fig. 7 using only the points of the vectors to avoid confusion.

The points shown represent ordinary differences of phase in objects like diatoms mounted in Canada balsam, and bacteria in water. The average light coming through the microscope at the point in the field under consideration can be represented by *ob*, the construction being by drawing *xy* (Fig. 7) through the average density position of the group of points representing the different phase angles of arriving waves. Clearly if the object is very near its mounting medium in refractive index the point *b* will approach complete transparency without phase shift which results in no image at all.

This is all normal microscopy, and by stopping certain diffracted beams (shutting down the apertures) some sort of effect can be seen which may or may not be like the object. We must now examine the system for making an object visible by using these differences of phase directly to produce a 'black and white' image.

If two sine waves (so-called because the amplitude of the wave is proportional to the sine of the angle of rotation of the producing vector) are added together in phase, the brightness will be equal to the algebraic addition of the quantities. If the waves are 180° out of phase, they are in opposition, and the result will be darkness. If waves which are coherent, but of intermediate number of degrees of phase difference, are added, intermediate results will be obtained dependent upon the laws of vector addition. If we chose 90° as the phase difference between waves, one will be at its maximum while the other is at its minimum, therefore they will have maximum effect upon the total when this careful arrangement of angles is altered.

To make this clear in the subject under consideration let us again examine Fig. 7, redrawn as Fig. 8. The points shown on the circle are dependent upon diffracted beams which cause an average brightness *ob*. If there were no diffraction (i.e. no object), the brightness would be *oa*. If it were possible to remove the causal direct beams from the system, the light from the diffracted beams would be represented by *ba*. This occurs in dark ground illumination and indicates why delicate objects emit such faint light under these conditions. If, however, we can arrange that instead of having the direct beams and diffracted beams near each other in phase and adding up to a uniformly bright field, we can move the direct beams to be 90° out of phase with these diffracted beams, the two sets of beams will be in a position to have the maximum effect upon each other. The argument here can be restated thus: if one quantity is zero, an infinitely small addition to this will be observable against this first zero. If, however, we have a large quantity, an infinitely small addition to this will not be detectable, and the system becomes much less sensitive.

Shifting the direct beams by 90° will cause Fig. 8 to have its origin at *o'* (vector *ob* is added to point *b* at angle of 90°). It will be seen now that when the 'brightness' vectors are redrawn they will be shortest near this point. This means that objects will be seen the darkest which cause this phase angle compared with the mounting medium in light which passes through them. This is fundamentally the operation of the phase microscope.

It would appear from Fig. 8 that the phase angle which will give the best contrast is considerably removed from the reference point, and may be equally visible by methods of critical microscopy. This, however, is not the case because if all the diffracted beams from a transparent object enter into the final image, then the image must be correct and consequently invisible when in exact focus. The phase microscope, however, will give an image when in focus. This is a very important distinction.

An objection still remains, however, that the place of best contrast is for objects which might be seen by dark ground methods. If, however, we can reduce the amplitude of the dioptric beam without interfering with its phase in relation to its diffracted beams, the point *o'* can be brought nearer to *b* (Fig. 9), to a point *o''*.

The point of maximum contrast is now much nearer that of complete transparency (*a*), but the actual contrast obtaining must be found by comparing *bo"* with *o" a*, hence it will be seen that if the direct beam is too much attenuated a wrong type of dark ground effect will follow due to excessive interference with the back lens aperture. An optimum condition can therefore be found by examination of the diagrams, and in general it will be found that attenuation of the order of 8 : 1 of the direct beam will give the best results (see later).

From the above reasoning it is apparent that each refractive index in an object has its own optimum conditions of attenuation of the direct beam for maximum contrast. When objects are near complete transparency and zero phase shift compared with the mounting medium it is most desirable that the attenuation be variable. This has been attempted with success by using polarized light as the illuminant and inscribing phase shifting devices upon a piece of polarizing material. Rotation of the plate will then cause variation of intensity over the part of the aperture covered by polarizing material.

Phase-shifting Devices

When it is required to alter the phase of a beam of light relative to another beam, the simplest method is to cause the beam under consideration to pass through a greater thickness of material. The thickness of material necessary may be found from the following calculation.

Let n=refractive index of material to be added to the surface, and λ=wave-length of light. Let d=distance of phase shift (in this case=$\frac{1}{4}\lambda$=$\frac{1}{4}$ wave-length), then

$$\tfrac{1}{4}\lambda=(n-1)d \text{ and } d=\frac{\lambda}{4(n-1)}.$$

suppose we take n=1·5 as an illustration, we see that

$$d=\frac{\lambda}{4\times \cdot 5}=\tfrac{1}{2} \text{ wave-length added thickness.}$$

If the added material has a refractive index of less than 1·5, it will be seen that a greater thickness must be provided.

If material is etched away, the same result obtains except that the phase will be advanced in traversing the thinner section.

A petrologists' $\frac{1}{4}$ wave plate is said to produce a difference in phase of 90° due to the different speeds of transmission along the axes of mica. It does not matter which physical mechanism is used so long as a 90° phase shift is produced between the central and diffracted beams.

It is not possible to use the whole of the back lens of an objective in the ordinary way in a phase-contrast microscope, but no serious loss of resolution should take place in consequence. It is necessary to separate the direct beam and the diffracted beams in the objective. If it is required to see fine objects, plenty of room must be allocated for diffracted beams clear of the direct beam or else both will experience phase shift and the result will be impaired.

If a small central area only is used, it is easier to separate the direct and diffracted beams, but aperture is seriously reduced. We may use therefore a dimetral 'strip' of the back lens so that resolution is not impaired in one direction, and its width is found by experiment to be about one-fifteenth of the back lens diameter. The phase plate in this case looks in plan like Fig 10 A and in section like Fig. 10 C. This disk should be just about the size of the back lens of the objective and should be placed near this back lens so that it intercepts as much light as possible and fills as much of the eyepiece field as possible, but there is no point in engraving this phase-shifting device upon a component lens surface within an objective.

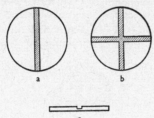

Fig. 10. The shaded areas in a and b represent the phase shift mechanism in two types of phase plates. c represents a side elevation of a, showing the depression $\frac{1}{4}\lambda$ deep in plane glass plate.

This 'strip' phase plate has the disadvantage that the effects surrounding the out-of-focus image are irregular, and it is not possible to vary the amount of attenuation of the dioptric beam by a mechanism involving rotating the plate. Resolution, however, is good, and can be better if a cross is used (Fig. 10 B).

After theoretical treatment, it has been decided that an annulus is the best shape to use as it covers a fair area of the objective aperture at a good intermediate position and will give symmetrical effects above and below focus, together with contrast effects covering most of the eyepiece field. Fig. 11 shows the area covered by a phase annulus. It should be about half way between the centre of the system and the outside edge of the back lens, and still only about one-fifteenth of the back lens diameter in width. It will be seen that this phase plate can be rotated, and so a variable attenuation of the 'central' beam can be arranged by polarization means.

Fig. 11. The shaded area represents the area covered by the phase annulus.

It now remains to be seen how a beam of light can be projected through the object on to the objective so that it covers an area exactly equal in size and position to the phase plate. This is very easily done, and needs only a cardboard or adjustable plate with the necessary apertures cut in it and placed in the stop carrier of the condenser. A good achromatic condenser should be used, as a fair image is required of this plate.

The annulus type of phase plate requires an object under the condenser or in front of the light source which will form a 'shape' on the back lens of the objective equal in size and shape to the phase plate. This will be as Fig. 12. This may be painted on plane glass or built up by any of the well-known methods.

Fig. 12. Illuminator plate. The double shaded areas represent the opaque portions. The clear annulus shows the area for the free transmission of light.

Some trials will have to be made as to size, but great accuracy is not required as the condenser may be moved up and down to arrange the final co-ordination. This movement will not interfere with conditions of critical light as the latter means a back lens *evenly* filled with light over the area required, and does not mean the illuminant must be focused upon the object, although this is the easiest way to obtain critical light.

Accurate centring of these components is necessary, but can usually be 'touched up' with the substage centring screws.

Mention has been made of variable attenuation of the direct beam in order to obtain maximum contrast with objects of different refractive indices. This can be arranged by covering the illuminator plate

(Fig. 12) with Polaroid so that light coming through the annulus into the condenser is polarized. If now the annulus of the phase plate is made of polarizing material, it is clear that almost total blackout may be obtained if one plate is rotated relative to the other. Of course this must be arranged without interfering with the phase-shifting properties of this annulus.

Fixed attenuation phase plates are made commercially by sputtering or evaporating metal on to the phase plate phase-shifting area. This area receives previously a correct depth of etching for a final reduction of $\frac{1}{2}$ wave-length of thickness (see preceding calculations).

In some cases the $\frac{1}{2}$ wave-length is built up by evaporation processes. Experiments have been made with indian ink, but it is very difficult to obtain a correct thickness of even deposit. It will be seen from Fig. 9 that it does not matter whether the $\frac{1}{2}$ wave-length thickness is built up or taken away from the phase plate, it only removes the point o' or o'' into the lower sector of the circle. $o'b$ is added to ob in the opposite direction and so will give opposite contrast, the darkest region (o' to the circumference of the circle) will correspond to an object with a higher refractive index than the mounting medium instead of to one lower, and therefore of quicker transmission, as shown, page 408.

It is suggested by the author that in many cases, due largely to failure to understand the theory of the apparatus and partially to prejudice against apparatus behind the back lens of the objective, microscopists are failing to make full use of a very powerful physical method of research.

Correct phase-contrast microscopy is a much more critical study than is critical or 'Nelsonian' microscopy because it requires the latter as a basis before any proper understanding of the new method can be obtained.

To Set up a Phase Microscope

A number of phase contrast outfits are now on the market and each manufacturer includes instructions relating to his particular instrument. The following notes deal with the matter in principle.

The most important requirement in the system is that the image of the phase diaphragm in the illuminating system shall be arranged to coincide with the phase shifting area on the plate in the rear focal plane of the objective and this coincidence must occur in both position and size.

(1) Remove the diaphragm plate from in front of the lamp or from the substage condenser and set up the microscope as for ordinary critical work upon transparent objects. Observe that

the objective back lens is evenly filled with light; there is no need to remove the normal phase contrast objective or phase plate if this happens to be separate

(2) Use as an object a thin layer of saliva scraped from the inside of the cheek, placed between a slide and cover glass. This preparation contains epithelial cells which make excellent test objects for the phase and interference microscopes.

(3) Insert the diaphragm plate in the illumination system.

(4) Examine the objective back lens with the naked eye and observe that the correct shape of light pattern appears there, i.e. for the cross type of phase plate it will of course be a cross and for an annulus, a ring. To obtain this, adjustment of the substage condenser focus or any auxiliary lenses present may be necessary, but it is usual only to raise the substage condenser in order to cause it to project an image into the objective rear focal plane.

(5) Insert into the draw-tube any telescopic viewing device there may be and look at the objective phase plate carefully, then adjust the position of the diaphragm plate and the centration of the system so that exact coincidence occurs between the image of the plate and the phase shifting area of the objective phase plate. The size of the diaphragm image may be altered to obtain coincidence of size by adjusting within limits the focus of the substage condenser. If it is a cross, the diaphragm plate must also be revolved until coincidence is found, and in any case the plate must be of such a size that great changes of substage condenser focus are not called for. If it is an annulus system, the centring of the substage may be used to obtain coincidence.

(6) Study the specimen and it should be found that it exhibits strong dark or light contrast where it was transparent by ordinary transmitted light.

(7) If the system has variable attenuation of the direct beam, the attenuation must be varied to obtain optimum contrast for the specimen under examination and the less the difference in refractive index between the specimen and surrounding medium, the more the control beam must be attenuated. It is usual to have incorporated in the system a brightness adjustment to the lamp so off-set variations in phase plate absorption.

Workers are recommended to try phase contrast microscopes upon a number of specimens; good examples are epithelial cells, large bacteria in water and blood cells in plasma.

Limitations of the Phase Contrast Method

The phase contrast method will not show objects in contrast when the difference in refractive power between them and the surrounding medium is very small. (See 'Interference Microscopy', page 422.) It also will not work well on thick or on confused objects; it will not, for example, work through a pond life trough nor upon an unprepared section of a leaf; one is too thick and the other too confused. However, the system works very well upon thin layers of organisms like bacteria, cell cultures, spermatozoa and laminated structures sometimes studied in the physics laboratory.

It is possible to calibrate a phase system to measure thicknesses and refractive index differences but for this work the interference microscope is superior for the reasons there given, and the matter will not be further treated in this chapter.

CHAPTER 24

Interference Microscopy

IF TWO nearly flat surfaces of transparent material are rested upon each other and illuminated, preferably from above, interference patterns known as Newton's rings will probably be seen. These are due to interference between light reflected from the two surfaces which are a small distance apart; the light has a slightly greater distance to travel to and from the second surface than from the first, consequently it becomes out of phase with light reflected from the first and some wave-lengths are cancelled out. If the distance apart of the surfaces is such that some wave-lengths (which happen to be 180° out of phase) are cancelled, the complementary colours appear in the light reflected out of the system and we have the colours shown by an oil film on water so often observed in the streets on a wet day. If the incident light is monochromatic, black and single colour reflection effects occur because there is present only one colour to cancel out. If the surfaces causing the interference are silvered so that about 90 per cent of the light is reflected backwards and forwards between them, a large number of times, the fringes or Newton's rings become very sharp and may be fine lines; this is known as multiple beam interference. If the source of light is composed of spectrum lines, each line of the source produces its own interference lines and in the case of multiple reflection these lines are quite sharp and separate from each other. The distance of one dark line to the next in one colour of light is $\frac{1}{2}$ wave-length of light, and when the surfaces between which interference is taking place are nearly parallel the lines may be about 10 mm. apart. If we take the wave-length of light as $\frac{1}{4}$ micron, it can be seen that a separation of lines of 10 mm. represents a change of level of $\frac{1}{8}$ micron. Over the area where the lines have this spacing a kink in a line of, say, 1 mm. deviation, represents a feature which at that point has a change of level of

$$\frac{1}{6}\mu \div \frac{10}{1} = \frac{1}{60} \text{ micron}$$

This figure shows that a very high degree of resolution can be obtained by the method.

In order to become familiar with the method, the following

macroscopical apparatus may be used. Obtain a sheet of typical plate glass about 80 mm. in diameter and 10 mm. thick, and half aluminize one side in an evaporation chamber. Obtain a microscope cover glass and treat this in the same way. Rest the cover upon the plate glass, the aluminized surfaces together, and illuminate the whole with a sodium lamp. If the system is examined by means of a telescope or with the naked eye, Newton's rings will be seen very clearly between the glasses. As the cover glass is moved about or is pressed lightly with a needle, the position and number of the rings will be seen to change rapidly. If white light is used, the rings appear coloured but are less distinct.

The apparatus as applied to the microscope is very similar. A specially good transparent optical flat is fixed below a medium power objective as shown in Fig. 1A.

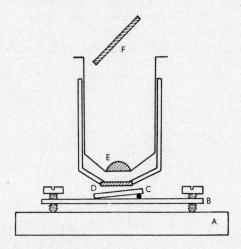

Fig. 1 (a). A. Stage.
            B. Stage levelling device.
            C. Specimen.
            D. Optical flat coated with rhodium giving 50 per cent reflection from lower face.
            E. Objective.
            F. Vertical illuminator.

Several arrangements are possible, for example, in some the optical flat is borne on levelling screws and some older objective changer slides have squaring-on screws which may be used for levelling purposes. On the whole it is better to use an independent levelling stage mounted upon the ordinary stage because this is a much more robust job. The idea in all cases is to present the surface to be studied to the reference flat in such a way that the surface is slightly inclined with respect to the flat, but before making this adjustment the microscope tube must be substantially vertical with respect to the specimen and this is normally arranged by means of the levelling stage. When these adjustments are made, a controllable number of interference bands will be seen crossing the field of view, the number being controlled by means of the fine adjustment to the levelling device.

The method has some unexpected uses; if, for instance, a culture of diatoms is allowed to live on a slide with a cover glass and the object is illuminated with a vertical illuminator, interference patterns can develop between the gelantinous coat of the diatoms and the underside of the cover glass when the diatoms are moving in contact with it. The cover is in this experiment taking the place of the reference flat, and this method was used by the Author to study the mechanism of diatom movement which can clearly be seen to originate in a streaming of protoplasm along the outside of the diatom. The direction of streaming changes at regular intervals and is probably related to the phenomenon of cyclosis seen in most plant cells.

Commercial Apparatus

The Hilger Watts surface finish microscope is an attachment for an ordinary engineer's instrument; it can be used to study any surface configuration so long as the appropriate optical flat is obtainable. It is not usually necessary to apply a linear magnification of more than 200 times to develop the fringes clearly.

In many branches of industry it is necessary to examine surface finish to very fine limits. Piston rings, ball bearings, ball races, injection moulds, etc., should have a finish good to a thousanth of a millimetre. This goes beyond the resolving power of the optical microscope and these ultra fine finishes can only be examined optically by using interference techniques.

For some time, interference fringes have been used for high precision measurements of length but it is only recently that instruments have been devised which can be used for the examination of metal finish by a mechanic who has no previous knowledge of the laws of physical optics. Hilger & Watts have produced a model which has the advantage of small size, adaptability and low price so that it is suited for use by the mechanic either on the bench or by using the special V base, directly on the machine without dismantling the work. It can be used for the examination of flat, spherical or cylindrical work pieces by employing the standard comparison flat or for concave work surfaces using the comparison hemisphere. The magnification is ×125 which is sufficient for workshop purposes. This instrument does not take the place of the larger two-beam interference microscopes with higher magnifications but has its own particular function as a workshop tool.

The image seen in the microscope can be interpreted easily and requires no calculations or reference to tables and it is repeatable. The measurements of surface roughness in terms of light waves are absolute and not comparative but for convenience, sample pieces with the required finishes can be provided as standards, thus enabling comparisons to be made with specimens of known CLA or RMS values.

If required, photographic records on 35 mm. film can be made for record purposes with a camera attachment.

Not only in the engineering shop but also in the science laboratory is this instrument of value where the measurements of small variations in depth are required as, for instance, in the measurements of thickness of metallic or dielectric films deposited in high vacuum processes, the finish of diamond tools, etc.

In its standard form this instrument is not intended for evaluation of surface roughness of coarser machine surfaces, say, with surface irregularities of about 20 microns, but with a replica technique its use can be extended.

The attachment is self contained and simply screws into the tube. Illumination is provided by a miniature high intensity mercury vapour lamp which is connected with a capacitor and switch and which plugs directly into the main (210–240 V.). Light from this mercury lamp (A) passes through a green filter (B) and adjustable iris diaphragm (C) and is focused by condensers (D) after reflection at a 50 per cent reflecting surface (E) on the back focal plane of the objective (F) which is especially designed to work with parallel light. In front of the objective is the spring-loaded reference flat (G) coated with a surface film to render it reflecting and at the same time non-absorbing so that none of the incident light is wasted. The surface of this scratch proof reference flat, is pre-adjusted to be in focus when viewed through the eyepiece (H) and when the tube is lowered so that the flat rests on the work, the interference fringes are visible. The flat itself can be given a slight tilt by rotating a knob; this varies the spacing between the fringes. When the interference fringes appear straight across the field of view from left to right the tube is perpendicular to the work and the slight tilt of the reference flat is all that is required to separate the fringes. Fig. 1b.

The standard reference flat has a reflectivity of about 50 per cent which will give sharp fringes with most of the metallic surfaces met with in the workshop. For examination of surfaces of higher reflectivity an extra comparison plate with a higher reflectivity is provided. These flats screw out and are easily interchangeable. Also for examination of concave surfaces a comparison hemisphere is provided.

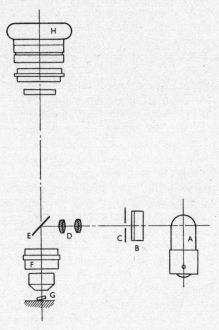

Fig. 1 (*b*). Diagram of surface finish interference miscroscope.

The standard stage of a microscope may be fitted with a dovetail slide to take cross traverse micrometer tables and circular tables if required and centres for small cylindrical work pieces up to 25 mm. in diameter. By using a specially designed V stage with centre clearance the microscope may be used on cylindrical or flat work during machining without any necessity for removing the work from the machine.

Interpretation of Fringe Pattern
When the fringe pattern is in view in the microscope the reference

flat is making a very small angle with the work and the fringes represent contours. It can thus be seen that if the work is quite flat these contours will be straight lines and if the angle of tilt is reduced the fringes will appear to be further apart.

If, however, the work is not flat, the fringes will appear like contour lines on a map. Any irregularity in the surface of the work equal to half the wave-length of the light used will displace a contour line by one fringe spacing (as explained above), in this case approximately 10 microns. Since these fringes are clear and sharp it is easily possible to read irregularities to one-tenth of a fringe spacing, i.e. to 1 micron.

If the work is spherical the invisible layers will take the form of concentric spheres and the inter-sections with the reference plane will be circles like a section through an onion, and any faults in smoothness will appear as irregularities on these circular fringes and faults in a sphericity as departure of line circular fringes. It can be seen also that a cylinder, if examined with a flat parallel to its axis, will give straight line fringes.

For examining concave surfaces a comparison hemisphere is substituted for the flat. The shape of the fringes will then be curved but the roughness will be given in every case by the irregularities in the fringes whatever their general shape.

Interference microscopy, in the present day meaning of the word, is a method for examining phase changing specimens which has been known and discussed for a considerable time, but a general appreciation of its practical value has arisen only of recent years, largely as a result of experience with the phase contrast technique of Zernike.

Phase contrast and interference microscopy both provide intensity contrasted images of phase changing specimens by means of optical interference phenomena, but the important characteristic of the interference method is that the mutually interfering beams which produce the contrast are generated by an interferometer system incorporated into the microscope itself, thus avoiding dependence upon diffraction by the object structure.

Two advantages follow immediately from this independence. First, that contrast can be obtained for features causing phase changes which are too gradual to diffract an adequate proportion of the light outside the phase step of a phase contrast microscope, thus enabling phase gradients as well as abrupt changes to be perceived. This results in an image which brings out the general morphology of the specimen better than phase contrast does, and which bears a resemblance to that obtained by staining methods. Secondly, the direct illumination need no longer be restricted to a narrow portion of the aperture, thus avoiding the artifacts that result from severe stopping-down, such as the confusing images of the substage stop formed by features which are partly out of focus.

The foregoing are of advantage from the observational stand-point, but of *greater importance* are the facilities that the instrument affords for measuring the precise amount of phase change produced. It is possible to see phase changes of less than 1/120 wave-length, and under good conditions changes of 1/300 wave-length have been observed.

The interference microscope thus permits both observation and measurement of the various phase changes produced by transparent objects and this alone is often valuable; but this value is enhanced if the phase changes produced by objects can be converted into, for example, biochemical information about the total protein content of a living cell, or its water content. Formulae are given below for converting the measured phase relationships into refractive indices, and in addition the reader is referred to the papers listed in the Bibliography.

Description of the Instrument

The optical design of the Baker interference microscope was devel-oped by Mr F. H. Smith in the firm's laboratories; while in mechan-ical design it closely follows the Baker Series 4 instruments, with important modifications to meet the specialized requirements of an interference microscope (Plate 11).

Coarse and fine focussing movements operate on the stage, while the body-tube remains stationary, although its position in relation to the stage of the instrument can be changed by releasing the lever on the right hand side of the limb and sliding the tube in its dovetails. The lever operates a cam and locking pad which hold the tube firmly. The body should be firmly supported by the left hand whilst this adjustment is being made. This arrangement of fixed body and focussing stage has distinct advantages, particularly in photo-

---

Fig. 2. A. Swing-out polarizer. The rotation of this element controls the intensity relationship between the double-refracted beams, permitting the out-of-focus image to be extin-guished for normal transmitted-light conditions.
   B. Double-refracting plano-concave lens.
   C. The double-refracted rays entering the Abbe condenser.
   D. Double-refracting plate cemented to the front lens of the condenser rendering it bi-focal.
   E. Double-refracting plate rendering the objective bi-focal.
   F. The re-combined double-refracted rays.

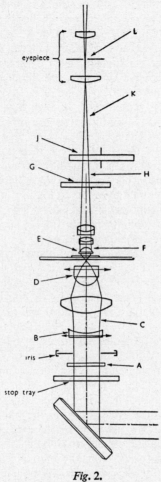

Fig. 2.

G. Quarter-wave plate.

H. The re-combined rays circularly polarized in opposite directions by the quarter-wave plate.

J. Rotatable analyser, with swing-out section, calibrated in degrees.

K. The phase relationship between the circularly polarized rays is adjusted by the analyser.

L. Final image exhibiting interference between the in-focus image of the object superimposed upon the out-of-focus image.

micrography. Focussing controls are in a convenient, low position, enabling the hands to be rested on the bench; whilst the milled heads for coarse and fine adjustment are placed in close proximity, so that both can be reached by a single movement of the hand.

In the interference microscope, each objective requires its own special condensing system. To facilitate the easy and rapid change of condensers the sliding type of changer (after Akehurst) is provided, in which each unit is mounted on its own dovetailed plate, which slides into its corresponding fitting on the substage. (See 'To Line up a Microscope for Critical Work', page 369.)

Plate eleven shows the general appearance of the interference microscope and the positions of the different parts.

Below the condenser (C) is situated the polarizing plate (P) with a lever to control its orientation, and mounted on a hinged bracket to enable it to be swung out of the axis. Above the polarizer is the iris diaphragm and below it the tray for colour filters, as found in conventional substage arrangements.

The condensers are pre-centred and placed in their correct orientation in relation to their changer slides. *No attempt should be made to unscrew any part of them* because they are lined up with great accuracy on a special bench. The pairs of knurled screws (SC) on the upper surfaces of the plates control the tilt of the systems, as explained below.

The objectives are mounted on the usual type of rotating nose-piece changer (N). Unlike normal objectives, *these must never be removed* from their nosepieces and are, in fact, pinned in position to prevent this accident from happening. In order that other objectives may be used on the stand, provision is made for the entire nosepiece with its objectives to be removed from the body-tube on dovetail slides and replaced by other nosepieces containing alternative interference objectives or normal objectives.

Situated in the body-tube above the nosepiece is the tube-length correcting lens (T) for the binocular tubes. This is mounted in a swing-aside fitting controlled by a lever; on the left side of the body and above it is the quarter-wave retardation plate (RP) on a slide marked with the 'in' and 'out' positions. Above the quarter-wave plate is the rotating analyser (RA) graduated in degrees and with its 'out' position marked by a red line on the graduation circle.

Interchangeable monocular and binocular eyepiece fittings screw into the upper end of the body-tube; when the former fitment is in use the tube-length compensating lens *must be swung aside*.

The mirror bracket plugs into the tail of the limb by a three-pin fitting and is thus readily removable if it is desired to use the microscope in a horizontal position. For photomicrography the Projecto-lux camera unit is available.

Choice of Objective System
The *Shearing System* objectives are designed for making sensitive measurements upon separated features, e.g. single cells; but they are not so well suited for continuous specimens such as sections: for these the *Double Focus* systems are more suitable, but measurements with these will not be quite so accurate as with the shearing system. The following notes explain the essential optical characteristics of these alternatives.

The Optical Systems
(a) *Shearing System.* Figure 3 illustrates this form of the instrument. The same portion of the luminous source is imaged by the condenser in two laterally separated areas in the region of the object, while the subsequent effect of the special form of objective brings the images in these two separated areas into coincidence in the image plane, which consequently contains a correctly focused image of the specimen superimposed upon a mutually coherent laterally displaced reference area. These interfere in a manner which is controlled by the double-refracting phase-shifting system, which also permits the interference contrast to be varied between positive, central dark field and negative contrast.

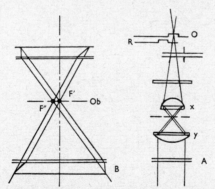

Fig. 3. Paths of rays through shearing system.
    A.   Basic parts of instrument.
    O.   Object wave with focus profile.
    R.   Reference wave with sheared profile.
    B.   Enlargement of space xy.
    F'.   Object focus.
    F".   Reference focus.
    Ob. Object plane.

The *Reference Area* with which any given feature of the object is interferometrically compared lies in a laterally displaced region of the field, isolated from the object feature, see Fig. 4. Thus, the shearing system permits a complete spatial segregation between the object features and the reference area, when the object feature is below a size dependent upon the character of the objective in use.

In the case of the ×10 shearing objective, the centre of the reference area is separated from the optic axis by a distance of approximately 330 microns; for the ×40 system, this distance is about 160 microns; and for the water immersion ×100 system, the corresponding distance is about 27 microns. These dimensions signify that for the three systems, objects of up to 330μ, 160μ, and 27μ respectively in diameter can be viewed *without any overlap* between the object and its reference area. Complete freedom from overlap,

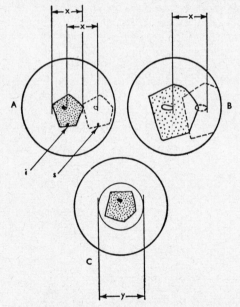

Fig. **4.** A. Shearing system showing complete separation of true image i and sheared one s. X is separation between optic axis and centre of reference area=maximum size of cell which can be viewed without overlap.
   B. Shearing system showing some overlap between i and s. Area overlapped invalid for measurements.
   C. Double focus system. Y is diameter of reference area available for measurement of cytoplasm of cell shown. Size of cell must be less than diameter Y.

however, is not as a rule necessary, and measurements can usually be effected in *any area free from overlap*, Fig. 4 (A, B).

(*b*) *Double-Focus System.* Fig. 5 illustrates the optical layout of the microscope for this system, in which the optical properties of the matched condenser and objective pairs impart double-focus effects.

The double-focus systems differ from those already described in that reference area for any given feature of the object, instead of being located in a laterally displaced region, surrounds the feature, Fig. 4 (C).

The resulting images interfere in a manner which again is controlled by the double-refracting phase-shifting system.

Instructions for setting up and using the interference microscope are give below, but the following section, explains some basic theoretical principles.

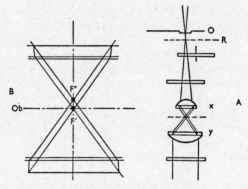

Fig. 5. Double focus system.

A. Basic parts.
O. Object wave focus profile.
R. Background reference wave.
B. Enlargement of space xy.

F'. Object focus.
F''. Reference focus.
Ob. Object plane.

Theoretical Treatment

Phase changes exhibited by objects made visible by interference microscopy can be measured, and from the quantitative knowledge of these phase changes, much useful information can be derived (see References).

Interference phenomena can be used to determine phase relationships between mutually interfering beams of light and it will be seen later how these can be converted into useful information concerning microscopic objects. The essential step in this process is to cause the object to modify the phase relationship between the beams, so that the measured phase relationships can be directly related to the optical path-length-changing properties (optical thickness) of the object. The interference microscope achieves this selective action on the part of the object by obtaining the mutually interfering wave trains in the combined beams from different portions of the object area, whereby a wave train proceeding from an object feature becomes combined with a wave train from another portion of the object area, conveniently referred to as the *comparison area* or *reference area*. Consequently, the final image is a compound one comprising two superimposed mutually different views of the object area which are interferometrically compared. This image-doubling is achieved by double-refraction due to anisotropic crystalline material in the objective system.

Pairs of wave trains can mutually interfere only if they originate from precisely the same portion of the light source, which means that the two superimposed different images of the object area must be illuminated by identical images of the light source in perfect point-to-point registration upon each other. In other words, the image-doubling power of the objective must be exactly compensated out (for the light source) by introducing a precisely complementary form of image-doubling between the source and the object space, i.e. in the condenser system. This is carried out by the incorporation of double-refracting crystalline elements in this system as well as in the objective. The essential optical systems are diagrammatically shown in Figs. 2, 3, 4.

Owing to the wave-like behaviour of light, the intensity which results from the combination of two beams originating from precisely the same portion of a luminous source depends not only upon the individual intensities of the beams, but also upon the distance between a crest of one wave train and a corresponding crest of the other. This principle is diagrammatically illustrated in Fig. 6 (*a-d*), which shows how two entirely different wave *amplitudes* ($A+B$ and A^1+B^1) can arise from the super-position of pairs of identical wave trains, only differing by the distance between successive crests in the two pairs (45° or $\frac{1}{8}$ wave-length in $A+B$; 135° or $\frac{3}{8}$ wave-length in A^1+B^1). This crest-to-crest distance is conventionally known as the *phase relationship* or *phase difference*.

Vector Diagrams
Although the wave diagrams of Fig. 6 clearly illustrate the dependence of the resulting intensity upon the phase relationship between the combined beams, a more convenient and exact means of repre-

sentation is the vector diagram. Two such diagrams, illustrating the two conditions in Fig. 6 (A, C), are shown in Fig. 6 (B, D). Each vector is a straight line whose length is proportional to the amplitude (crest-to-trough displacement distance) of one wave train and whose direction represents the position of a crest of that train for one moment of time. Since this moment is arbitrary, all that is really relevant is the directional relationship (conventionally known as *phase angle*) between the two vectors representing the two wave trains.

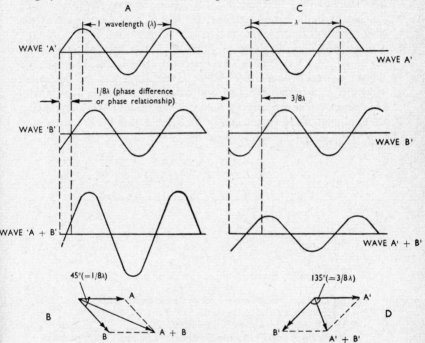

Fig. 6. A, B. Wave and vector diagrams showing combinations of wave trains with difference of $\frac{1}{8}\lambda$.

C, D. Diagrams showing combination of two waves with difference of $\frac{3}{8}\lambda$.

It will be seen that the vector diagram achieves considerable geometrical simplification and this facilitates the derivation of a simple formula establishing the resultant amplitude obtained by the superimposition of two identical wave trains. For the special case, where the two trains have same amplitude, 'A', this formula is

$$RA = 2A \cos\frac{\theta}{2}$$

where RA is the resultant amplitude and θ is the phase angle between the trains. In general it is only relative amplitude which is of interest, so that 'A' can be omitted from the equation. In practical double-beam interferometry this special case of amplitude equality obtains to a sufficiently close approximation.

Minimum Illumination Condition

An example of particular importance is where the phase angle θ is 180°, for then $\theta/2$ becomes 90°, for which the corresponding cosine value is zero. The amplitude for this special condition is, therefore, also zero, so that the illumination is at a minimum.

This condition of minimum illumination can be used to establish a known phase relationship between two wave trains, to serve as a datum against which other phase relationships can be measured. How this can be done is vectorially illustrated in Fig. 7. A phase-changing system for varying the phase relationship by any known amount is adjusted to give minimum illumination for the portion of

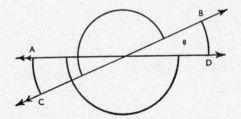

Fig. 7. Vector diagram illustrating transference method of phase measurement.
- A. Background beam adjusted to give minimum illumination at the reference portion of the field.
- B. Object beam at feature to be measured.
- C. Background beam readjusted to transfer minimum illumination to the feature.
- D. Object beam at reference portion of field.

the visual field to which it has been decided to relate quantitatively another portion where the phase relationship is different. The 180° phase relationship for the reference portion of the field having been thus established, the phase changing system is again adjusted, this time to establish the same minimum illumination condition for the other portion of the field. The amount by which the phase relationship has to be changed to move the minimum illumination condition from one portion to the other is clearly the difference in phase relationship between these two portions.

Direction of Measurement

From the same figure, however, it can be seen that there is an ambiguity as to whether the phase angle is to be measured in a clockwise or anti-clockwise direction. The correct direction for the angular phase measurement corresponds with the actual direction of the phase modification being measured. That is to say, that a phase advance must be measured in the phase advance direction and a phase retardation in the corresponding retardation direction. In practice, the direction of the phase modification is usually known, or can be ascertained in a manner to be explained later.

Design of the Present Instrument

There are various methods for manually changing the phase relationship by known amounts, but many of these are excessively coarse for the very small phase changes which are frequently encountered, for example, with cytological material. However, with interferometer systems of the double-refracting type, such as that employed in the Baker interference microscope, the circumstance that the two interfering wave trains are polarized in mutually perpendicular planes enables one to employ a birefringent compensator to effect the phase change. Various forms of these are familiar to users of normal polarizing microscopes. The chosen compensator is very simple and has the additional merit that the measuring principle involved is absolute, so that the uncertainties of calibration are avoided.

The mutually perpendicular transverse vibration directions of the two polarized wave trains are converted into corresponding circular vibrations by transmission through a birefringent quarter-wave retardation plate whose vibrations are inclined at 45° to those of the incident wave trains. Because the plane polarized vibrations of the incident wave trains are mutually perpendicular, the circular vibrations into which they are converted by the quarter-wave plate, occur in opposite directions, the one clockwise and the other anti-clockwise. The advantage thereby gained is that the resultant of the two combined circular vibrations of equal amplitude occurring in opposite directions is always a plane polarized vibration, inclined in precisely the same direction as that of the vector-resultant of the vector diagram for the phase relationship between the two same trains. How this happens is diagrammatically explained in Fig. 8, from which it can be seen that the inclination of the plane polarized resultant of the two oppositely circular-polarized wave trains is directly proportional to $\theta/2$, where θ is the angular phase relationship between the wave trains. The angular inclinations of the plane polarized resultants of the various phase relationships can be read

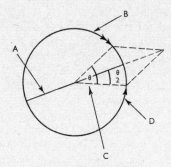

Fig. 8. The plane polarized resultant of two circularly polarized
components.
A. Plane polarized resultant of two circular components.
B. Right hand circularly polarized component.
C. θ=phase relationship between the two circular components.
D. Left hand circularly polarized component.

off directly from a goniometer analyser set to minimize in turn the
intensity of each resultant required (Plate 14, A, B, C).

Matching Method

Phase measurements obtained by transferring the minimum illumination condition from the reference portion of the field to tne
position being measured by means of the goniometer analyser, are
extremely useful but tend to become uncertain when the features
being measured are too small for the minimum illumination condition
to be precisely determined by the eye. For such cases a useful procedure is to adjust the goniometer analyser to obtain equality of
illumination between the reference field and the minute feature. This
is a highly sensitive condition which tends to render the feature
invisible against its reference background. The vector diagram for
this method is shown in Fig. 9. It will be seen that the corresponding
phase adjustment is precisely half that required to transfer the
datum minimum illumination condition to the feature and that the
corresponding goniometer analyser rotation is consequently only
$\theta/4$ where θ is the required angular phase relationship. This method
is applicable to isolated granules, particles and filaments.

Half-shade Eyepiece

This is an additional piece of apparatus which greatly enhances the
accuracy with which phase measurements can be made. The eyepiece
field contains a small area where there is a uniform instrumentally
produced phase change surrounded by a sharply defined boundary.

As a result the two portions of the field generally assumed different intensities which, however, become equal for a very sharply defined position of the analyser. The feature to be measured is made to straddle the boundary between the two portions of the field, the condition of equality being obtained for the reference area and then for the feature.

Integral Wavelengths

The foregoing account of phase measurement has assumed that the phase relationships are less than one wave-length, that is less than 360°. The measurement of phase angles in excess of 360° is slightly complicated by the circumstance that the vector diagrams for phase relationships which differ by integral multiples of 360° are identical. In other words, the above methods are not directly applicable to the determination of integral numbers of wave-lengths. Although phase modifications exceeding 360° are rarely encountered in biological work, it may be necessary to know the modified techniques for dealing with them.

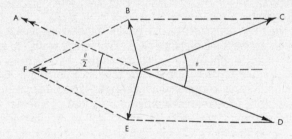

Fig. 9. Vector diagram of matching method for measuring very small features.
A. Background reference beam adjusted to darkest condition in part of field under observation.
B. Resultant for feature.
C. Object beam retarded by feature.
D. Object beam in reference region of field.
E. Resultant for reference field.
F. Background reference beam readjusted to make feature match reference field.

Since the phase modification introduced by an object feature is merely the result of a displacement of the associated wave train along its direction of propagation, the magnitude of the phase modification is virtually inversely proportional to the wave-length of the wave train, a given linear displacement (path difference) containing more short-wave crests than long-wave ones. Consequently, illumination by white light, or light including a wide range of wave-lengths, gives rise to a mixture of different phase relationships. Only for the special case where there is no path difference between the pairs of wave trains will there be one phase relationship for all the different wave-lengths simultaneously. For a polarizing system, therefore, this is the one unique condition for which a neutral coloured minimum illumination condition can be obtained. This neutral minimum, conveniently referred to as 'black', therefore serves as a reference datum for the determination of path differences for any number of wave-lengths. All that is required is a device which will directly modify the actual linear, path-length relationship.

In its simplest form such a device consists of a wedge made from birefrigent material such as crystalline quartz, cut parallel to the optical axis. Sliding such a wedge across the optical aperture provides a variable thickness of birefringent material and therefore serves to vary the path difference between the wave trains, provided that its own vibration directions are approximately parallel to those of the wave trains.

Fringe Eyepiece
Perhaps the most convenient arrangement employing such a wedge is the fringe eyepiece (see page 441). The wedge with its scale is pushed through a slot in the eyepiece mount to positions which give, say, first the black datum for the reference portion of the field and secondly for the feature being measured. The wedge calibrations may be in terms of linear path difference instead of angular phase, since this simplifies the calculation of the refractive indices of features whose thicknesses are known. It is not, of course, essential to select the black interference colour as the datum, any other of the sequence of interference colours may be chosen, but black is probably the most easily recognized. Wedges can also be used in monochromatic light to produce a system of fringes for estimating phase changes by the method of fringe deformation. (Plate 18.)

Refractive Index Calculations
The refractive index, μ, can be readily found if the thickness 't', of the object is known ('t' can be directly measured for objects which can be assumed to be spherical or cylindrical by measuring their diameter):

$$\mu = \frac{\theta \lambda}{360t} + n$$

where θ is the change in phase relationship (in degrees) produced by the object, λ is the effective illuminating wave-length expressed in the same units as 't', and 'n' is the refractive index of the reference portion of the object space, usually the fluid in which the material is immersed. If the actual linear path difference (p.d.) is directly known, for example, by employing a p.d.-calibrated wedge in heterochromatic light, then this formula simplifies to

$$\mu = \frac{p.d.}{t} + n$$

If 't' is not known but it is possible to change the medium in which the material is immersed, then the refractive index can be found from the following equation, where 'A' is the phase difference produced by the object immersed in the medium whose refractive index is 'n_1', and 'B' is the corresponding phase difference for the changed medium of refractive index 'n_2'.

$$\mu = A \frac{(n_2 - n_1)}{A - B} + n_1$$

A helpful feature of this equation is that the phase differences 'A' and 'B' need not be expressed in any particular form, so all that is necessary is merely to use the actual reading differences taken direct from the phase control of the microscope.

Volume and Dry-Mass Calculations
Refractive index values are often very useful in many fields of research, but it is only recently that this has become true for cytology. The important discovery that the increase in refractive index resulting from a one per cent increase in concentration of the solid substances contained in cells is 0.0018, to within 10 per cent for all such substances, has made it possible to calculate the concentration when the refractive index is known. Thus, if zero concentration is taken as equivalent to a refractive index of 1.334, the concentration of cytological substances is

$$C = \frac{\mu - 1.334}{.0018}$$

where C is the concentration in terms of percentage and μ is the refractive index as measured by one of the above methods.

If one is interested only in the combined amount of cytological substances, it is not necessary to know the refractive index because the equivalent thickness of the combined dry substances is proportional to path difference. Consequently, the equivalent volume of dry, concentrated substances in solution is proportional to the product of the cell's mean path difference (m.p.d.) and its area.

Since the increase in refractive index due to a one per cent increase in concentration is 0·0018 to a close approximation, the solid substances contained in the cell, and considered in isolation from the water component, are optically equivalent to a substance having a refractive index equal to $1·334 + ·0·180 = 1·514$, provided that the surrounding medium is water. So the average equivalent thickness of concentrated substances $(^t)_m$ is

$$t_m = \frac{m.p.d.}{1·514 - n}$$

where m.p.d. (mean parth difference) $=_m\theta\lambda/360$, and n=refractive index of the surrounding water, usually about $1·334$. Consequently, the total equivalent volume of dry substances is

$$V = A_c \left(\frac{m.p.d.}{1·514 - n} \right)$$

where A_c is the area of the cell in the same units squared as m.p.d. and V in the same units cubed.

The dry volume may be converted into dry mass for an assumed value of specific weight. For the majority of cases a value of $1·25$ grammes per ml. is a close enough approximation. More detailed information is available in the appropriate references.

To Set up the Interference Microscope: Illumination

The usual light sources are satisfactory on account of the high light transmission of the optical system. Excellent results can be obtained with Köhler illumination. For precise phase measurement a mercury green filter is advisable and a neutral for obtaining interference colours in white light. For preliminary setting-up, however, the filament structure of the lamp in the back focal plane of the objective which results from Köhler illumination should be destroyed by introducing a diffusing screen between the lamp and the substage condenser, since this structure mars the visibility of the interference figure in the back focal plane of the objective.

When available, the ideal source is probably a mercury vapour lamp, with its entire visible region used for obtaining interference colours and a mercury green filter for precise phase measurements.

The substage iris diaphragm should be used in a manner similar to that which is customary in normal bright field microscopy, i.e. it should be opened as wide as is compatible with adequate contrast. This applies particularly to the double-focus type of system. The shearing interference systems do not suffer in contrast from severe stopping down. When using the ×100 double-focus water immersion objective, the substage iris should be completely opened. Only *distilled* water should be used as the immersion medium for both types of ×100 objective. Neither immersion oils nor organic solvents should be allowed to come into contact with the fronts of any of the objectives supplied with the interference microscope.

Adjustment of the Instrument

(Refer to Plate 11 for the positions of the components and controls named in the following paragraphs.)

1. Insert and lock into the substage slides the condenser unit (C) appropriate to the objective being used and rack up to stage level.

2. Rotate the lever of the substage polarizer (P) to the 'off' position to eliminate the second interference beam and thus produce normal bright field conditions.

3. *Focus on a preparation* and select a reasonably clear area. Adjust the condenser focus if necessary.

4. Rotate the substage polarizer lever to the 'on' position.

5. With the quarter-wave plate (RP) 'in' (i.e. with '$\lambda/4$ in' visible on the slide) and the goniometer analyser (RA) in the body on the operating portion of its scale (the 'out' position is marked by a red line interrupting the scale), remove the eyepiece and inspect the back focal plane of the objective by the use of the telescope provided. In diffused white light this should exhibit an interference pattern, which can be traversed across the aperture by turning the goniometer analyser. By this process a fringe can be brought into the centre of the aperture and subsequently broadened to fill it as uniformly as possible by adjustment of the two condenser levelling screws (SC) either simultaneously or alternately.

Typical appearances in the back focal plane of a ×40 objective are reproduced on Plate 15. It will be found of assistance to compare these illustrations with the figures actually seen, until one becomes quite familiar with the working of the instrument.

Shearing Systems

When white light is used, rotation of one substage screw only, or rotation of both in opposite directions, will cause a series of interference fringes to move across the objective aperture, thus permitting selection of the required order of interference. When this order, or fringe, has been brought to the centre of the objective aperture, both screws should be simultaneously operated in the *same* direction to spread this selected fringe over the major portion of the aperture. Whether the rotation of the two screws should be clockwise or anticlockwise to bring this about is at once apparent because the incorrect direction will make the fringe narrower instead of broader. It should be noted that for each shearing system there is only one fringe which can be spread over the major portion of the aperture. Peripheral portions of the aperture which may be slightly beyond the spread of the chosen fringe can be masked out by adjustment of the substage iris diaphragm and this method can be employed when it is desired to use orders other than the optimum one.

While making these adjustments slight irregularities can sometimes be seen in the objective aperture and these are usually due to striae in one or other of the crystal optics. Care is always taken, however, to ensure that such irregularities are too small to impair the performance of the instrument.

Double-Focus System

When the double-focus system is in use a series of fringes in the aperture of the objective indicates that the microscope is out of adjustment, as above; but in this instance one does not have to select the order in which the instrument is to work. The goniometer is rotated until a fringe lies across the centre of the aperture, then one of the substage screws is employed to broaden this fringe, and the other to check any resultant rotation of the pattern.

Need to Check Adjustment

Each time the object or the object-slide is changed, the adjustment of the microscope must be checked, whether the shearing or the double-focus system is being used.

When the most uniform illumination in the objective aperture has been obtained, the instrument is correctly adjusted and it remains only to remove the viewing telescope and to replace the eyepiece and examine the specimen. This should exhibit strong interference contrast, which can be varied by rotation of the goniometer analyser (see Plates, 12, 13 and 14).

Phase Measurements

The most direct method of measuring the phase differences introduced by various features of the specimen is to rotate the goniometer analyser to a position which causes a clear reference portion of the field adjacent to the selected feature to assume a clearly recognizable hue or density, and then again to rotate the goniometer analyser to a second position which causes the selected feature to assume the same appearance as was previously obtained for the reference portion of the field.

Twice the angle through which the goniometer is rotated from the

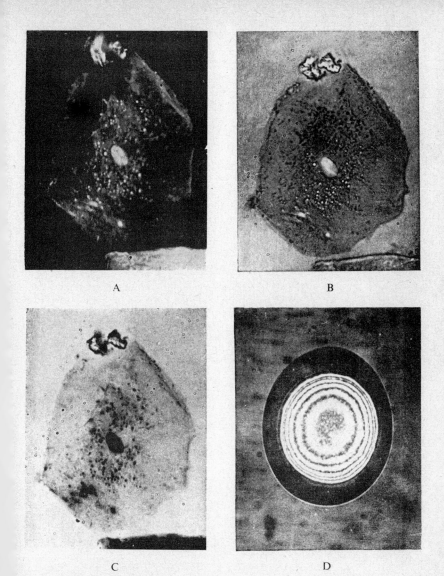

(A) Analyser set to darken background (223 deg.).
(B) Analyser set to darken cytoplasm (185 deg.).
(C) Analyser set to darken nuclear region (121 deg.). The analyser was being rotated in an anti-clockwise direction, indicating progressively increasing retardation.
(D) Sea Urchin Egg, with fertilization membrane, mounted without pressure. This preparation exhibits six fringes, denoting a phase difference of 6λ.

Courtesy of Dr J. M. Mitchison.

PLATE 14

A, B, C: Fresh Epithelial Cell, taken with $\times 40$ shearing objective.

D: Sea Urchin Egg, taken with $\times 10$ double focus objective.

1

2

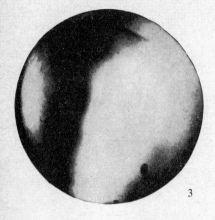

Plate 15

Illustrating the progressive broadening of the fringes seen in the rear focal plane of a double focus objective when the levelling screws are appropriately adjusted.

1. System completely out of adjustment.
2. System still out of adjustment, but fringes broader.
3. One fringe commencing to fill the aperture. Ideal adjustment is achieved when one fringe has completely filled the aperture.

3

The Half-shade Eyepiece.   Plate 16   The Fringe-field Eyepiece.

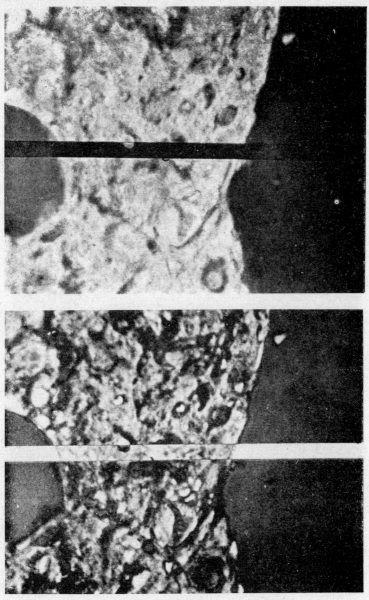

Upper: *Right-hand* side, analyser setting giving a close match between half-shade strip and background (177 deg.).

Lower: Area immediately to the *left* of the very prominent particle on the boundary shows matching condition (41 deg.).

PLATE 17

Typical appearance using the half-shade eyepiece.
Taken with ×40 shearing objective, mercury green line.

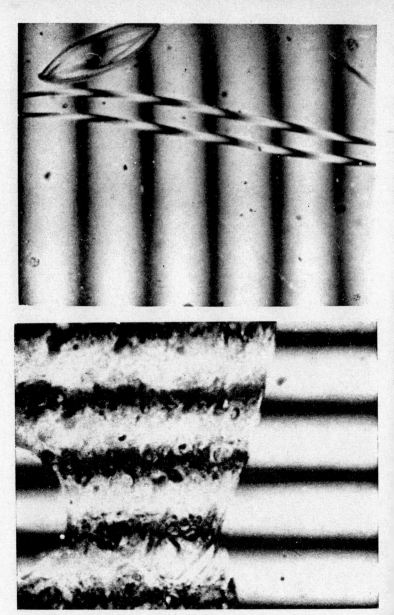

Upper: Shows displacement of the fringes by the siliceous girdle of a diatom.
Lower: Fixed, unstained section of cat pituitary to demonstrate a specimen in the fringe field. Mounted in liquid paraffin (refractive index $1\cdot47$). Optical thickness, $0\cdot28\lambda$: $\lambda = 0\cdot546\mu$.

PLATE 18
Typical appearances using the fringe eyepiece.
Taken with × 40 shearing objective, mercury green line.

first to the second position is the fractional part of a wave-length, expressed in degrees, by which the selected feature has changed the phase relative to the adjacent reference area (see Fig. 10).

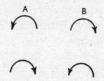

Fig. 10. Chart showing the direction in which the Goniometer Analyser has to be rotated to secure advance or retardation of the reference beam with different objectives.
A. Double-focus × 100 Water immersion objective.
B. All other types and powers of objective.

The correct direction in which to rotate the goniometer analyser is determined by the sense of the phase changing property of the features being measured. The phase changing system of the instrument is so designed that a phase retarding feature (one having a refractive index higher than the surrounding medium) requires an anti-clockwise rotation for all the objectives except the × 100 double focus system. For a phase advancing feature the corresponding rotations are, of course, reversed.

If the sense of the phase change to be measured is not already known from the nature of the preparation, this can be found by, for example, the Becke line test.[1]

Before attempting to apply the Becke test the second beam should be suppressed by rotating the polarizer situated below the condenser to the 'off' position.

When the phase change is too small to provide adequate Becke line conditions, then it can be assumed that it is less than 180°, in which case the correct rotation direction is the one which causes the feature to change towards the original reference appearance as soon as rotation starts.

The recommended reference appearance is the darkest condition obtainable in monochromatic light, preferably mercury green. It is,

[1] When light impinges upon the interface between two media whose refractive indices differ by a substantial amount, the light is reflected and refracted towards the one having the higher refractive index and results in the appearance of a fine white boundary line. It is usually necessary to reduce the cone of illumination to see this effect clearly. The line moves *inside* the object upon *increasing* the distance between the preparation and the objective (lowering the stage of this particular instrument) when the object retards the light, and vice versa. A good deal of biological material is geometrically too amorphous to produce a definite 'line', but the object either brightens or darkens relative to its surround, and these appearances correspond to the line moving 'in' and the line moving 'out'.

however, possible to use such a readily recognizable colour as purple, obtained by the use of unfiltered light.

The phase changes due to features which are too small for their densities to be readily recognized can be measured by the matching method instead (see Fig. 9 and page 432), in which case the goniometer analyser is rotated from the reference position only far enough for the small feature to match its surround. The required phase change is then *four times* the angle through which the analyser has been rotated. This method is often useful for such small isolated features as bacteria and fine filaments. It is often useful also for such features as granular inclusions in cells: the cytoplasm surrounding the inclusion then being used as the selected reference portion of the field.

Practical Examples

(1) The Becke test indicated a feature showing a *retardation* (brightening of the feature upon increasing the distance between the objective and the slide), therefore a rotation of the analyser in an anti-clockwise direction using a ×40 objective was required to darken the object. Twice the angle between the reading for the darkness of the surround and darkness of the feature gave the angular measure of the amount by which the feature was *retarding* the light relative to the surrounding material.

(2) The Becke test was negative, indicating a phase change of less than 180°. Rotation of the goniometer analyser in a clockwise direction with the ×100 objective from the 'dark surround' position darkened the object, thus again indicating a *retardation*, and twice the angular difference between the readings indicated the amount of retardation, as before.

Phase Changes Greater than one Wave-length

(*a*) *Gradual Changes*. The above methods of measurement suffices provided it is known that the phase changes being measured are less than one wave-length. With living biological material this is usually the case, and even in the rare cases when it is not, the maximum phase change is practically always approached by a gradient of phase change which starts at less than one wave-length, resulting in the formation of a fringe system upon the object (see Plate 14, Fig. D). The whole number of wave-lengths associated with the maximum change is then merely the number of dark fringes between the thin edge of the gradient and the region being measured. Any reversal of the sense of the gradient can be detected by rotation of the goniometer analyser, when the fringe or fringes in the reversed region will move in a direction opposed to the motion of the fringes in the other region of the gradient.

(b) *Abrupt Changes.* There is, however, a special case to which the fringe-counting method for determining integral numbers of wave-lengths does not apply. This is when the boundary of the phase change is substantially vertical, resulting in an abrupt transition of phase change. There are no fringes to be counted and a more elaborate technique is therefore required.

Fringe Eyepiece

The best method for determining integral numbers of wave-lengths is to employ a fringe eyepiece system in place of the standard one; and to adjust the quartz wedge to transfer the blackest appearance of the reference portion of the field to the region being measured, white (unfiltered) light being used. The number of fringes which traverse the region while this adjustment is being made corresponds to the required number of integral wave-lengths. When using this ocular, both the goniometer analyser and the quarter-wave slide should be in their respective 'out' positions.

A method which has the advantage of not calling for instrumental elaboration is to obtain the integral number of wave-lengths from two phase readings instead of one, the first being made with light of one mean wave-length and second with light of a different mean wave-length. It is, of course, essential that the two readings are obtained from two separate settings for the reference portion of the field, corresponding to the two wave-lengths.

A formula for this method is as follows:

$$\text{Phase change} = (A_1 - A_2)\frac{\lambda_2}{\lambda_2 - \lambda_1}$$

where 'A_1' is the phase difference with light of shorter wave-length λ_1, and 'A_2' is the corresponding phase difference with light of longer wave-length λ_2. It is advisable to ignore the fractional components of the result, since this has already been accurately obtained in the form of A_1. The difference between the two wave-lengths should not exceed λ_2/N, where N is the number of wave-lengths being measured. This raises no practical difficulties, because abrupt changes exceeding five wave-lengths are extremely improbable, especially with biological material.

Good quality commercial filters will usually be found adequate, even in white light, and the mean visual wave-lengths are usually obtainable from the manufacturers. It is recommended that one of the two filters should be the mercury green one previously mentioned (see page 436). The second filter could then be blue. If a mercury vapour source is used, then the green filter will provide the single green line, the second filter can be one designed to transmit only the blue line.

Practical Notes on the Interference Microscope

(1) The ×100 objectives are corrected for *water* immersion and must *never* be oil immersed, or cleaned with organic solvents.

(2) It is advisable to ensure that No. 1 cover glasses of a *measured* thickness of 0·18 mm. are employed for all preparations. This applies equally to both dry and water immersion systems.

(3) More than usual cleanliness in respect of condenser and objective surfaces, slides and covers, is necessary with the interference microscope.

(4) The objectives are permanently fixed in the nosepiece and no attempt should be made to remove them.

(5) The front component of the ×100 objective is nearly flush with the mount, so that particular care should be taken not to press it on the slide. *This objective is fragile.*

(6) If a fringe eyepiece is employed both the quarter-wave plate and the analyser must be placed in their respective 'out' positions.

(7) When using Köhler illumination it is particularly important to see that the filament image is centred on the aperture of the substage condenser; this can be checked either by direct inspection of the condenser iris diaphragm or by examination of the back focal plane of the objective.

(8) When obtaining the interference figures in the back focal plane of the objective for setting-up purposes, insert a ground screen in front of the lamp to destroy the Köhler image of the filament.

(9) If an attempt to obtain a setting of the goniometer analyser is frustrated simply by the 'out' position of the scale appearing in the window, this condition is immediately rectified by turning to the diametrically opposite side of the scale.

(10) Make sure the sliding changer of the substage is pushed completely 'home' when changing condensers, and is properly locked in position.

Interference Colours

In general, the intensity variations associated with the phase variations produced by an interferometer system when illuminated by monochromatic light are transformed into colour variations when the light contains a mixture of widely differing wave-lengths. This phenomenon results directly from the circumstance that any given linear optical path difference between the interfering beams corresponds to phase differences which are inversely proportional to wave-length. For example, if one considers an inteferometric field across which there is a uniform gradient of increasing optical path difference it is apparent that the corresponding gradients of phase difference are inversely proportional to wave-length. Consequently the field will be crossed by fringes whose spacing is proportional to wavelength, so that a mixture of wave-lengths results in a corresponding mixture of fringe spacings. For instance, red fringes will be more widely spaced than will be blue ones, resulting in a cyclic variation of the red-blue intensity relationship. This variation in intensity relationship applies, of course, to all the colours present in the illuminating beam so that there is a corresponding progressive variation in the colours seen in the interferometric field.

The colours associated with given optical path differences for white-light illumination when the analyser is set to make the zero path difference appear black, are set forth in the table (see page 447). Orders higher than 3 are not included because their hue saturation is insufficient for the estimation of path differences.

Half-Shade Eyepiece

The accuracy of visual phase difference measurement with a normal interference microscope is limited by the precision with which a convenient condition, such as that of minimum brightness, can be recognized. Not only is this particular condition often marred by flare and slight errors of adjustment, but it can be very difficult to recognize when the object to which it has been transferred by instrumental phase adjustment is very small in relation to the field of view, owing to the disturbing effect of glare from the rest of the field.

The half-shade eyepiece has been introduced to increase greatly the accuracy of phase measurement by exploiting the fact that the eye is far more precise in its estimation of equality of illumination between two adjoining areas in the field, than it is in recognizing a selected level of illumination at any given region. The Baker half-shade eyepiece carries a 45° inclined viewing prism having a totally reflective face which receives the primary image of the object formed by the microscope objective. A central horizontal strip of the reflecting face is coated with an aluminium film. Owing to the different character of metallic reflection from total reflection at a dielectric/air interface, the two perpendicularly plane polarized beams of the

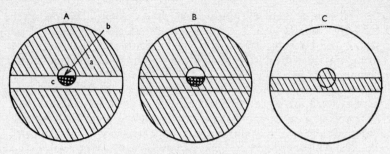

Fig. 11. Appearances seen with half-shade eyepiece.
A. Non-balance condition.
B. Balance for reference area (Reading 1).
C. Balance for object (Reading 2).
     a. Reference area.
     b. Object.
     c. Half-shade strip intersecting field.

double-refracting interference microscope system experience a relative phase shift at this metallized strip. In other words, the strip is exactly equivalent to an artificial object producing a constant and uniform phase change with extremely clean-cut edges and consequently changes its relative luminosity when the phase changing system is varied. There are two such phase settings for which the strip matches the surrounding field. One of these two positions is extremely sensitive, and is clearly recognized because it yields a markedly *lower over-all brightness* than does the other, insensitive condition. The normal measuring procedure consists in adjusting the phase system to yield the sensitive balance condition for the reference portion of the field and then to make a further adjustment to transfer this same sensitive balance condition to the feature whose phase shift is being measured, the feature having been previously brought into a position where it is straddled by at lease one edge of the strip. In general, the accuracy is more than four times greater than with the normal, direct method.

The prismatic, locally metallized face, together with the primary image of the object, are viewed through a low-power auxiliary microscope provided with a screw-in form of focussing adjustment. This arrangement permits the metallized strip to be focussed with the necessary accuracy, and also allows the use of alternative eyepieces.

The metallized prism is carried in a horizontal slide, which also carries an adjacent prism without a metallized strip, and which can be pushed into the optical system when it is desired to return to normal viewing.

The half-shade eyepiece unit screws on to the microscope body in place of the normal monocular tube or binocular head and is locked in place when it is squared-on to the microscope limb. The unit contains its own phase-changing system, so it is necessary to ensure that the analyser and quarter-wave slide in the body are both in their respective 'out' positions.

Fringe-Field Eyepiece
Since the introduction of the Baker interference microscope, experience has shown that special circumstances may exist for which the uniform field provided by this instrument is less useful than would be a field crossed by a system of straight interference fringes. For the uniform field condition, phase changes in the specimen can be measured by adjustments of the analyser, but when it is preferred to make rapid visual estimates, one has to rely upon the information provided by the hues of the interference colours seen in white light. This method depends upon an uncommonly precise acquaintance with interference colours and is not usually favoured.

If, however, relevant optical conditions can be modified in such a way as to produce a field which is crossed by straight interference fringes, visual estimates of phase changes in the object can be made by observing the extent to which the object deforms the straightness of these fringes.

The straight fringes are produced by introducing an optical path difference between the object beam and reference beam which linearly increases across the field. In the absence of localized phase changing features, this uniform gradient of optical path-difference produces, in monochromatic light, a system of evenly spaced interference fringes so that the corresponding distribution of intensity can be used to determine the phase-change produced by some localized feature. This method is useful when photographic densitometric apparatus is available.

The apparatus for producing the fringes consists of a Ramsden eyepiece with focussing eye-lens at the focal plane of which is a quartz wedge supported in a metal carrier which is inserted into a slot provided with a rotation movement through 90°. Between the two eyepiece lenses is an analyser plate in a slide to allow the analyser to be placed in or out of the optical system. The eyepiece unit is pushed into the top of the monocular tube and then secured by the knurled screws after having been orientated with the analyser slide in a diagonal position. The fringes can then be placed either in a vertical or horizontal position by revolving the carrier in the slot into a position which is respectively horizontal or vertical. A shortened monocular tube is provided to compensate for the additional 'tube-length' occupied by the slow below the eyepiece. Both the quarter wave-plate compensator and revolving analyser in the body of the microscope must, of course, be in their respective 'out' positions.

To permit viewing of the back of the objective, the whole eyepiece is easily removed by loosening the securing screws.

If it is desired to change over to normal uniform field conditions without exchanging the eyepiece for a standard one, it is only necessary to push the eyepiece analyser to its 'out' position and move both the quarter wave compensator and analyser in the body to their respective 'in' positions.

Normal quartz wedge compensators can also be provided in metal carriers for this eyepiece in place of the steeper wedges required for the fringe field condition. Such compensators are also available with a calibrated scale for which use the eyepiece is provided with a cross-lined graticule. In this form, the eyepiece can, of course, be used with a normal polarizing microscope for the estimation and measurement of birefringent retardations.

Subjects suitable for study with the Interference Microscope

In the previous pages several comments have been made concerning objects which can best be studied by interference methods but some extra remarks by way of conclusion may be useful.

The light beams which are employed to produce the contrast must, in the system described, be taken from parts of the object area which are covered by the objective field of view. It follows that the objects to be studied must not be entangled in dense masses of other or similar material, and particles must not be closely packed on the slide or else diffraction rays from them will interfere with the wanted effects and cause errors in interpretation. If the area of the object is large and it is of an even clear structure the effect of stray diffraction is less important because the object forms an even background to the observations against which interference effects from included structures can be well seen and measured. If cells are being studied they should be isolated. Such objects as spherical particles of very different refractive index from their mountant interfere with the contrast because they act as lenses and cause errors in the microscope set-up which cannot be overcome except by using a mountant of nearly the same refractive index as the object and depending upon interference contrast to find it.

Any object which has the property of bi-refringence will upset the working of the microscope because so many parts of it depend upon polarization phenomena for their operation and as the expression

No.	Retardation $\lambda = 589m\mu$	Order	Interference Colours between Crossed Nicols
1	0	0	Black
2	40		Iron-grey
3	97		Lavender-grey
4	158	$\frac{1}{4}$	Greyish blue
5	218		Clearer grey
6	234		Greenish white
7	259		Almost pure white
8	267		Yellowish white
9	275		Pale straw-yellow
10	281		Straw-yellow
11	306	$\frac{1}{2}$	Light yellow
12	332		Bright yellow
13	430		Brownish yellow
14	505	$\frac{3}{4}$	Reddish orange
15	536		Red
16	551		Deep red
17	565		Purple
18	575		Violet
19	589	1	Indigo
20	664		Sky-blue
21	728		Greenish blue
22	747		Green
23	826		Lighter green
24	843		Yellowish green
25	866		Greenish yellow
26	910	$1\frac{1}{2}$	Pure yellow
27	948		Orange
28	998		Bright orange-red
29	1101		Dark violet-red
30	1128		Light bluish-violet
31	1151	2	Indigo
32	1258		Greenish blue
33	1334		Sea-green
34	1376		Brilliant green
35	1426	$2\frac{1}{2}$	Greenish yellow
36	1495		Flesh-colour
37	1534		Carmine
38	1621		Dull purple
39	1652		Violet-grey
40	1682		Greyish blue
41	1711		Dull sea-green
42	1744	3	Bluish green
43	1811		Light green
44	1927		Light greenish grey
45	2007		Whitish grey
46	2048		Flesh-red

TABLE 14

'bi-refringence' means by definition 'two refractive indices' it will be seen from the description of the operation of the interference system that incalculable mixed effects must occur when the object puts its own finger in the pie. In general bi-refringent substances cannot be properly studied in an interference microscope.

Materials such as replicas of all degrees of coarseness, transparent films of all kinds (except bi-refringent ones) and most biological sections if thin and clear make suitable objects. Great progress in cylology has been made by employing the measuring properties of the interference system to sort out cell constituents.

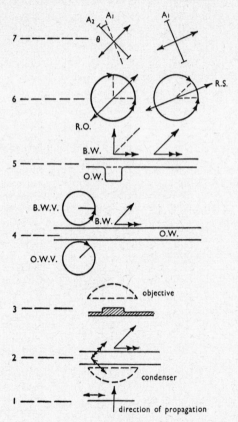

Fig. 12. Diagrammatic summary of the basic phase measuring system.

7. At A_1 the analyser is set to darken surrounding portion of field, and at A_2 to darken the object element. Twice the angular difference between these settings (θ) is the angular phase difference required.

The Interference Microscope in Quantitative Cytology

One of the most important applications of interference microscopy is to the determination of the dry mass of living cells. In certain cases much other useful cytological information such as the concentration of solids and of water, cell thickness and volume can also be deduced.

The basic principle underlying all such determinations is that there is a direct relationship between the concentration of a solution and its refractive index. Many experimental observations have shown that this relationship can be expressed in the form

$$\mu = \mu^\circ + a\ C$$

where μ is the refractive index of the solution, μ° is that of the pure solvent and C is the concentration of the dissolved substances. a is a characteristic constant known as the *specific refraction increment*. It is important to note that C must be expressed in terms of weight per unit volume, or the linear relationship between refractive index and concentration will not hold. C is most conveniently given as grammes per 100 ml. of solution, in which case a— becomes the increase in refractive index per 1 per cent increase in concentration.

a has been measured experimentally for many substances. From the cytological point of view proteins are of the greatest interest since they usually constitute the major solid component of living cells. Nearly all soluble unpigmented proteins have values of a which do not deviate from ·00185 by more than ±2 per cent. Values for nucleic acids and nucleoproteins are less certain, but are evidently very close to those for proteins. Carbohydrates generally have lower values (about ·00143), but carbohydrate-protein complexes have

---

6. R.S. is the vibration direction resulting from combination of the circular vibrations shown at (4), and R.O. is the corresponding direction for the combined vibrations at the image of the phase retarding element in the preparation.

5. Both waves reach the image plane, but only the object wave (O.W.) reproduces the profile modification due to the phase retarding object element, background wave (B.W.) having been de-focused by double refracting objective. The two phase relations which result are shown by the vectors.

4. Both waves leaving the quarter-wave plate circularly polarized in opposite directions as shown by circle diagram. The positions of the rotating phase relationship shown by vector diagram.

3. Represents phase retarding element in the preparation.

2. Two waves derived from a single wave by the double-refracting plate. The resulting vibration directions shown by double arrows. Phase relationship between waves indicated by vector diagram.

1. Single wave front, double arrow indicating plane polarization after leaving polarizer.

values similar to those for proteins. It is not possible to discuss the refraction increment of ordinary lipids since these are normally insoluble in water, but lipoproteins have values between ·00170 and ·00185. Taking all these facts into consideration and remembering the predominance of proteins, a mean value of 0·0018 for the refraction increment of protoplasm seems reasonable, and it is highly improbable that this value would be in error by more than 10 per cent.

Dry Mass

Let us now consider the case of a living cell of refractive index μ, examined in a dilute salt medium of refractive index μ_o which can be taken as approximately 1·334. We have

$$\mu = \mu_o + a\ C$$

Hence $$\mu - \mu_o = a\ C$$

Multiply each side by the cell volume, written as At, where A is the projected area and t the thickness

thus $$A\ (\mu - \mu_o)\ t = a\ ACt$$

But by definition the product of thickness and refractive index difference between object and medium is simply the optical path difference, p.d.

so that $$A \times p.d. = a\ ACt$$

The quantity ACt is the concentration multiplied by the volume, in other words the dry mass. If C is expressed in grammes per 100 ml. of protoplasm, the dry mass in grammes is $\dfrac{ACt}{100}$

i.e. total dry mass $= \dfrac{A \times p.d.}{100a}$

or dry mass per unit area $= \dfrac{p.d.}{100a} = \dfrac{p.d.}{·18}$

This remarkably simple result shows that the path difference as measured by interference microscopy is directly proportional to the dry mass at the point measured. Although in deriving this formula it has been assumed that the cell is homogeneous, the total dry mass of a heterogeneous cell can be found by dividing the cell into a number of homogeneous elementary areas and summing or integrating the product of $\dfrac{p.d.}{·18}$ and the elementary area. This may be

a very tedius process, and it is usually simpler, though less informative, to determine the dry mass per unit area at different regions. It will be seen that the determination of dry mass does not require a knowledge of the cell thickness t, which may be extremely difficult to measure.

Concentration of Solids

As already shown p.d.$=a$ Ct. Hence the solid concentration C at any point $=\dfrac{\text{p.d.}}{at}$ If t is known or can be measured, C can be found. The measurement of the thickness of irregularly shaped living cells is extremely difficult, but t can be found with reasonable accuracy in the case of spherical or cylindrical cells.

An alternative method applicable to cells of any shape is to determine the refractive index of the cytoplasm by immersion refractometry. In this the cell is immersed in an isotonic protein solution (for full details see references) and observed by phase-contrast or interference microscopy. When the refractive index of the cytoplasm matches that of the immersion medium, the cytoplasm becomes virtually invisible. The concentration of solids C is then found from the equation

$$C=\frac{\mu-1\cdot334}{\cdot0018}$$

This method has the advantage of not requiring the measurement of t, but precautions are necessary in order to ensure that the cell does not shrink or swell in the protein medium.

Cell Thickness and Volume

Since

$$\text{p.d.}=aCt$$

$$t=\frac{\text{p.d.}}{aC}$$

Thus t can be calculated if both p.d. and C are measured. Similarly, the volume

$$V=At=\frac{A\times\text{p.d.}}{aC}$$

Concentration of Water, Mass of Water and Total Wet Mass

If the concentration of solids is C, it might be thought that the concentration of water would be $100-C$. In general, however, C grammes of solids do not occupy C ml. in solution. One gramme of protein actually occupies about $\cdot75$ ml. in aqueous solution. Hence,

100 ml. of a protein solution containing C grammes of protein would contain $(100 - \cdot 75C)$ ml. of water, i.e. the concentration of water is $(100 - \cdot 75C)$ grammes per 100 ml. of protoplasm. Since concentration is mass per unit volume, it follows that

$$\text{Mass of water} = \frac{100 - \cdot 75C}{100} \times At$$

$$= \left(1 - \frac{3C}{400}\right)At$$

or, mass of water
per unit area $= \left(1 - \frac{3C}{400}\right)t$

This result can be expressed in a number of different forms which can be conveniently used according to circumstances.

Thus, mass of water
per unit area $= t - \frac{3}{400}\frac{\cdot \text{p.d.}}{a}$

$$= t - \frac{\text{p.d.}}{\cdot 24}$$

$$= \frac{\text{p.d.}}{a}\left(\frac{1}{C} - \frac{3}{400}\right)$$

$$= \frac{\text{p.d.}}{100a}\left(\frac{100}{C} - \frac{3}{4}\right)$$

$$= \text{dry mass per unit area} \times \left(\frac{100}{C} - \frac{3}{4}\right)$$

The total wet mass is simply the sum of the mass of water and the dry mass.

Hence total wet mass $= \left(1 - \frac{3C}{400}\right)At + \frac{A\,Ct}{100}$

$$= \left(1 + \frac{C}{400}\right)At$$

or, total wet mass
per unit area $= \left(1 + \frac{C}{400}\right)t$

$$= t + \frac{\text{p.d.}}{400a}$$

$$= \frac{\text{p.d.}}{100a}\left(\frac{100}{C} + \frac{1}{4}\right)$$

$$= \text{dry mass per unit area} \times \left(\frac{100}{C} + \frac{1}{4}\right)$$

Density

As shown above,
$$\text{total wet mass} = \left(1 + \frac{C}{400}\right)At$$

hence density of protoplasm (=wet mass per unit volume)
$$= 1 + \frac{C}{400}$$

In some cases it is more convenient to express the results in grammes per unit wet weight rather than volume. This is easily done by dividing the concentration in grammes per unit volume by the density.

Thus concentration of solids in grams per 100 grammes of
$$\text{protoplasm} = \frac{C}{1 + \dfrac{C}{400}}$$

$$= C - \frac{C^2}{400} \text{ approx.}$$

Similarly, the concentration of water in grammes per 100 grammes of
$$\text{protoplasm} = 100 - C + \frac{C^2}{400} \text{approx.}$$

In many cases the term $\dfrac{C^2}{400}$ is small compared with C and can be neglected.

Combined Refractometry and Interferometry

A method which is capable of yielding much information in certain cases is to measure the optical path difference produced by the same object in two media of different refractive indices. Let μ be the refractive index of the object and t its thickness. Let μ_1, μ_2 be the refractive indices of the mounting media and p.d.$_1$, p.d.$_2$ the corresponding path differences measured in these media.

Then
$$\text{p.d.}_1 = (\mu - \mu_1)t$$
$$\text{p.d.}_2 = (\mu - \mu_2)t$$

so that
$$\mu = \frac{\mu_2 \text{ p.d.}_1 - \mu_1 \text{ p.d.}_2}{\text{p.d.}_1 - \text{p.d.}_2}$$

and
$$t = \frac{\text{p.d.}_1 - \text{p.d.}_2}{\mu_2 - \mu_1}$$

In the case of living cells it is necessary to use isotonic protein immersion media. In this case we can write

$$\mu = 1 \cdot 334 + aC$$

and our equations become

$$C = \frac{C_2 \text{ p.d.}_1 - C_1 \text{ p.d.}_2}{\text{p.d.}_1 - \text{p.d.}_2}$$

and

$$t = \frac{\text{p.d.}_1 - \text{p.d.}_2}{a\,(C_2 - C_1)}$$

where C_1, C_2 are the concentrations of protein in the media. In general it is convenient to carry out one measurement in a simple saline medium, in which ase $C_1 = 0$ and the equations reduce to

$$C = \frac{C_2 \text{ p.d.}_1}{\text{p.d.}_1 - \text{p.d.}_2}$$

and

$$t = \frac{\text{p.d.}_1 - \text{p.d.}_2}{aC_2} = \frac{\text{p.d.}_1}{aC}$$

This method, though potentially very valuable, may be difficult to carry out in practice in the case of living cells, since it is essential that these should not move or alter in shape or volume during the measurements.

Measurements on Fixed Tissues
After the usual processes of fixation and dehydration the cell solids can be regarded as existing as a network of precipitated particles or fibrils. The spaces between the particles will generally be filled by the mounting medium, but the medium will not usually penetrate into the dried particles or fibrils themselves. Let us imagine that the specimen is composed of a dried tissue of overall thickness t containing particles of total thickness t_p and refractive index μ_p. The thickness of the empty spaces will thus be $t - t_p$. When the tissue is mounted in a medium of refractive index μ_m the total optical path through it will be

$$\mu_p t_p + \mu_m\,(t - t_p)$$

The optical path through an adjacent clear region containing the medium alone is $\mu_m \cdot t$. Hence the path difference measured between the specimen and the adjacent clear region will be

$$\begin{aligned}
\text{p.d.} &= \mu_p t_p + \mu_m\,(t - t_p) - \mu_m t \\
&= (\mu_p - \mu_m)\,t_p
\end{aligned}$$

PLATE 19
Engineering Inspection Microscope: The Brinell Test Microscope.

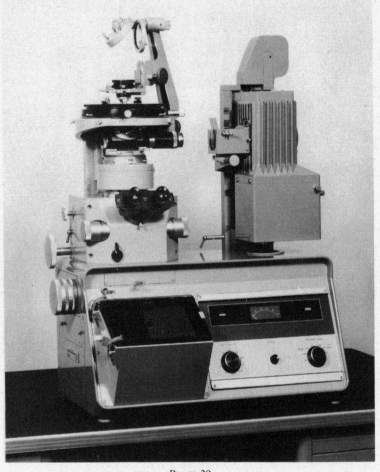

PLATE 20
Vickers Projection Microscope by Cooke, Troughton & Simms set up
for the Micro Examination of a Metallurgical Specimen.

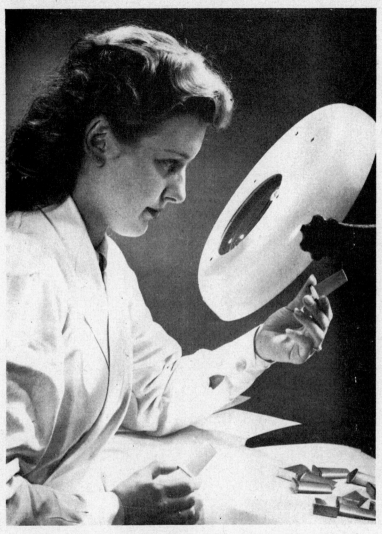

PLATE 21
The Bench Inspection Microscope (see page 468) by Engineering Developments
Ltd., Hemel Hempstead.

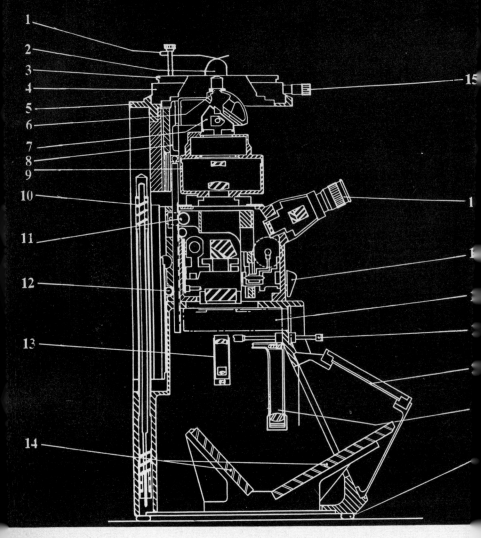

PLATE 22

Sectional Diagram of the Vickers Projection Microscope.

1 Stage clips.
2 Specimen.
3 Gliding stage.
4 Micrometer stage.
5 Stage support.
6 Sextuple objective changer.
7 Illumination box.
8 Slow motion carriage.
9 Magnification changer.
10 Weight relieving spring.
11 Slow motion transfer gear.
12 Rack and pinion (coarse motion).

13 Zoom projection eyepiece.
14 Mirrors (2).
15 Stage traverse micrometers.
16 Binocular eyepiece.
17 Selector switch (Visual only—35 mm. photo visual—macro).
18 Micro shutter.
19 Selector rod (zoom eyepiece—macro—35 mm. corrector lens).
20 Focusing screen.
21 35 mm. corrector lens.
22 Anti-vibration mountings.

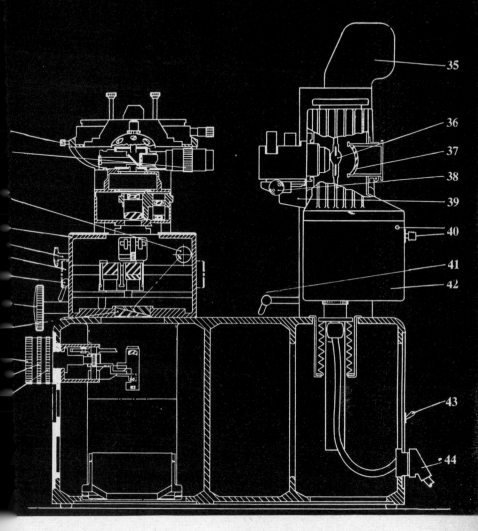

23 Stage clamp screw.
24 Field iris control
25 Photomultiplier cell.
26 Microscope Block.
27 Subsidiary coarse motion pinion control.
28 Fine motion heads.
29 Clamp for coarse motion.
30 Coarse motion milled head.
31 Pick-off prism for photomultiplier.
32 Objective setting scale.
33 Final magnification scale.

34 Magnification changer scale.
35 Weight relieving spring for xenon lamp.
36 Xenon lamp.
37 Reflector unit.
38 Condenser focusing control.
39. Condenser bracket.
40 Lamp centring control heads.
41 Clamp lever for lamp slideway.
42 Xenon lamp casing.
43 Lamp mains switch.
44 Multi-pin plug.

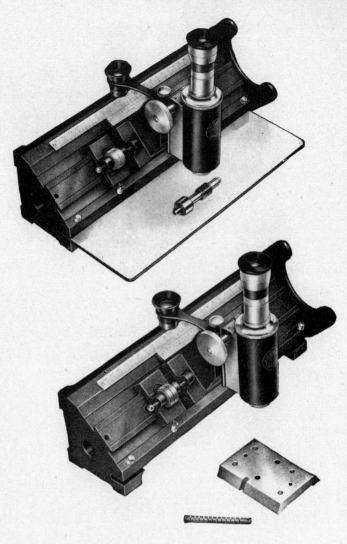

PLATE 23
Vernier Measuring Microscope.

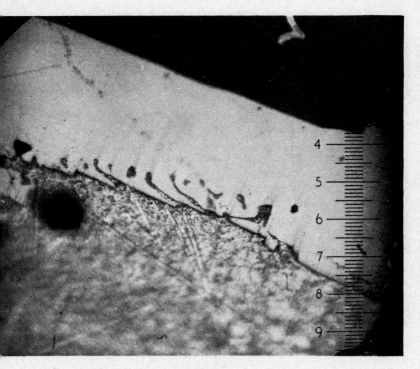

PLATE 24

Sectioned and Polished Junction of Indium and Germanium in a Transistor.
1 scale div. = 1 micron. 4 mm. apo. and approx. × 10 micrometer scale eyepiece;
standard vertical illumination, apparatus as in Plate 2.

PLATE 25

Hydrofluoric Acid Etched Surface of Germanium Crystal.
2 mm. apo. objective magnification 1 micron = 1 mm. standard vertical illumination in mercury green light with apparatus as in Plate 2.

As shown in the previous section, by taking two measurements of p.d. in two media of different refractive indices it is possible to calculate both μ_p and t_p. It must be stressed that t_p is the *effective* thickness of solids, i.e. the thickness which they would occupy if all the interstices were eliminated. The true thickness t of the section cannot be measured by interference microscopy unless the specimen can be mounted in a medium which does not penetrate either the interstices or the solid particles.

The exact interpretation of μ_p is somewhat obscure. The mean refractive index of tissue sections is generally about $1 \cdot 54$ and similar values are found for dry protein fibres such as casein, leather and wool. For proteins in solution the formula $\mu = 1 \cdot 334 + aC$ holds. If this formula is extrapolated for a solid 'solution' of protein, we must put $C = 133$, since 133 grammes of protein would occupy 100 ml. in solution. We then find $\mu = 1 \cdot 574$ (for $a = 0 \cdot 0018$). This is much higher than the experimental value. It may be that tissue sections and protein fibres are not usually fully dehydrated and more adequate drying or precipitation might raise the refractive index. This requires further investigation. According to the formula a refractive index of $1 \cdot 54$ corresponds to a solid content of 114 grammes per 100 ml. On the other hand, it is equally possible that the experimental values are correct, but that a does not remain constant at very high values of C, and indeed the concept of a solid 'solution' is a very theoretical one. For the same reasons it seems scarcely possible to define clearly what μ_p means in the case of dry substances. Davies *et al.* (1959) have suggested that a should be taken as between $\cdot 0019$ (for dilute protein solutions) and $\cdot 0015$ (for dry protein) and that the upper and lower limits for the concentration can be calculated from the refractive index μ_p. It is doubtful, however, whether this procedure gives much more than an indication of the degree of dehydration or precipitation of the specimen solids.

The derivation of absolute values for dry mass is also made difficult by uncertainties in a. If measurements are made on a fixed sections mounted in water, we have, as in the case of living cells,

$$\text{dry mass per unit area} = \frac{\text{p.d.}}{100a}$$

If the path difference is measured instead, in a medium of refractive index μ_m, we have

$$\text{dry mass per unit area} = \frac{(\mu_p - 1 \cdot 334)t_p}{100a}$$

As shown at the beginning of this section

$$p.d. = (\mu_p - \mu_m)t_p$$

Hence dry mass per unit area $= \dfrac{p.d. + (\mu_m - 1 \cdot 334)t_p}{100\alpha}$

In each case it is assumed that the water or other medium has no chemical action on the specimen and does not destroy or dissolve out any solid substance. Despite the uncertainties in the absolute values for dry mass these formulae can be used for comparative studies, for example, to investigate the variations in the distribution of dry mass in different parts of the same or similar specimens. The removal of solid substances by enzymes or solvents can also be followed in this way.

REFERENCES

BARER, R., and ROSS, K. F. A. (1952). 'Refractometry of Living Cells', *J. Physiol*, *118*, 38P.

BARER, R. (1952). 'Interference Microscopy and Mass Determination', *Nature*, *169*, 366.

BARER, R., HOWIE, J. B., ROSS, K. F. A., and TKACZYK, S. (1953). 'Applications of Refractometry in Haematology', *J. Physiol.*, *120*, 67P.

BARER, R., ROSS, K. F. A., and TKACZYK, S. (1953). 'Refractometry of Living Cells', *Nature*, *171*, 720.

BARER, R. (1953). 'Determination of Dry Mass, Thickness, Solid and Water Concentration in Living Cells', *Nature*, *172*, 1098.

BARER, R. (1955). 'Phase-Contrast, Interference Contrast and Polarizing Microscopy', in *Analytical Cytology*. Ed. R. C. Mellors, New York, McGraw-Hill.

BARER, R. (1956). 'Phase-Contrast and Interference Microscopy', in *Physical Techniques in Biological Research*, Vol. III. Ed. G. Oster and A. W. Pollister, New York, Academic Press.

BARER, R., and DICK, D. A. T. (1955). 'Mass, Concentration and Thickness of Living Cells in Tissue Culture', *J. Physiol.*, *128*, 25P.

BARER, R., and JOSEPH, S. (1954). 'Refractometry of Living Cells—I', *Quart J. mic. Sci.*, *95*, 399.

BARER, R., and JOSEPH, S. (1955). 'Refractometry of Living Cells—II', *Quart. J. mic. Sci.*, *96*, 1.

BARER, R., and JOSEPH, S. (1955). 'Refractometry of Living Cells—III', *Quart. J. mic. Sci.*, *96*, 423.

BARTER, R., DANIELLI, J. F., and DAVIES, H. G. (1955). 'A Quantitative Cytochemical Method for Estimating Alkaline Phosphatase Activity', *Proc. Roy. Soc. B.*, *144*, 412.

DAVIES, H. G., and WILKINS, M. F. H. (1952). 'Physical Aspects of Cytochemical Methods', *Nature*, *169*, 541.

DAVIES, H. G., WILKINS, M. F. H., CHAYEN, L., and LA COUR, H. (1953). 'The Use of the Interference Microscope to Determine Dry Mass in Living Cells and as a Quantitative Cytochemical Method', *Quart. J. mic. Sci.*, *95*, 271.

DAVIES, H. G., ENGSTRÖM, A., and LINSTRÖM, B. (1953). 'A Comparison between the X-ray Absorption and Optical Interference Methods for the Mass Determination of Biological Structures', *Nature*, *172*, 1041.

DAVIES, H. G., and ENGSTRÖM, A. (1954). 'Interferometric and X-ray Absorption Studies of Bone Tissue', *Exp. Cell. Res.*, *7*, 243.

FRANÇON, M. (1954). *Le Microscope à Contraste de Phase et le Microscope Interferentiel*. Paris, Editions du CNRS.

HALE, A. J. (1956). 'A Quantitative Study of the Colloid in the Thyroid Gland of the Guinea Pig', *Exp. Cell Res.*, *10*, 132.

MELLORS, R. C., KUPFER, A., and HOLLANDER, A. (1953). 'Quantitative Cytology and Cytopathology', *Cancer*, *6*, 372.

MELLORS, R. C., STOHOLSKI, A., and BEYER, H. (1954). 'Measurement of the Organic Mass of Sets of Chromosomes in Germinal Cells of the Mouse', *Cancer*, *7*, 873.

MELLORS, R. C., and HLINKA, J. (1955). 'Interferometric Measurements of the Dry Weight of Genetic Materials in Sperm Nuclei', *Exp. Cell Res.*, *9*, 128.

MERTON, T, (1947). 'On a Method of Increasing Contrast in Microscopy', *Proc. Roy. Soc.*, A, *189*, 309.

MITCHISON, J. M., and SWANN, M. M. (1953). 'Measurements on Sea Urchin Eggs with an Interference Microscope', *Quart. J. mic. Sci.*, *94*, 381.

OSTERBERG, H. (1955). 'Phase and Interference Microscopy', in *Physical Techniques in Biological Research*, Vol. I. Ed. G. Oster and A. W. Pollister, New York, Academic Press.

ROSS, K. F. A. (1954). 'Measurement of the Refractive Index of Cytoplasmic Inclusions in Living Cells by the Interference Microscope', *Nature*, *174*, 836.

ROSS, K. F. A. (1955). 'A Critical Method for Measuring the Diameter of Living Bacteria with the Interference Microscope', *Nature*, *176*, 1076.

SMITH, F. H. (1954). 'Two Half-Shade Devices for Optical Polarizing Instruments', *Nature*, *173*, 362.

SMITH, F. H. (1955). 'Microscopical Interferometry', *Research*, *8*, 385.

SMITH, F. H. British Patent Specificat on 639014.

SMITH, F. H. US Patent Specification 2601175.

Earlier references may be found in BAKER INTERFERENCE MICROSCOPE, *Second Edition, 1955.*

CHAPTER 25

The Reflecting Microscope

THE ADVANTAGE to be had from a reflecting microscope system is that there is no chromatic aberration over a range of wave-lengths extending from the ultra-violet to the infra-red, therefore with such an instrument microspectroscopy can be performed over a very wide range of wave-lengths; absorption characteristics of materials can be compared in wave-lengths far apart in the spectrum without any change in the microscopical arrangements.

A practically useful reflecting microscope cannot be made with spherical surfaces, therefore the figuring of the objective to the correct curvature is a difficult task. A successful reflecting microscope was made in 1939 by Burch of Bristol and has since been much improved. The description following is taken largely from a paper[1] upon the construction of this and later instruments by Dr Keohane, also of the University of Bristol, who took an active part in the work.

The first Burch objective employing non-spherical surfaces was demonstrated in 1939 (Fig. 1); it was of NA 0·58 and was the only one of all that have been constructed to date with both concave and convex mirrors aspheric. It was manufactured from speculum metal (an alloy of copper and tin of approximately 60:40 composition) for several reasons. Firstly, in the process of making the mirrors aspheric, frequent testing of the surface was necessary and this was greatly facilitated by their retaining a high reflectivity throughout the polishing operations. The second reason for the choice was that, as the thermal diffusivity of a metal is very much higher than that of glass, thermal disturbances, created during the polishing and final figuring of the surface, are rapidly dissipated, so that the time interval between successive workings and testings could be quite small. Thirdly, of the metallic materials known to the optician, speculum has a 'flowing' quality and readily takes up a very high polish free from 'lemon-peel' effects. The original primary mirrors of this first microscope were cast with the central hole 'in situ' which resulted in there being a very large number of pits in the castings and incidentally made them extremely liable to fracture. From the rough castings the mirrors were first ground spherical on a quite crude spindle using cast iron laps (later it was found to be better to use brass laps as

[1] Journal of Quekett Mic. Club. Series 4, Vol. IV, No. 2.

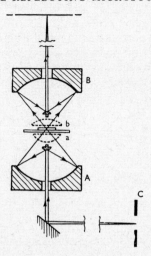

Fig. 1. Burch reflecting microscope.
A. Condenser.
B. Objective.
C. Lampfield diaphragm.
a. and b. Normal incidence oil components (dotted).
  With them in use, N.A.≃0·95.
  Without them (dry system), N.A.≃0·65.

crystals of speculum were not 'picked' out of the surface so easily) and then brought to a polish by conventional methods using pitch laps with rouge. The final shape of the surface required about 200 microns of asphericity (large when it is considered that only about 1/100th of this is required for most telescope reflectors) and this was polished in by hand with a series of pitch-covered ebonite ring laps that were worked in turn over successive zones of the mirror until the required depth was reached. The resulting surface was tested by knife-edge methods and besides having raised cusp-shaped zones at the positions of overlap of the polishing laps, the mirrors frequently lost revolution symmetry. Interferometric testing methods were later devised which, although highly sensitive, were extremely tedious. Each test necessitated the removal of the primary mirror from its cradle on the polishing machine and its mounting with the convex mirror, before the wavefront from the mirror pair could be compared with a spherical wavefront from a comparison mirror. The procedure adopted was to place the primary aspherized concave with respect to the convex mirror so that the wavefront from the mirror pair was free from spherical aberration. Then, the edge of the convex was polished down until the off-axis coma introduced by the setting of the mirrors was annulled. This testing procedure was repeated many tens of times between figurings of the surface, a process performed by means of rouge and a cob of pitch on the tip of a knitting needle. Patience brought success and the final wave-front

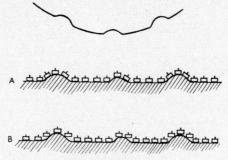

Fig. 2. Angle control.
A. Polishing pad following surface freely.
B. Pad faces not following inclinations freely but free in vertical plane only.

from the objective had a residual error of only $\frac{1}{8}$th of an interference fringe. The potentialities of the instrument appeared to be tremendous—an objective completely achromatic over the entire spectral range made photography in the ultra-violet of tissues and cultures possible. Quite apart from the increased resolution gained by nearly halving the wave-length of the illumination, the property of differential absorption exhibited by some compounds of cytochemical importance made 'natural staining' a possibility. Thus, it could be hoped that some of these substances could be located with precision in cells at various stages of their growth or metabolism.

The methods employed in the manufacture of the first, left little likelihood of the production of a succession of these instruments. In an attempt to acquire mastery of the aspherizing technique, rather than to produce more microscopes, Burch devised a polishing method now known as 'Angle Control'. The essence of the method is that the polishing pad smooths the surface that it has already generated. Fig. 2 shows the mechanism of the process. Suppose the non-spherical surface with its errors to be developed on a plane. When, as in diagram A, the polishing pad is unconstrained and allowed freely to follow the surface, as much time is spent by the pad on that part of the surface it is intended to leave unchanged as on the areas of error; whereas in B the hills are polished away more rapidly than the rest of the surface and the ultimate figure tends to conform more precisely to the ideal surface that is required. This is 'Angle Control'.

An angle-control machine was constructed with its own interferometer and with this a second microscope of NA 0·65 was completed. These objectives were given to laboratories engaged in work on which it was hoped they would be of use, and, as a result of some promising results, the Nuffield Foundation generously supported a project for the development and manufacture of a number of NA 0·65 Burch type reflecting microscopes which were to be made available to a few chosen laboratories. A small group was formed by C. R. Burch which include Mr M. J. Willcocks of Clevedon, who was to be responsible for the manufacture and commercial transactions related to the instruments.

The microscope primary mirrors are still manufactured in speculum metal, a method having been found by which the mirror blanks can be cast free from strain and are no longer so fragile; the convex mirrors are now made in stainless steel. An attempt has been made to produce an aspheric concave mirror in this metal but polishing proved far too slow for it to be a manufacturing proposition.

It is possible to increase the numerical aperture of the microscope to 0·95 by fitting hemi-spherical glass or quartz-oil components. These are used at normal incidence, as no refractions then occur when rays enter or leave the component and therefore no chromatic aberration is introduced. The NA 0·65 objective when used dry, has an effective working distance of about 13 mm. compared with the fraction of a mm. possessed by the equivalent refractor; even when used with oil the clearance between the specimen and the lower surface of the oil component can be made as much as $\frac{1}{2}$–1 mm. This incidental property of long working distance possessed by the Burch type objective is of advantage in the microscopic examination of hot metal surfaces and also makes micro-dissection a much less difficult task.

The design is such that phase contrast can be applied to the instrument either by etching a phase ring on the oil components or, by placing a phase disk in the system, as near as possible to the surface of the convex mirror. The obscuration of the aperture of the objective by the shadow of the convex mirror is only 2 per cent of area, in addition to which there is a very small assymmetric obscuration caused by the arm which supports the secondary mirror.

The wavefront emerging from the microscope objective is corrected to about 1/10th of a double incidence reflection interference fringe (this corresponds to a surface error in the primary mirror of only λ/40). The substage condenser is a mirror pair nominally identical with that of the objective and corrected to about the same degree of accuracy. A useful field of about 300 μ over which there is no appreciable error other than field curvature enables the Burch type reflector to be compared favourably with the best equivalent aperture refractors when used for visible microscopy, whilst it possesses the advantages of complete achromacy, over the entire spectral range.

The difficulty of producing accurately aspherized surfaces has led in recent years to a revival of interest in bi-spherical reflecting objectives. Burch showed that such a system could be aplanatic up to NA 0·5 at the expense of an obscuration of about 25 per cent of aperture area. One of the applications of reflecting microscopy has been to microspectrography—a technique whereby absorption measurements are made at various wave-lengths on small areas of cells or tissues. Various theoretical considerations suggest that, due to a redistribution of energy in the Airy pattern of the image caused by the obscuration, the latter should not exceed 20 per cent if the absorption measurements on small areas are to be reliable. With this in mind attempts have been made to design objectives with lower obscurations than the bi-spherical aplanat suggested by Burch, and also with the numerical aperture of the objective increased to a value 0·6–0·7. If the system is to be spherically corrected and there is also to be reduction in obscuration at a higher NA, it can only be achieved by tolerating considerable off-axis aberrations— coma and astigmatism are introduced which seriously limit the useful field diameter of the objective.

Norris, Seeds and Wilkins designed and constructed a series of such objectives of small corrected field diameter for use in micro-spectrographic systems in which only a small area in the central field is used. It has been suggested by Grey that the conventional tolerances on spherical aberration alone are exceeded at the field centre of these objectives, but presumably this would mainly affect the limit of area to which a particular absorption could be attributed. It is questionable what the limiting factors are in obtaining accurate quantitative results (scatter, refraction, fixation methods, etc.); it may well be that the Norris type objectives have limited use in such techniques.

More recently a corrected three-mirror spherical system has been devised by Steel, whilst Drew has manufactured an aspheric mirror pair of Gregorian design and of NA 0·3. Using semi-automatic ring lap polishing methods he constructed a solid all-reflecting objective of similar design and of NA 1·0 (Fig. 3). In addition to these purely

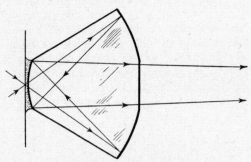

Fig. 3. Drew pattern solid reflecting objective.

reflecting systems, a series of catadioptric objectives have been designed and constructed, more particularly in the United States of America, in which most of the convergence is obtained by the reflecting surfaces, whilst the errors introduced as a result of the low obscuration of the designs, are annulled by refracting elements. These elements are designed to be of relatively low power so that the achromacy of the system is affected as little as possible. Now that fused silica, fluorite, and lithium fluoride can be obtained in suitable sizes of optical quality, these objectives can be constructed substantially achromatic over considerable spectral ranges. Many variants of these systems have been devised of which Johnson's in this country was amongst the first, but it is felt that such objectives do not rightly belong to the category 'Reflecting Microscopes' and it is not proposed to deal with them at greater length here.

It will be gathered from the above account of the construction of

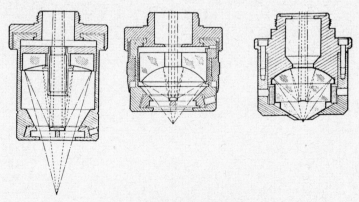

Fig. 4. Reflecting objective systems.

reflecting objectives that this kind of micrscope is very much of the one-off sort and at the time of writing is not a normal commercial proposition. Aspherical grinding techniques are being developed rapidly and the great value of cell analysis by means of 'physical' rather than 'optical' microscopy, employing interference, phase contrast and spectrographic methods has become so apparent that reflecting microscopes are inevitable in the next few years.

CHAPTER 26

Engineering and Metallurgy

A GREAT PERCENTAGE of the work done upon the micro-scopical structure of metals is performed by etching the polished surfaces and examining them with top light. Because of this, many metallurgical microscopes are built without substages and often without an axial hole through the stage.

Metallurgical Stands

When a microscope is set up for vertical illumination it is usual to find the lamp and various pieces of accessory apparatus mounted on separate stands with or without an optical bench to unite them and if the microscope body is moved along its axis to focus a speci-men, the vertical illumination light input hole moves at a right angle with the axis of the illuminating system (see 'Simple Vertical Illumin-ator', page 210). To avoid this interference with the illumination it is usual to make the microscope coarse adjustment pre-set and arrange the stage to rack the specimen along the microscope optical axis for focussing purposes. The fine adjustment, which has only a very small travel, is usually left in the conventional place behind the coarse adjustment. The system is good and practical but care must be taken that the stage dovetail slides are sufficiently heavy to bear the load of the stage and a piece of metal under examination which may weigh half a pound. The parallel type of slide is better in wearing qualities than the dovetail type and is very much easier to adjust and repair. A rotating stage is useful when work with incident polarized light is contemplated.

The metallurgical microscope is best used for convenience in sup-porting specimens with its stage horizontal but it should not be used visually in this position, therefore an angle eyepiece is desirable (see 'Instruments and Use of the Eyes', page 18).

Most metallurgical work can be carried out with instruments of normal size and these should be mounted on an optical bench (Plate 2) and must have a solidly mounted camera always at hand. Owing to the large amount of work sometimes undertaken some large rack mounted microscopes have appeared (Plate 20). These are really no different from the ordinary kind but they have all the usual accessories built-in in such a way that they may be rapidly engaged and disengaged by means of levers and knobs. Illumination is also built-in and follows conventional lines. The camera is permanently mounted above the microscope and a large viewing screen also is fitted, it is then often called a 'Projection' microscope but must not be confused with a 'Microprojector'. When the instrument is not of the inverted kind the weight of stage and camera are counterbalanced, thus making for very smooth and light controls.

The very large instruments are often inverted, that is the objective looks upwards at the object. (See 'The Inverted Microscope', page 165.) This arrangement has many advantages in any microscope but in this case it is particularly useful because a weighty specimen of any physical size can be placed upon the stage, and the relative position of objective and specimen-underside is always the same. Once an objective is focussed on the plane specimen, face downwards, it need not be again adjusted except for fine focussing, through a series of examinations.

It is desirable to possess a metallurgical microscope with the features described above, i.e. a racking and rotating stage and a fine adjustment preferably acting upon the nosepiece only, but with the addition of a normal though simple substage. The replica techniques described and mentioned in principle under 'Replicas' require transmitted light for their study. There is no difficulty in designing such a microscope and several are advertised by the manufacturers.

Optical Systems

Metallurgical microscopes may be monocular or binocular but for research and experimental work the monocular is to be preferred, mainly because accessories are much easier to fit into the simple body tube.

Metallurgical objectives are always corrected to work upon uncovered objects while oil immersion objectives are of the same correction whether a cover is used or not. Metallurgists usually prefer a tube length of 180 mm. for their objectives as this length allows greater space for accessories and when the microscope is closed down to about 150 mm. it is possible to see a recognizable image when a cover glass is present.

Most work upon etched specimens is best performed with high power apochromatic dry objectives because with these the contrast in the image of the etching marks is greater than when an immersion fluid is present, but a water immersion lens is very useful because under it may be performed etching both chemical and electrolytic in the immersion medium itself at high power of magnification. (See page 476.)

The eyepieces for use in metallurgy may be normal Huygenian and one must be fitted with an eyepiece micrometer scale. A monochromatic light source should be provided together with the high power illuminating system described under 'Illuminants' and 'Vertical Illumination'.

A full description of vertical and other opaque object illuminations appears under the heading 'The Study of Surfaces'.

Engineering Techniques

This Section must inevitably be a collection of facts which may be associated with the foregoing detailed descriptions of methods of using microscopical apparatus.

Brinell Test Microscopes

The Brinell and similar tests for hardness of metals require the accurate measurement of an indentation in the specimen, made by a tool driven by measured pressure. The instrument is low powered, magnifying about 50 times, requires no special illumination and stands upon the job being tested. It has fixed tube length and an eyepiece scale which is usually graduated in hardness units direct. Focussing is by sliding body tube. This is a small specialist instrument requiring no particular skill to use and not liable to any troubles.

The Brinell test impression covers a relatively large area of the material, say 1 mm. and is used only as a general hardness test, not a test of the hardness of single crystals or small areas of the metal substance. Single crystal tests are carried out under an ordinary microscope with the Micro Hardness Tester described on page 471.

Toolmakers' Inspection Microscope

This is a similar device to the Brinell instrument but is meant to be held in the hand after the fashion of a fountain pen. The lens system is non-standard and the shell of the body protrudes beyond the objective to form a rest for the instrument upon the specimen. The inside of this extension is plated to form a reflecting illuminator. Fig. 1. The instrument magnifies about 25 times and is usually not fitted with a scale.

The inspection magnifier is often a simple lens arranged in a stand with a brass scale in the front of the instrument which lies on the job in the focus of the magnifying lens, it thus forms a magnifying ruler of considerable accuracy. The device appears to have been invented by Mr P. Langabeer and when constructed with the

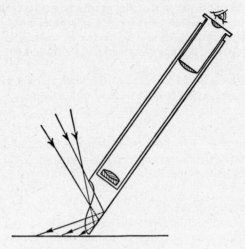

Fig. 1. Pocket Inspection Microscope.

modern photographically-reproduced scales and a built-in lamp, is capable of any elaboration and accuracy. Such direct reading devices are of the greater assistance because they do not depend upon calibration through an optical system, are very robust for use in machine shops and are easily carried in the pocket. Fig. 2.

Fig. 2. A. Simple lens.
         B. Stiff wire frame.
         C. Scale on diametrical strip (strip cut from high quality steel rule).

The Bench Inspection Microscope
This consists of a 150 mm. diameter, 300 mm. focal length lens mounted in a disk of plastic about 300 mm. in diameter which forms

a rim round the lens in which is mounted a circular fluorescent tube. The lens is large enough for both eyes to be used in comfort thus allowing the advantage of binocular vision for manipulations. The whole is mounted in a gimbal in a large fork which is counter-balanced and can be adjusted to and from the bench for focus. The device as purchased is refreshingly simple and robust and may be easily made up in the laboratory is desired. There is no particular reason for using a fluorescent tube but with it, the light is soft, evenly distributed and cool.

The greatest use for the bench magnifier is in delicate fitting jobs, instrument repair and inspection, and in some turning operations. Owing to the long working distance the magnification is only about two but owing to the fact that both eyes can be used to wander over a large field, the assistance to vision appears much greater. (Pl. 21.)

The Greenough Binocular

The instrument is fully described in the section devoted to it. In its lowest power form it has a working distance of about 100 mm. and magnifies about 15 times. It is of the greatest use when cutting fine threads in a lathe. A heavy stand can be provided with a stout arm which extends over the lathe. The focussing arrangements are in the Greenough body-part and so the stand need only be a pre-set arrangement.

The Greenough instrument takes standard eyepieces but not standard objectives. It may have an eyepiece scale fitted and can be calibrated from a ruler or stage micrometer. Only one eyepiece, the non-adjustable one, should have the scale fitted.

The uses to which the Greenough binocular can be put are infinite. It is however, a microscope in the proper sense of the word and cannot take the place of the toolmakers magnifier or bench magnifier. Its working distance is too short for many simple operations to be carried out under it and it cannot be hand held while in use.

Measuring and Travelling Microscopes

Any monocular microscope may be mounted upon a stand and caused to move rectilinearly in various directions and if scales are fitted to record the movement the whole becomes a measuring microscope.

In practice the microscope is mounted on a stout vertical pillar up and down which it slides for focal adjustment. The vertical pillar is mounted on a saddle and slides horizontally along a stouter block which forms the foot of the instrument. Both motions are carefully calibrated in tenths of a millimetre vernier read, and the total horizontal travel is about 250 mm. but may be made any length for special purposes. Both slides are accurately surface ground to ensure accurate travel for the microscope.

A normal eyepiece scale is fitted which interpolates the slide graduations and allows readings to a few tens of microns to be taken. The instrument is used to study and measure such objects as screw threads, the spacing of features or objects and thicknesses of parts into which micrometers cannot be inserted. Usually these microscopes take standard objectives and eyepieces, the usual objective is a 50 mm. In the eyepiece is fitted a graticule, pointer, or a scale.

The travelling microscope is practically essential when long pieces of metal or other large objects have to be made accurate to, say, a few tens of microns in length. The microscope may be used to step along the object with considerable accuracy, the method being as follows. A stop is set up on a flat bench against which the object to be measured can be returned accurately. Carefully measured gauge lengths of material are placed end to end against the stop, the measuring microscope being the last and containing within its base length the end of the object to be measured. Tiny adjustments in length of the object are studied under the microscope. Once set up the microscope may be clamped in place.

Profile Microscopes

These are ordinary monocular microscopes with low power objectives arranged to project an image of the specimen on to a screen of some kind. Screws and the like are placed on the stage of the microscope and are illuminated with transmitted light. The result is a black shadow of (say) the screw, on the screen which can be examined, traced and measured easily. It should be remembered when using a profile microscope to examine a screw or similar object that the instrument presents a shadow image and this is not a true representation of the optical section of the screw. Fig. 3.

The light does not produce a shadow of only the points a, b, c, d, on the axis XY, but a composite shadow including the points a' a'', f' f'', c' c'', d' d'' owing to the slope of the threads. The shadow therefore is not the exact profile of the thread. If the microscope has high resolution, it has a much better chance of forming an image of a thin section only, within its depth of focus. Unfortunately these instruments do not have high resolution, the numerical aperture being usually about $0 \cdot 1$ in order to obtain great depth of focus. When flat objects, or any without a sloping structure are examined in this way, no inherent errors are present.

When screw threads are to be examined carefully in a profile

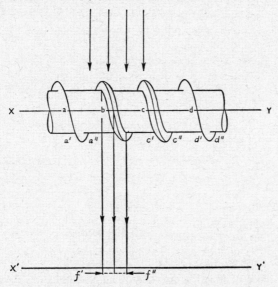

Fig. 3. XY.   Section plane required.
X'Y'.   Shadow obtained.

microscope they should be turned about until the profile of the thread at its root is the narrowest that it can be shown on the screen. This is the position of least error but it cannot be quite correct and the only way to do the job accurately is to grind the screw until a thin longitudinal section is formed then the profile will be substantially free from shadows from other parts of the thread. A sectioned replica may be used, page 223.

The Micro Hardness Tester

This apparatus consists of a sharp chip of diamond mounted upon the front lens of a high power dry objective. The angle of the point being about 30° (Fig. 4). The point may be mounted upon a cover slip which is oiled on to the front lens of the objective for the experiment.

The specimen is mounted upon a pivoted platform which can give under the pressure of the point on the objective when the objective is lowered. The platform is loaded with gramme weights so that the pressure applied by the objective is directly measured in grammes. The size of the micro-indentation is measured directly by means of an eyepiece micrometer but each point used on an instrument requires calibration upon a specimen of known hardness. The instrument can be used to measure the hardness of a single crystal in a metal or other substance and is now commercially available.

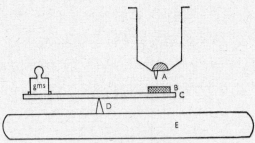

Fig. 4. A. Diamond point.
B. Specimen.
C. Platform.
D. Knife-edge bearing.

Preparation of Surfaces of Metal Specimens
When a piece of metal is cut, smoothed and polished, a very small amount of detail can be seen upon it in a microscope because the joints between one crystal and another are rendered invisible by the fusing of the metal in the polishing process. The flowing and smearing also covers over some of the scratches formed in previous stages of the preparation. In order to show the structure of a metal it is necessary to etch the surface either chemically or electrolytically to remove the artifacts which follow polishing. The etch may do two things, it may alter the level of the surface thus throwing the crystals composing the metal into relief or it may attack the crystal boundaries and so outline each crystal. Different etches produce different results which also vary considerably with the strength and length of time the etch is used. Each worker on a problem works out his own etch and the number published is formidable, however a few typical ones are published under 'Etching', page 476.

The sample of metal may be detached from the parent piece by sawing or chipping and chipping where this is possible, is likely to be the better process because less heating is involved than when a saw is used. Metal must be kept very cool if portions are detached by sawing as the structure can be altered for hundreds of microns on each side of the cut if the saw is allowed to become hot. Very hard metals are more affected than soft ones owing to the fact that they tend to crack. Specimens are best cut with a hand saw used normally or with a machine, pump-fed with cutting compound. It is convenient to detach pieces about 1 cm. diameter and about $\frac{1}{2}$ cm.

thick as these are large enough to polish evenly and not so thick that they tend to roll over in the fingers while being rubbed. If thin pieces have to be used they should be mounted upon plates of glass about 40 mm. square and 3 mm. thick, with Araldite.

Sawing Sections for Metallurgical Examination
Soft metals like silver, brass, lead and mild steel may be sawn in the ordinary way with a hack-saw. Little distortion of structure is likely because the mass is soft and conducts heat away from the cut easily, and any troubles encountered are likely to be during the polishing processes.

Harder metals like steels and the semi-conductors like germanium must be cut with great care. Damage is caused to the structure in two ways, the first being cracks due to expansion and contraction of the specimen near the heated cut face, and the second being cracks and shivers due to hard pieces of material either from the saw or the specimen jamming between the blade and the job. Such jams cause damage by fracture and it appears as fine cracks running into the specimen a distance of several hundred microns often not visible except when brought up by etching under high power examination.

Naturally the type of saw which is charged with diamond dust or carborundum is the worst for doing damage of the fracture kind but the toothed saw generates most heat. When sawing with a mechanically driven toothed saw a jet of cutting compound, say paraffin or cotton seed oil, must always be directed into the cut and on to the saw in an attempt to cool both job and saw and to remove cut particles. The essence of ordinary polishing is to move the polisher as lightly and rapidly as possible over the job so that maximum temperatures and even surface fusion takes place, but when cutting for microscope work, the opposite conditions are required, the saw must move slowly so that the minimum fusion takes place and all must be kept as cool as possible.

For hard materials which cannot be cut with a toothed saw the choice lies between the diamond-charged disk and the disk or wire charged with loose diamond or carborundum sludge. In general the diamond-charged disk is best so long as the job is not held against the cutter by a positive feed. A limited pressure spring holder should be interposed between the saw and the feed mechanism so that should hard particles enter the cut the job can give until they are washed out by the coolant.

The alternative device, the wire fed with cutting sludge has much to recommend it in practice for such delicate tasks as cutting crystals and small pieces of germanium. The sludge is loose in the cut but the wire gives so easily that damage due to particles jamming does not result. The sludge is fed in mixed with the coolant. The best results are obtained with a continuous belt of wire travelling in one direction only, wiped clean automatically at the leaving end of the cut by a pad and continuously fed at the input end with filtered sludge. The wire is very difficult to join into a belt as it must be spliced and soldered so that no lump is present. In view of this difficulty a reciprocating wire may be used with success but it cannot be easily fed with sludge in a regular manner and so becomes uncertain in action. All wire cutting methods are slow and are apt to produce ridged surfaces.

Flattening the Specimen

Having sawn or chipped off a piece about 1 cm. cube in size it must be prepared for etching. Very small metallurgical specimens should be embedded in a synthetic resin like Epikote until the block is about 1 cm. cube in size.

The ordinary soft metals may be filed flat and often the best results are obtained by resting the file on the bench, flat side upwards (both sides of a 'flat' file are not flat) and rubbing the specimen upon the file. Only sharp new files should be used, the finer the better.

The hard materials cannot be filed and so must be ground but there is a basic matter often overlooked with respect to grinding which has vital bearing upon metallurgical techniques. An abrasive material when it gets between rubbing surfaces always sticks in the softer surface and cuts the harder, therefore it is always the harder material which is ground away quicker. If it is decided that a loose abrasive must be used to grind a specimen flat, then the lap must be of a softer material than the specimen. The lap must of course be sufficiently firm to hold its shape and may require a stout backing plate.

The best lap for almost all metallurgical purposes is a lead one charged with diamond dust in watch oil and when materials as soft as copper are lapped on lead no particles become embedded in them. Ordinary emery paper is not a good material for finishing purposes because it is too uncertain in its composition and action. If it must be used it should be placed upon a flat plate of glass and stuck down with wax. An India hone used under a running tap is excellent for preparing the harder metals but tends to chip brittle substances. The running tap carries away stone particles before they damage the job.

Polishing the Specimen

A large amount of damage to the micro structure always occurs during polishing and so the damaged layer must be etched away afterwards. Polishing as applied to a metallurgical specimen is really little different from grinding, the abrasive particles in the polishing material are merely smaller and are supported by a soft backing substance so that they may follow the contours left by the previous process and not change the shape of the surface unduly. Ordinary metals and hard brittle substances are best polished with diamond powder upon a rotating iron disk covered with Selvyt cloth and cooled with kerosene; alumina and magnesium oxide may also be used for polishing softer metals. The size of the diamond abrasive particle should be 1 micron. The cloth, a woollen structure with special pile, is charged with a pinch of diamond powder, about one carat to a 200 mm. diameter wheel. The wheel is kept wet with kerosene from a bottle and the whole will work continuously for several months before needing recharging with diamond.

Soft metals like lead cannot be polished in this way because their surfaces fuse and collect diamond particles. Such surfaces are best prepared with a sharp knife but if they have to be polished the pumped sludge method must be employed as follows. A roller must be made up about 100 mm. in diameter and as long as the width of the specimen, covered with smooth chamois leather stuck down with a waterproof glue so as not to show any ridges at the joint. The roller is mounted horizontally and driven at about 500 revolutions per minute. Sludge is made up from No. 1000 carborundum suspended in water and is pumped between the roller and specimen on the input side. The specimen must be mounted upon a sort of carriage adjustable in height like the table of a milling machine and the apparatus is operated after the style of a horizontal miller.

It will be found that all types of specimens can be polished by one of these methods or a minor modification of it. Devices made up of a mixture of metals are always difficult but the aim should be to lap with a material softer than the softest component. In this connection experiments may be made with the material of the roller described above; wax has been used with success when charged with diamond powder also plastic coverings look as if they will be useful especially for the harder metals. The great enemy of polishing even when considering small areas for microscopy is particles of abrasive dust from the atmosphere or from other works, therefore the greatest cleanliness must always be observed and laps should be frequently washed down and kept covered when not in use. During grinding a specimen should often be examined under a low power microscope to see whether or not particles are becoming embedded in it and if they are the lap must be made of softer material. The particles may be got rid of by etching the layer away and starting again.

Etching: Chemical Method

The polished specimen is immersed in a liquid which attacks it, removing the damaged or non-typical structure. If the metal is an alloy an etch may be chosen to attack certain components more than others, so differentiating the structures. Most etching solutions attack the crystal boundaries quicker than the exposed faces of the crystal thus making the margins of the crystals visible.

The effect of an etch varies with its strength in a way not necessarily directly related to the length of time it acts. A weak solution of nitric acid about 1 part in 100 acting for many hours will show structure which strong acid acting for a few minutes will not, this fact is well known to artist etchers though they are not seeking micro-structure. In general experiments with strength of solution, temperature and time of exposure must be made on each particular job. The action may be watched under the inverted microscope and this method is to be recommended because etching processes often first reveal a structure and then destroy it by further action.

A list of typical etching solutions follows:

1. Iron and steel: 3 c.c. nitric acid + 100 c.c. alcohol.
2. Alloys of iron: 2 gms. picric acid + 100 c.c. of 25 per cent solution of caustic soda.
3. Copper alloys: 20 gms. chromic acid, 15 gms. sodium sulphate, 100 gms. water.
4. Aluminium alloys: 10 gms. caustic soda, 100 c.c. water.
5. Tin and lead alloys: 6 gms. silver nitrate, 100 c.c. water.
6. Germanium: Hydrofluoric acid used in degree of dilution suited to the structure sought.

Electrolytic Etching

In this process the specimen is placed in an electrolyte and voltage is applied between the specimen and the liquid, the polarity being dependent upon the experiment but usually for ordinary etching the specimen is made negative. Almost any liquid may be used depending upon the results sought but caustic soda in dilute solution is often effective. Chemical etches usually produce results when used electrolytically but different results are obtained with electrolytic etches depending upon whether they are used with weak currents ($20\mu A$ for many hours) or with heavy currents, examined in the light of the requirements.

In all etching work it is usually necessary to watch what is happening under the microscope and a suitable apparatus is shown in Fig. 5.

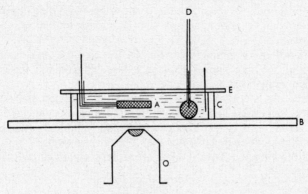

Fig. 5. A. Specimen.
        B. Slide (1 mm.).
        C. Deep cell.
        D. Thermometer.
        E. Perspex lid with electrodes.
        O. Inverted microscope objective.

The deep cell must be deep and wide enough to hold enough etching liquid to make sure that its strength does not change appreciably during the experiment. Because an inverted microscope is used, any quantity of liquid may be placed on the stage. The function of the lid is to support the electrodes upon which hangs the specimen which can be placed near the thin bottom of the cell for examination under high power. (Page 165, 'The Inverted Microscope'.) The electrolyte should be circulated with a pipette during the experiment. Etching liquids often contain hydrofluoric acid and other unpleasant materials which attack glass therefore the cells for microscopical etching studies should be made of polythene. As this substance is not transparent, a lens cannot work through it so usually a window made from a cover glass is inserted with wax as a sealing compound, which will last long enough for an observed etch to be carried out. An alternative is to look into the top of the cell but with a protecting cover glass stuck to the objective front lens with cedar oil.

When performing long experiments at low current density care should be taken to observe whether or not the set-up is light-sensitive. Germanium is an example of a material which is sensitive and a different current flows in the dark from that which flows when the microscope light is on.

When examining specimens like transistors it is possible to show the junctions within the germanium base-plate by means of electrolytic etching where both transistor electrodes are connected to one supply pole and the germanium to the other with the liquid left unconnected. The method is capable of an almost infinite number of variations.

For work of this kind the Beck Reflecting Objective is useful because of its long working distance (see 'Objectives: Reflecting Objective', page 70).

All etching methods show the crystal boundaries of fatigued specimens very well and approaching failure can be detected long before mechanical breakdown occurs because the etch attacks the fatigued boundaries much faster than the normal ones.

The following is a list of typical electrolytic etches or polishes:

1. For Steels: Acetic acid 760 c.c.
Perchloric acid 180 c.c.c of 65 per cent solution
Water 50 c.c.

Apply about 50V d.c. between the specimen and the bath making the specimen negative.

2. Tin and Aluminium: Perchloric acid 190 c.c.
Acetic anhydroxide 800 c.c.
Apply 10V d.c.

3. Copper, Brass and Cobalt: Orthophosphoric acid
Apply 2V d.c.

4. Zinc: Potassium hydroxide 25 per cent solution
Apply 16V d.c.

5. Lead and Alloys: Acetic acid 700 c.c.
Perchloric acid 300 c.c.
Apply 6V d.c.

6. Phosphor Bronze: Methyl alcohol 2 parts
Concentrated nitric acid 1 part
Apply 50V d.c.

7. Deep Etches for Iron and Steel: Strong nitric acid 2 ml.*
Ethyl alcohol 99 ml.
Used without electrolysis

8. Oxalic acid 10 gms.: Oxalic acid 10 gms.
Water 100 ml.
Apply 10V d.c.

9. Germanium: Dilute solution of caustic soda or hydro-fluoric acid
Apply sufficient voltage to pass a few milliamperes according to the experiment

* Use dil. nitric acid in alc. to show iron grain boundaries.

CHAPTER 27

Special Methods of Microscopy

Ultra Microscopy

IN EARLIER chapters of this work it has been stated that a particle of matter must be about $\frac{1}{2}$ the wave-length of light in diameter in order to be big enough to disturb a passing wave of light sufficiently for it to be seen in that light, and for an image of it to be formed.

This does not mean that smaller particles cannot be seen but it does mean that images of their shapes cannot be obtained.

If a powerful beam of light is applied to tiny particles they will scatter some of the light and may be seen in an optical instrument or with the naked eye as bright spots of light or as a haze, depending upon the size and number of the particles present. The smaller the particles the brighter the light required to make them reflect enough to indicate their presence. Examples of this scattering are motes in a sun-beam and the blue of the sky. Smoke in the beam of the projector in a cinema is an example of haze. Though the particles of smoke may be resolvable in a microscope they are not to the naked eye.

If microscopical objects are illuminated in this way and if their numbers are known an estimate of size can be obtained.

Raleigh used the following formulae in connection with scattering:

$$r = \frac{3M}{4}$$ where r is average radius of the particles

M is total mass of substance in the volume counted. To obtain M a volume of the field of view must be measured off with the eyepiece micrometer and calibrated fine adjustment. The number of particles visible in this volume must be counted. The ordinary mass of the substance must be obtained by normal dilution methods.

Raleigh's formulae can be taken further and used to obtain the amount (percentage) of light scattered by regular particles of any degree of fineness down to atomic dimensions.

In the formulae $s = \frac{kav}{r}$ where

a is the amplitude of the incident light,

s the amplitude of the diffracted light at a given distance r,
r the volume of the particle, $s\,a$ and v are proportional and inversely proportional respectively to r,

k is a constant. $\dfrac{s}{a}$ is only a ratio and v is (length)^2 therefore the

dimensions of k are (length)$^{-2}$. Then k must vary as (wave-length)$^{-2}$, wave-length being the only quantity left. From this it may be seen that,

$\dfrac{s^2}{a^2}$ is proportion to (wave-length)$^{-4}$. This gives a figure of about

16 times the amount of scatter for components of light in the blue end of the spectrum compared with those at the red end. Any particles in the atmosphere (even atoms) which cause scattering, scatter much more light in the blue than the red, therefore when the system is viewed from the side it is the blue which is predominant. When the particles are thickly distributed or when light has to travel a great distance, the blue components are scattered away and red and other colours only remain to be reflected from cloud etc. as at sunset.

In the Ultra microscope such colouring of particles is often seen. There is for instance a critical size of particle which reflects most light of a certain colour. Tyndall's Blue Sky tube was an example of this.

This illumination will show tracks of bubbles in an atomic ray chamber where true resolution is not possible because of small bubble size and time limitation. The method may be used to show such things as filaments or whiskers crossing gaps between surfaces when the size of the object is such that a low power, long working distance objective must be used. The author used the method recently to study the growth of whiskers on indium, tin and other materials across a gap of only 25 microns wide, but of about 3 mm. deep.

The apparatus is as follows:

The lamp must be bright (it is brightness that matters not quantity of light) and the condensing system must be arranged to suit the shape of cavity or liquid holder in use. Usually a slit must be used as the final source shaper and an image of this is projected into the object holder by means of a microscope objective. An image of suitable small size can be obtained from a 12 mm. objective. The function of the slit is to form a flat beam of great thinness so as to illuminate only a very thin layer of the specimen. If a gap is being studied the light must not touch the sides or the resulting flood of light drowns the low level of light reflected from the ultra microscopic particles.

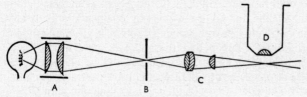

Fig. 1. Ultra microscope.

A. Lamp condenser.
B. Slit.
C. Micro objective used as auxiliary condenser.
D. Microscope.

Fair results can be obtained with a bunched filament low voltage lamp but a carbon arc must be used for the finest results with sub-microscopic particles. A flash tube may be used with photography.

The amount of light reflected from ultra microscopical particles depends much upon their shape as well as upon their volumes. The method cannot be considered a good quantitative one but is very useful as an indicator. When dark ground illumination is being used many of the twinkling effects visible are due to scatter from particles too small to produce an image of themselves yet their irregularity of shape shows in the changing amount of reflected light.

Fluorescent Microscopy

The technique of staining tissue to improve its visibility and differentiate its components is well known but many materials are damaged by the techniques. Some objects, however, and many common dyes, are fluorescent in the presence of ultra-violet light. If then the specimen is illuminated by ultra-violet light only, but is viewed through the microscope in the fluorescent light, differentiation of structure often results.

The method is as follows:

A convenient source of ultra-violet light is the high pressure mercury vapour lamp. A convenient type is the Osira made by the General Electric Company. This is composed of a quartz tube in which the discharge takes place, surrounded by an ordinary lead-glass envelope. The envelope should be removed by knicking the base with a file and separating by touching the knick with a red hot blob of glass on the end of a glass rod. The exposed quartz tube is a powerful radiator of ultra-violet light and the eyes must be protected from the light. The lamp should be housed in a light-tight metal can with an opening for a condenser lens about 30 mm. diameter. The can should not be ventilated. If a draught of air passes outside the discharge tube, instability of position of the arc is likely to result.

The hole must have a screen of Wood's glass or any other glass advertised by Chance Bros designed to pass ultra-violet but not visible light. The lamp condenser should be of quartz but a glass condenser passes a considerable amount of ultra-violet light and can be used. The microscope condenser and slide also should be of quartz but again, results can be obtained with glass ones. Often good results are obtained without condensers.

The eyes looking through the microscope are not in danger from the ultra-violet light mainly because of the presence of the glass in the magnifying system but also because of the proportional reduction of intensity of light in the magnified image.

The system must be set up in a dark room. Most objects are feebly fluorescent when magnified and they must not be mounted in the ordinary media because these are also fluorescent. Cedar oil, balsam and paraffin will all drown the fluorescence of the specimens. The mountants should be glycerine or polyvinyl alcohol. Most of the usual dyes, for example, methylene blue and neutral red, are fluorescent.

The method may be employed upon all sections, animal tissues living and fixed, wood and vegetable sections and morbid products. It is also valuable as an aid to analysis. Should an attempt be made to sort out adulterants by this means it is essential that test slides of the pure products be prepared as a reference. Different sources of ultra-violet light and different microscope lens systems give different appearances, and descriptions are rendered valueless.

The method of Fluorescent Microscopy is useful in connection with the study of Micro-incinerations where the mineral content of a section is shown by raising the section to red heat on a hard slide and studying the residue left in place. With care in heating and careful handling the method shows the distribution of mineral particles in a structure very accurately. Flinty and calcarous skeletons of minute animals are also well shown. It is usually not possible to mount the specimens permanently because of their great fragility, but a reasonably successful attempt was made by building a cell round the incineration on a heavy turn-table and slowly filling it up by stages with a very dilute solution of hardened balsam in zylol, allowing the balsam to harden off at each stage of the filling. When the cell is full it may be gently heated to allow the cover to settle down with the minimum disturbance of the incineration.

The following stains fluoresce under the influence of ultra-violet light:

1. (*a*) Water 97 ml., phenol 3 ml., auramine 0·1 gm. Apply this mixture for 2 to 3 minutes then wash in water.

(*b*) Decolourizing solution: 0·5 ml. hydrochloric acid, 10 ml. alcohol, 0·5 gm. sodium chloride. Dip the material in this solution for a few seconds only.

2. Fluorescent stain for Trypanosomes.

Acridine-orange 1:1000 solution in water with 0·9 per cent sodium chloride. Mix the blood containing the trypanosomes with this solution and spread upon a slide and dry naturally. When the slide is illuminated with strong blue or violet light and an orange filter in the eyepiece, the trypanosomes show up green.

3. Fluorescent stain for Dry Blood.

Mix the blood dust in methyl alcohol and stain in auramine 1:100 solution plus 5 per cent phenol for 4 minutes. Wash in water and keep in the dark. Mount in liquid paraffin.

The red cells will fluoresce dark green, and the trypanosomes golden.

In general the usual aniline dyes used for biological purposes have very sharply marked absorption bands and so are not satisfactory for fluorescent microscopy.

X-ray Microscopy

It is not possible to refract X-rays for discovering micro-structure in the way that light is refracted by a lens system and caused to form an image by the mechanism of interference.

Some interesting detail can be obtained by exposing specimens to soft X-rays and receiving the shadow image on a photograph plate in the usual way. It is essential to use nearly monochromatic X-rays and to receive the picture on a special fine-grain film such as the Lipmann type by Kodak. Such a film will withstand a subsequent optical enlargement of about 300 times.

The kind of X-ray apparatus is important. For sections of typical substances like wood, insect parts and fabrics about 30,000 volts is needed on the tube and the most suitable near monochromatic radiation must be obtained by using a target of either copper, iron, cobalt or molybdenum giving radiation to suit the specimen. The radiation should be of the order of 2 A.U. in wave-length and exposures are about 20 secs. A steady d.c. machine is best.

In general, when the exposure is correct, the density of most tissue or other substance provides differentiation of structure. Staining or soaking with materials like barium sulphate may be arranged to cause increased absorption of X-rays. The injection technique described on page 273 is useful.

The great advantage in the industrial field of the X-ray method is its simplicity. Standard medical X-ray apparatus may be used and the only other matter of importance is that the specimen shall be very near the photographic film. The film may be wrapped in black paper. Mounted specimens if of a dense kind like insect parts, photograph satisfactorily in their natural state, but in general the raw material should be taken and flattened against the black paper covering the film or plate. Ribbon sections may be photographed without dissolving out the wax.

The limiting factors in the method are as follows:

(1) The grain size of the film. If magnification after development is pushed above, say 300 times, even with special film some very convincing artifacts appear. (See 'Photomicrograpy', page 190.)

(2) The scattering of the X-radiation by the specimen and mountant. The X-radiation picture being only a shadow, cannot differentiate between layers in the specimen in the way that an image-forming system can.

(3) The resolution of the system is decided by factors (1) and (2). Although the X-radiation being of very short wave-length compared with visible light will pick out fine detail, it cannot subsequently be made full use of in an image by means of magnification. If the X-rays could form an image, the finess of detail visible would still be limited by the film grain size. In practice the grain of the film is the important limiting factor and the fact that the X-rays do not form an image reduces the resolution still further depending upon the thickness of the specimen. The best results are obtained with thin sections and soft X-rays, and the results can be roughly compared with those obtainable with a 16 mm. objective. The type of detail shown will of course be different.

Calcined specimens may be studied in this way and excellent photographs obtained, but it is doubtful whether or not any really useful information about structure as opposed to composition can be obtained by high temperature processes.

Ultra-Violet Light Microscopy

It has been explained in earlier Sections that microscopical resolution of detail is dependent ultimately upon the wave-length of the light used. The shorter the waves the smaller the diffraction patterns and the easier the separation of them with a given magnification, also looked at in another way, a particle too small to disturb a wave of long wave-length light will disturb one of short wave-length and allow an image to be formed.

Successful attempts were made during the last century to obtain microscopical images using ultra-violet light, and for photographic presentation employing a fused quartz lens system. If the frequency of the light is doubled (i.e. the wave-length is halved) twice the resolution results, and it is possible with transmission optical systems to obtain this result. Little modern work has been done with the ultra-violet light microscope because it has been eclipsed by the electron microscope which has carried the method of using high frequency light to its limit. However, the perfection and production of the Burch reflecting microscope has opened the way again for interesting spectroscopical work over the whole range of infra-red, visible and ultra-violet radiation with the same instrument, the reflecting system being perfectly achromatic.

When work with ordinary transmitted ultra-violet light is contemplated many unusual factors must be considered, for instance the microscope chosen must have a fine adjustment of the very highest class and it is usually necessary to employ a horizontal instrument of massive construction with a special slow speed micrometer fine adjustment acting on the nosepiece. An optical bench construction is essential because objectives, condensers and lamps have to be interchanged frequently. The lens systems of fused quartz must be assembled without cement because cements are either opaque to ultra-violet light in the region usually used (2,700 AU) or they are fluorescent. The immersion medium must be a mixture of glycerine and distilled water set to have a refractive index suitable to the quartz system in use determined by the lens designer, other media are opaque or fluorescent, the slides and cover glasses must be of fused quartz.

Normal focussing of the microscope cannot be performed when ultra-violet light is being used because the light is invisible to the eye and if the image is received on a fluorescent screen it cannot be focussed sufficiently accurately for photography of fine detail. There is no satisfactory alternative to making the microscope a precise piece of engineering equipped with a sliding objective changer of great accuracy, and arranging two objectives, one apochromatic and the other for ultra violet, to be parfocal. It follows that two condensers or a concentric condenser are necessary in connection with these objectives and two light sources also. The method of working these is as follows:

A test specimen is chosen which presents a sharp clean outline in both ultra-violet and visible light, such a specimen being the chips of

Dutch metal described as 'Avenue Gold' in 'Objectives', page 62, but it must be mounted in glycerine and water on an example of the quartz slides to be used on the apparatus. The quartz and glass objectives are mounted on their respective changers and set up using visual and photographic methods to give images in perfect focus so that the specimen can be set up with visual light and the change made to ultra-violet knowing that the focus will be perfect. Several trials with the photographic plate are necessary before the objectives can finally be made parfocal and it must be remembered that the setting is correct only for one wave-length of ultra-violet light.

Illumination of the specimen is always by means of transmitted light therefore a quartz condenser of ordinary construction is used. Visible light can be passed through with the aid of a mirror from a source at the side of the optical bench, the focus of the condenser in visual and ultra-violet light being sufficiently near for an illuminating system to be set up with visual light. When the ultra-violet exposure is to be made the mirror is swung aside, this also has the great advantage that lengthy setting-up procedures can be undertaken in visible light, the lethal ultra-violet being kept out until the moment of change of objective and exposure of the plate.

If great care is demanded in the illumination system a concentric dark-ground condenser can be modified to have a quartz transmission system in the centre zones leaving the ordinary visible light dark-ground part in place at the periphery, it then becomes possible to make the visual dark-ground part and the ultra-violet transmission part parfocal, the dark-ground part permitting a very accurate focus to be made in visible light.

An ultra-violet light objective was computed by the British Instrument Research Association to work on the cadmium line 2,753 AU and it was manufactured by Messrs R. & J. Beck, but it is not known whether or not any examples still exist, they are not now manufactured. The immersion medium of glycerine and water is carefully adjusted in proportions to have a refractive index the same as fused quartz for the 2,753 AU line. The refractive index of this medium is actually fairly low but this does matter in the ultra-violet system because the high resolution is obtained through the high frequency of the light as is also the case in the electron microscope which has in reality a mere pin-hole of aperture. The immersion system is used, however, in order to obtain any increase in aperture possible by this means but mainly to improve the correction of spherical aberration (see 'The Immersion System', page 37) and to allow the tube length to be fixed.

The transmitted ultra-violet light image-forming microscope is not likely to be much used except for special purposes because ultra-violet light of sufficiently high frequency to make possible major improvements in resolution cannot be passed through any lens system of the optical kind nor can ultra-violet light be focussed by electromagnetic means as can electrons. It appears that ultra-violet work in the future will be by means of reflection systems with the metallic mirrors coated with a material which will reflect a large proportion of ultra-violet light of short wave-length. Useful work was done and more can be done with short-wave-length light as an aid to resolution now that focussing and transmission difficulties associated with transmitted light lens systems have been overcome by using reflectors. The great advantage over electron apparatus is that living objects may be studied within time limits of minutes and there is no need for the specimens to be specially prepared except as described above.

Devices Employing Infra-Red Light

Light having a lower frequency than visible red will pass through an ordinary microscope so long as neither the mountant nor the cover glass are of the exceptionally thick sort. Objects mounted in Euparel under standard No. 1. covers generally show up satisfactorily. Infra-red has considerably greater penetrating power than has ordinary light and where resolution is not the main aim much can be learned about dense structure.

A filter which passes infra-red but not visible light can be pur-chased from the manufacturing photographers, the special plates and suitable developer are also provided. Any hot light source is suitable and the ordinary incandescent opal electric bulb used as for critical light may be employed as the infra-red source. Exposures will be found to be about twice those for panchromatic plates but experiments must be made with the source in use. Camera attach-ments should be made of metal because infra-red light is very likely to leak through wood, cardboard and leather.

The method of infra-red photography does not vary much from ordinary photomicrography, the apparatus is set up with visible light of a definite colour, and an exposure of a thin black object taken (see chapter 'Objectives', page 61). Several exposures are then taken without changing the microscope adjustments but with the infra-red filter in place, and depending upon the construction of the objective, some error in focus will be found. Trial alone will give the correct amount of shift of the focal adjustment to put this right. The amount of focal change necessary must be recorded on the fine adjustment milled head and when once determined for each objective the same kind of plates, and the same microscope and lamp should always be used.

The practical uses of the method are limited to penetration experiments, examination of differential reflections from metal surfaces, study of objects which cannot be readily stained and showing the internal structure of entomological specimens which may be heavily coated with chitin.

Infra-Red Image Converters

When infra-red-sensitive plates are used with a conventional microscope the worker cannot see what he is doing owing to the invisibility of the light. This is always a disadvantage and the alternative is to render the infra-red rays visible by an electrical method and either observe or photograph the results in visible light. The following circuit arrangements for doing this are due to B. K. Johnson of Imperial College of Science and Technology, London.

Image converter tubes are made by the electrical and applied physics industry generally and Fig. 2 shows diagrammatically the principle of what is in effect a modified form of photo-electric cell. The evacuated cell-chamber consists of a Pyrex glass cylinder 40 mm. in diameter and 40 mm. long, with plane end windows 2 mm. in thickness. A semi-transparent silver-caesium oxide layer is deposited on the inside of one of the end windows acting as the cathode, whilst

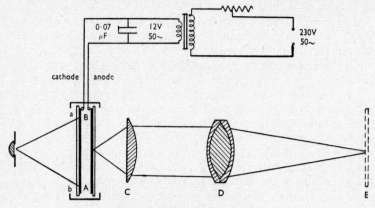

Fig. 2. Infra-red converter tube.
a–b. Image from microscope.
A.　Silver caesium oxide layer on glass.
B.　'Willemite' fluorescent screen.
C.　Lens for viewing screen B.
D.　Photographic lens (high aperture necessary).
E.　Panchromatic plate.

at a distance of about 5 mm.[1] from this and parallel to it is a thin glass plate coated with Willemite and superimposed with a small-meshed metallic grid, the latter acting as the anode.

If infra-red radiation is allowed to fall on the cathode surface, photons and electrons will be released, which may be accelerated (by application of a high voltage) to the anode where they form a fluorescent (green) image corresponding to the infra-red image. This fluorescent image may in turn be observed by a lens (45 mm. focus) cemented on to the window at the other end of the tube (Fig. 2).

The high voltage (4,400) applied to the cell can be supplied by comparatively simple electrical equipment, which is illustrated diagrammatically in the lower part of Fig. 2. Utilizing the mains supply of 220V a.c. and 50 cycles periodicity, a transformer (having a step-up ratio of 20 to 1) is interposed in the circuit and the high tension leads from the secondary windings are then taken directly to the cathode and anode of the photo-electric cell. A condenser (having a capacity of 0·07 μF) is placed in parallel across the secondary windings.

The arrangement for using the apparatus in conjunction with the microscope is shown in Fig. 3. A source (rich in near infra-red radiation) is situated at S; this may consist of either a carbon arc or a 'Pointolite' lamp (tungsten arc in argon). A quartz condenser

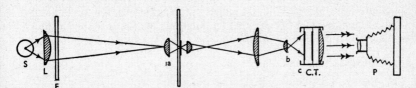

Fig. 3. Infra-red microscope.

S.   Source.
L.   Lamp condenser.
F.   Filter.
a–b. Microscope.
C.T. Converter tube assembly.
P.   Photographic or visual presentation.
b–c. Distance sufficient to obtain fairly large image, say 200 mm. with usual eyepieces.

lens L forms an enlarged image of the source on to the sub-stage condenser of the microscope. The latter is set up in the usual manner for visible light and the object focused; by using a projection eye-piece and the infra-red cell at a distance of about 150 mm. from the

[1] Some converter tubes may have a much greater distance between anode and cathode but with an electrostatic lens placed between them for focusing the electrons on to the fluorescent screen.

eyepiece, an image of the object may be seen on the cathode (which is white in appearance before the current is applied to the cell). A filter[1] (removing all the visible spectrum but transmitting the near infra-red) is then interposed at F, the current to the image-converter cell is switched on, and the fluorescent anode screen can be viewed by the eye E situated behind the magnifier M (already cemented to the cell end as mentioned earlier). It may be necessary to re-focus the microscope slightly (on account of the change in wave-length from the visible to infra-red radiation) in order to get a sharply defined image on the screen. It will be essential to use a panchromatic plate as the fluorescent image is of a green colour; but it will not be necessary of course, to use an infra-red sensitive photographic plate.

The thickness of glass contained in the lenses of the microscope objective, the eyepiece, and the sub-stage condenser may together amount to something of the order of from 10 to 15 mm.; but such a thickness will not reduce the intensity of the radiation $\lambda = 12,000$ A. very seriously. The condensing lens L, on the other hand, may be quite thick at the centre, and it may be advisable therefore to use a lens of quartz in place of a glass one. The use of the entire optical system made in quartz would enable transmission to be obtained up to 40,000 A., and indeed the gelatine infra-red filter mounted between quartz plates will transmit freely up to this wave-length, but the response of the silver-caesium layer in the converter tube is not so good in this region. Consequently, the region 10,000 to 20,000 A. might just as well be resorted to and thus enable the normal glass optics of the microscope to be used, with equally good results.

[1] Near infra-red filters may be made by employing: (i) a piece of Ilford infra-red dyed gelatine mounted between two pieces of thin glass; (ii) a piece of glass (2 mm. thick) containing nickel oxide as made by Messrs Chance Bros. Ltd. under their catalogue type No. OX$_5$; (iii) an opaque solution of iodine in carbon disulphide or alcohol in a glass-walled cell about 10 mm. thick. All these filters have a high transmission in the region 10,000 to 20,000 A.

CHAPTER 28

Laboratory Methods of Working Glass

IN THIS section, glass grinding and polishing only is described. For glass blowing as applied to the construction of chemical apparatus, works on chemistry must be consulted.

Optical glass is made by fusing a 'melt' of materials in a crucible in a gas-fired furnace. An earthenware pot must be used because hot glass very readily dissolves metals (except platinum) from any mixture which it may contact. Traces of metals produce the colour in common coloured glass. The selection of materials in the melt determines the refractive index and dispersion of the glass, and it may be very difficult to reproduce exactly a kind of glass once made. Messrs Chance Brothers will provide optical glass of all kinds and their lists should be consulted when design requirements are known.

One of the great arts in making optical glass is to keep the melt free from bubbles, and sufficiently well stirred to make the mass homogeneous, that is, free from stria and smears. After cooling, the pot is broken off the glass block and the block itself is broken up into lumps a few inches across. These lumps are examined in a parallel beam of light for stria and irregularities, and when satisfactory pieces are found, they are cut up for use as lens blanks.

Sheets of plate glass were made from a melt which was poured on to a plate and ground and polished as a sheet, but much glass is now continuously drawn by dipping a bar of earthenware the length of the width of glass required into a trough of molten glass and withdrawing it slowly up the side of a building about forty feet high. The glass follows the draw-bar after the manner of treacle, and if the speed and temperature are right, sheets about 12 metres square and more can be drawn. A flotation method also is used.

For laboratory purposes, all kinds of plate glass are suitable for making troughs, simple lenses and screens, and it is the construction of these devices that usually interests microscopists.

Sheets of glass may be cut with either a small hard wheel about 4 mm. diameter with a sharp rim, known as a wheel cutter, or with a fragment of diamond arranged to mark a groove across the sheet. There is very little difference in effectiveness between the wheel and diamond cutters, but the wheel is not good on very thin glass because the requisite pressure cannot safely be applied; the wheel is better for tracing shapes.

The aim when cutting a sheet of glass is to make a clearly defined groove across the glass so that when a strain is applied, the line of weakness is clearly defined and the break occurs with least effort exactly where wanted. Fig. 1.

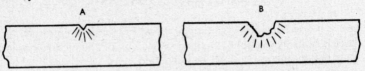

Fig. 1. A. Correct cut line of weakness clearly defined and intense.
B. Wrong cut (though deeper). No well defined line of weakness.

The wheel cutter is simple to use, the only requirements being a sharp wheel and practice. The surface may be marked dry or lubricated with a mixture of paraffin and turpentine, and in general the cutting compound is required when shapes are being cut. If the pieces required have straight edges, it is most important that the ruler which guides the cutter be really smooth and straight.

A diamond fragment for use as a glass cutter must be carefully selected for its shape. Its plan view should be something of the shape of Fig 2A. It is drawn along the sheet in the direction of the arrow, the cutting edge being shown up the centre of the figure. The leading slope of the fragment is less than the trailing slope, so that the diamond tends to dent the surface rather than plough it up. For this reason a writing diamond conical point cannot be used for cutting.

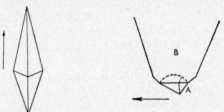

Fig. 2. a. View looking upwards at cutting diamond direction of cut
with arrow.
b. Side elevation of diamond in mount.
A. Diamond.
B. Steel mount.

More practice is called for in using a diamond than a wheel, for it can be seen from the diagram that it must be held in line with the cut.

Glass up to about 10 mm. thick can be cut with wheel or diamond cutters so long as (i) the glass is only a few months old, (ii) the surface is clean, (iii) it is of the plate kind, and (iv) the cut is correctly made. (Fig. 2A.) Old glass develops a skin and will not cut well. Scratched glass and blown glass have irregularities which interfere with the concentration of strain at the surface cut and so cannot be parted readily. When glass is irregular or very thick, it must be sawn.

To Cut Glass with a Glass Cutter

Select a piece of new clean glass, wipe the surface clean of dust, and rest it on a flat table covered with thin felt or baize but not thick felt like carpet underlay. Obtain a really straight ruler which had better be a metal one, of such thickness that it will guide the cutter but not foul the wheel. Draw the cutter firmly once from end to end of the line of the cut. If correct, the wheel or diamond should leave a clear fine line only. If it ploughs a groove which can be felt with the finger and seen to be rough, the cutter is blunt, the diamond not held in proper line, or too much pressure was applied.

Having secured a clean groove, take the glass in the two hands with the mark between them, the thumbs placed at A and B, and break it along the line. Fig. 3.

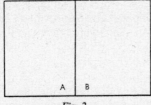

Fig. 3.

If the glass does not readily part, light tapping on the underside of the cut with the metal holder of the cutter will often start the crack. It can then be urged along by tapping or by pressing above the line of cut. After parting, a cut edge should be smoothed with emery paper to prevent splintering.

When it is required to part a piece from a tube or other tiny part, the part should not be broken by hand after marking with the cutter, because damage will probably occur to devices inside when the glass gives under pressure, also a tube may be too large in diameter readily to break off. In these cases, a glass rod about 2 mm. in diameter should be heated until a red-hot globule forms on the end, which is then pressed against the cut and usually the glass will part instantly.

To make out a sheet of glass for cutting, lay out the pattern on white paper and place this under the glass sheet to be cut.

To Grind Glass

The mechanism of grinding glass and other brittle materials has been described under 'Polarized Light', page 102. When glass is to be worked to shape it is usual to use a cast iron lap or one of brass, and a cast iron disk as described under 'Grinding Rock Sections' is quite satisfactory. Two iron blocks or laps are required for coarse and fine carborundum and the two should be well separated in the workroom to prevent coarse powder mixing with the fine. The grinding paste is a simple suspension of powder and water applied with brushes to the lap and the glass is rubbed on this and will be ground away.

For many purposes, including prism and lens making, it is necessary for the iron laps to be geometrically flat and this is arranged by having three identical plates of iron, preferably round, about 300 mm. diameter, and not less than 25 mm. thick, which are ground one upon the other in rotation in the following order: 1 on 3, 2 on 1, 3 on 2, 1 on 3, etc., the first mentioned being on top. This grinding rotation will flatten all plates as explained below.

If two equal diameter plates of any material are ground on each other, the upper becomes concave and the lower convex owing to the over-hang during grinding of the upper plate increasing the load on the peripheral zones of the lower. The system of rotation eliminates this effect by interchanging upper and lower plates. The use of three plates instead of two averages the results much better.

The coarsest carborundum used for grinding is about the grade of motor-car valve grinding paste but is much more evenly graded. The finest is known as 400 grade (some manufacturers use a mesh size, some a settling time) but its consistency is that of flour. This will grind a smooth silky appearance on the glass and is as near as one can get to the polishing stage. A job usually requires several grades of carborundum during grinding but it must always be remembered that a coarse grade of powder cannot be removed from a lap without the most elaborate operations, therefore a fine lap must only be used for fine powder.

If a piece of glass has to be ground flat, it must be backed up by several pieces of similar glass cemented or waxed close beside it on the handling plate to prevent the edges of the wanted piece from becoming rounded by the pile-up of loose abrasive along its front. (See below.)

Large pieces of glass for relatively rough work may be hand held and applied directly to the lap. If, however, the parts are small, they must be stuck to a block of glass of suitable size to handle, by means of pitch or beeswax. Opticians' pitch, consists of a mixture of pitch, beeswax and sawdust, so proportioned as to make a mixture which, when cold, holds glass firmly yet is brittle enough for the glass to be knocked off without danger. The proportions are usually about two of beeswax to fifty of pitch by weight. The proportion varies with the job and the temperature, and must be tried.

When the parts of a job are small and light, beeswax is a good cement, and it is always best to grind several parts together so that one backs up the other. This backing-up is most important when flat surfaces are being ground, a spherical surface tends to look after itself as described under 'Lens Making', page 496.

It is most important to wash the job under a running tap with a scrubbing brush before changing from a coarse grade of abrasive to a finer.

Anyone who can shape a piece of wood by rubbing it on a sheet of glass-paper can shape glass by rubbing it on a sheet of iron with an abrasive. It is not really necessary to have mechanical aids.

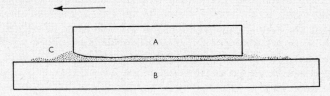

Fig. 4. A. Job.
      B. Lap.
      C. Pile-up of abrasive in front of job.

To Polish Glass

Polishing is not an extension of the grinding process. Grinding is a distinct breaking away of particles of a surface, while polishing is a fusing process. Polishing materials for glass are not abrasives but are materials chosen to generate heat on the surface and cause a flowing of the glass. Several materials can be used but rouge (iron oxide) and zinc oxide are commonly used. Naturally the polishing materials must be held on a soft matrix so that the flow of glass surface can follow the accurately ground curves of the glass, and a suitable medium is opticians' pitch, which can be moulded to suit the surface in question, is hard enough to keep its shape and cannot scratch the surface to be polished.

To make a polishing lap for flat surfaces, a layer of pitch about 3 mm. thick is spread when hot upon an iron sheet (a lap will do). The surface of the pitch is then pressed flat with another (clean) lap or a piece of plate glass. The firm smooth surface thus formed is charged with rouge and water and used in the manner of a lap. The pressure on the job must be even and light and the job must be inspected periodically so that polishing is not continued beyond polish or there is danger of distorting the surface.

It is most difficult to judge the quality of a polished surface until a large amount of experience has been accumulated. One way is to allow the reflection of a dark bar in a window frame to be seen on the surface of the polished article; the eye should look at the surface where the joint between dark and light occurs. In this region the slightest irregularities upon the surface of the glass can be seen and it is recommended that comparisons be made between the reflection from a top-quality commercial prism and a piece of partially polished ground glass, or a shiny bottle surface be made, when, after a little practice, the differences become very apparent.

Lens Making (See also 'Objectives')

Lenses are made with spherical surfaces because they form naturally to that curve, as surfaces of the same radius, one concave and the convex, are the only shapes which can fit each other exactly in all positions of rotation. For this reason two surfaces rubbed together at random will ultimately have spherical surfaces formed on them. If we take a roughly concave hole and work it on a roughly convex projection, the two will end with the same radius, exactly spherical surfaces, and this is the principle behind all lens grinding.

Two tools are required, male and female. These are first turned in a lathe to the correct mechanical curvature worked out from the focal length of the lens required. The tools are then worked one upon the other with abrasive until the mechanically perfect spherical form results on both. The glass blank is then put in place of one of them and the glass is ground by the tool to the same curvature. During the grinding the tools are ground together occasionally to preserve sphericity.

During the grinding, plenty of abrasive and water must be applied. The grade of abrasive may become finer as the job progresses but it is usual to use two sets of tools, one for coarse grinding and the other for fine. The coarse tools tend to change their curvature with continued use and have to be trued up to save too much work being thrown upon the fine tools.

To polish a lens, one of the tools is covered with pitch, and while warm and plastic, the pitch is pressed on to the lens surface, so taking up its curvature. A few furrows are cut at random across the pitch to make spaces for pieces of matter to lodge if any become detached during polishing. The polisher is charged with rouge and water and polishing proceeds after the manner of grinding, and if there is a tendency to polish much quicker at one place than another, a little of the polisher is cut away with a knife to check its action in that zone.

The actual set-up for lens making usually consists of a vertical rotating spindle which can revolve on a good bearing at about 300 revolutions per minute. On top of the spindle is a chuck which can hold pegs fitted into brass disks upon which are mounted in pitch lens blanks, etc. Grinding tools are usually made with spiggots or shanks to fit the chuck direct. It matters little whether the tool or the job rotates on the axis of the spindle, but it is usual to rotate the job.

Tools are made from solid cylinders of cast iron or brass, the shank is turned in a lathe in the normal way but the curvature on the end face has to be made by mounting a swinging tool holder on a tool post below the horizontal axis of the lathe. In this way the tool describes an arc. (Fig. 5.)

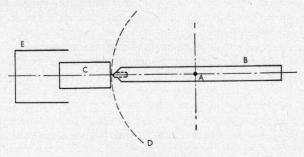

Fig. 5. A. Tool post.
         B. Tool arm.
         C. Job.
         D. Radius of cut.
         E. Lathe head.

If a convex tool has to be made, the tool post is moved under the job on the lathe axis and the tool is mounted elsewhere on the arm.

There is no very great accuracy required in this apparatus and a laboratory lash-up will give highly satisfactory results. Of course, the distance of the tool from the tool post must be correct because this distance determines the radius of curvature of the lens. However, this is a dimension which is not of the greatest importance and the grinding of the tools together and the polishing and grinding of the lens itself off-set great accuracy in the turning of the tools.

An easy way of making accurate tools has been mentioned under 'Objectives' but is repeated here for completeness. A series of balls for use in ball bearings should be obtained, as these are very accurately spherical and make excellent male tools. The female tools are cast round them from hard tinman's solder to any other hard metal.

The small piece of glass to be worked by this kind of tool is mounted on a short stick about 50 mm. long and 6 mm. diameter, by means of pitch. It is then worked in all directions on the tool, or the tool on it, depending upon convenience. A firm steady pressure is demanded, and in general it is not possible to make a good lens unless there is a rotating spindle available. A large astronomical mirror may however be made without rotating machinery so long as the worker takes the greatest care to move tool and job evenly over each other in the manner described in books on amateur astronomy.

For making tiny lenses the ball bearing method is the best and to make the tools, a cavity, the shape of the ball selected, is pressed in a block of solder in the plastic state. The two tools so made are ground together as usual and the lens worked in the hole. Polishing is performed in a hole in a block of pitch, made in the same way. The piece of glass from which the lens is made is mounted on a length of match-stick with pitch.

So much success in hand lens making is dependent upon the manual skill of the operator that descriptions are only useful to outline the methods, but one of the most important factors concerned in obtaining accurate spherical curves is the length of the stick or handle used to hold the lens. Nothing but the feel of the job can tell this, but generally large lenses, say 100 mm. in diameter, require short stubby blocks of pitch to hold them by, while small lenses about 3 to 12 mm. diameter required sticks about 25 mm. long which can be easily held and manipulated in the fingers.

How to Trepan Glass
Trepanning is the process of removing a disk of material from a sheet, thus obtaining either the disk or a large hole of the diameter required. A disk usually has to be prepared for lens making.

A trepanning tool consists of a piece of brass or iron tube of length greater than the thickness to be cut, with thin walls, say 1 mm. The tube is revolved on its axis in a drilling machine and the end of the tube is fed with abrasive and water. The abrasive sticks in the material of the tube and cuts the glass.

In actual construction, the tube is about 1 mm. in wall thickness and may be between 10 and 100 mm. in diameter. One end is blocked with a disk and a spiggot is fixed so that the instrument can be chucked, and at the other end an axial slit is made in the tube with a hacksaw blade. Fig. 6.

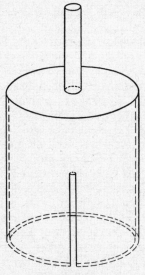

Fig. 6. Trepan.

In operation, the glass from which a disk is required is stuck upon a waste piece with beeswax so that when the trepan breaks through, the wanted disk within the trepan tube does not spin round with it and scratch itself. A wall of plasticine is then built round the position of the intended cut allowing about 6 mm. of clearance between the wall and the trepan. The miniature pond thus formed on the job is filled with a mixture of carborundum and water, the trepan is lowered into it, and cutting commenced. The slit in the side of the trepan allows the carborundum-water mixture to work under the cutting edge of the tube and so keep it continuously supplied and cooled. A speed of revolution of about 100 r.p.m. is right for a trepan of, say, 25 mm. diameter. Smaller trepans may be revolved faster in inverse proportion to their circumferences. If a power drill is used (and this only is really satisfactory), the control handle may be loaded to provide constant, even pressure on the tool. The only difficulty likely to be encountered with a trepan is in starting it, as it often wanders on the surface of the job before it cuts a groove in which to locate itself. If the cutting edge of the trepan is turned true and it rotates truly on its axis, and both job and trepan are properly clamped, real trouble should not ensue; it should be started gently but firmly.

How to Drill Glass

Trepanning is the only safe way of drilling glass when the hole required is over about 10 mm. diameter, because if the job is properly backed-up to take the strain when the glass below the cut becomes thin, a breakage is very rare and the cut may be taken as gently as necessary. When small holes, say between 8 mm. diameter and 12 mm. diameter are required, and when the job is strong like a bottle or sheet of plate glass, the tungsten carbide-tipped drill is the only tool worth considering. The drill takes the form of an ordinary twist drill (or shell bit in large sizes over 6 mm.), fitted with the tip and revolved at a speed normal for metal drilling. Such a drill will cut glass, steel, files, bricks and concrete, cleanly and quickly. Usually no lubricant is required, but for glass a mixture of equal parts of paraffin and turpentine should be used to prevent local heating. The drills require to be ground to a straight vertical front edge and a large clearance angle, and are usually correct as purchased except that the clearance angle is insufficient for drilling glass. Most examples of this type of drill can be ground on an ordinary emery wheel so long as plenty of time is taken on the job and the drill is well cooled frequently by dipping in water. The tungsten carbide itself is not likely to be much affected by the heat generated during grinding, but the tip which is only brazed on to the shank may be fused off.

The method of using the carbide-tipped drill is exactly the same as that for drilling metal. The drill will start iself without centre potting and should progress through the glass, leaving a dusty waste. If it squeaks a lot, it is too blunt or the clearance angle is insufficient. There is considerable danger of the drill breaking out flakes of glass on the undersurfaces of the job when it breaks through, but on the whole there is little trouble due to this when the rough nature of the cutting action is considered. If the job is flat, a waste piece of glass may be stuck to the back of the job with glue or beeswax to prevent flaking, but in any case the pressure upon the drill must be reduced when break-through is imminent. If possible the hole should be drilled towards the centre of the job from both sides, but if this is attempted the lining-up must be good or the job will split when the drill reaches the first hole on a different axis. Holes may be drilled relatively safely in bottle-shaped objects by first filling them with warm fluid beeswax and drilling the hole after the wax has hardened. The wax may afterwards be poured out after warming in an oven and the vessel washed clean with carbon tetrachloride (Thawpit).

Very small holes down to about 1 mm. diameter are drilled with fine chisel-pointed bits which operate in a reciprocating manner, driven usually by means of an electro-magnet and armature after the style of an electric bell but at a much greater reciprocating speed. These drills punch their way through the work slowly and the hole has to be cleaned out frequently. The modern reciprocating drill for glass has been developed from this but is now driven by apparatus working at supersonic frequencies, the reciprocating movement being obtained by magneto striction in a bar of nickel; the frequency by a value oscillator and amplifiers, and is of the order of 20 kilocycles. The drill bit need not now be a chisel-pointed affair, but can be flattened and of any cross-sectional shape, therefore holes of almost any shape can be drilled in brittle materials, of which glass is an example.

How to Saw Glass

The ordinary method of glass cutting is a somewhat chancy procedure and cannot be applied to glass of irregular quality and thickness nor to irregular objects. These should be sawed to shape by means of a diamond disk as mentioned on pages 104 and 473, or with a similar disk charged with carborundum and water. The proper device is the diamond-armed disk and the pieces of glass may be hand held against this in the same way that a circular saw is used. The apparatus consists of the following parts:

(1) A sheet iron disk about one foot diameter mounted on a horizontal axle and revolved at about one thousand revolutions per minute. The arming of this disk is described below.

(2) An iron table through which the disk protrudes exposing its upper third, with its axle below the table, or a moveable platform which can be carried past the disk under the control of a mechanism; in this case it is usual to mount the disk on the end of the axle so that it stands clear of the driving mechanism.

(3) A pumped or gravity supply of paraffin oil to the rim of the disk at such a rate that the cutting rim and job are always wet.

(4) A fence or other arrangement on the cutting table against which the job can be rested to ensure linear travel.

To Prepare the Cutting Disk

Suitable mild steel or plastic armed disks can be purchased from the lapidiaries but they can equally well be laboratory made. A disk of mild steel is cut from a flat sheet about 1 to 2 mm. thick by means of a guillotine or a fly cutter, but not by means of shears or nibblers because these warp the sheet by stretching. The centre hole must then be drilled and the rim trued in a lathe. For the best work the sides of the disk should be slightly undercut to allow clearance while operating but this is not essential as the arming procedure raises a considerable burr at the rim. If undercutting is decided upon, the disk must be stuck with pitch on to a lathe face-plate and turned down the required amount using the face-plate and pitch as the support and vibration damper.

The arming procedure consists of embedding in the edge of the disk fragments of diamond which act like the teeth of a circular saw. The fragments are of irregular shape but with dimensions of the order of fifties of microns. They are mixed with grease and rolled into the rim of the disk with a hard steel or agate roller applied to the disk in the manner of a knurling tool, the operation being allowed to go on for, say half an hour, to make sure that the rim is well filled with diamond fragments and the steel burred over them to hold them in place. When the saw is put into operation, the steel covering the fragments is soon rubbed away exposing the points which then prevent any further wear until after several months the wheel needs re-arming. The rate of cutting of such a wheel for a sheet of glass about 8 mm. thick is about 25 mm. in five minutes, and for quartz about one-third this speed.

How to Cut a Glass Disk

It is comparatively easy to cut a circular piece of glass when it is the disk shape which is wanted, as opposed to the hole left by the removal of a disk. When a circular crack is described by the cutting wheel on the glass sheet, the crack is started by briskly tapping the underside of the sheet below the cut and tracing the crack round the cut by tapping until the disk is loose. It will be found impossible, however, to remove the disk from the sheet because of the irregularity of the fracture in the direction of the thickness of the glass. When the disk only is wanted, the outer piece can be cut as shown in Fig. 7 and then broken away, but when the hole is the wanted part, the centre disk must be cut in four directions as shown at B, after which a smart tap at the centre will cause cracks to spread to the circle and pieces will fall out, thus freeing the rest.

The scribing apparatus consists of a metal post about five inches

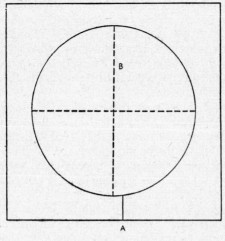

Fig. 7.

high, through which passes a cross arm with a clamp, carrying at its outer end an ordinary cutting wheel. The post rotates upon its base which is fitted with a rubber sole to cause it to remain steady on the glass while cutting arm is revolved. After cutting, both disk and hole, whichever was required, must be rubbed round with medium grade emery paper to remove the very dangerous sharp edges and prevent cracks spreading. This process is carried out commercially.

To Cement Glass

In works upon microscopy, very large numbers of different kinds of cement have been mentioned for such purposes as sticking cells to slides and making pond life troughs. Some of these are still useful and are mentioned in these pages, but on the whole, for making up larger laboratory apparatus, only two cements need be considered now and they are Araldite, made by Aero Research Limited, Cambridge, and marine glue, made by most London manufacturing opticians.

Successful use of these materials depends upon obtaining good flat surfaces to mate up before sticking, and upon the raising of the *whole* of the job to the temperature required to make the cement fluid. All the joints must be made while the whole job is hot and mechanical supports must be arranged to hold things in place until the whole has cooled down to nearly room temperature. If Araldite is used, the best form is that called thermo-setting, which requires about two hours cooking at about 200 ° C. and is sold in sticks which are applied to the hot job, fuse and leave a stiff liquid on the surfaces to be joined.

Araldite can be used to cement metal to glass but unequal contraction of the parts during cooling usually sets up severe strains at the surface of the glass, which usually result in the glass being pulled out in lumps at a later date.

Glass panels in larger aquaria and in windows must not be stuck in by such a hard cement as Araldite because mechanical shocks are then transmitted from the frame to the glass and the glass will break. Panels of this kind must be set behind a metal framework and be cushioned with common putty or marine glue. Ancient putty will yield when glass has to be replaced if it is coated liberally with soft soap.

Troughs and aquaria intended to hold sea-water are best constructed of seamless glass, though small ones for use on a microscope are satisfactory when built up with Araldite. Ordinary cements will not withstand salt water for a long period (say months) and metal work is attacked readily, thus changing the chemical constitution of the water. The common glass battery jar makes a good salt water aquarium but it must be stood outside for some weeks before use, with a dilute solution of soda in it to get rid of the acid. It is thought that such a jar never is rid of the influence of the acid, and delicate creatures may always be affected.

CHAPTER 29

Miscellaneous Pieces of Apparatus

A Method of Setting out the Fabric Shape for a Conical Object

THE DESIGN illustrated (Fig. 1) is for a conical-shaped pond-net 150 mm. diameter at the top, 25 mm. diameter at the lower end and 150 mm. deep.

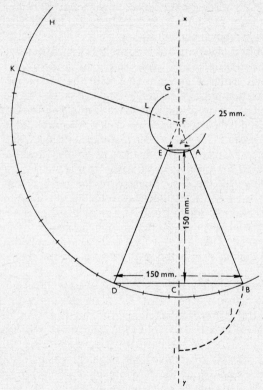

Fig. 1. Cutting diagram for cone (pond net).

Draw a full-size profile of the net—ABCDE about the axis line *xy*.

Extend lines BA and DE, both cutting *xy* at F.

With centre F, and radius FA draw the arc AEG, and with centre F and radius FB draw the arc BDH.

With centre C and radius CB draw the arc BI (which is the quadrant of the circle) meeting xy at I. Divide this quadrant into four parts (bisect twice). Each part, e.g. BJ, is therefore one-sixteenth of the required 150 mm. diameter circle.

Commencing at B and with radius BJ mark off 16 steps along the arc BDH to point K. Join KF, cutting the arc AEG at L.

Then the portion ABDKLE is the shape of the piece of material required.

Allow about 6 mm. extra on the line KL for sewing the net together. Also allow 6 mm. extra along the arc BDK for sewing the net to the ring. The bottom end AEL should be oversewn to stop it tearing away from the glass tube.

This method of setting out is entirely satisfactory for articles such as pond-nets, lamp shades, cones, etc., and leads to an appreciable economy in material as compared with the hit or miss methods sometimes attempted.

With reference to a suitable fabric for a pond-net, it has been found both difficult and expensive to get the usual bolting silk, and trials have been made with other materials. Yellow nylon parachute cloth, which has been on the market as surplus government stores, has answered the purpose most satisfactorily.

In bolting silk the clear opening of the mesh is 130 to 170 μ, whilst in the case of the yellow nylon the mesh is much finer giving square openings of roughly 100 μ. Further, it should be noted that many of the holes in the nylon are much smaller, as the weave is rather irregular. This material is quite suitable for collecting plankton down to medium size. A nylon stocking is a good filter.

Adjustable 'Cross' Aperture for Phase Contrast Apparatus

For phase contrast experiments often carried out with laboratory-made apparatus a cross-shaped aperture adjustable in all directions was found essential. The following apparatus was made out of 20 SWG tin plate and standard BA screws and nuts. (Fig. 2.)

When heads K are slacked off, segment pieces C fall together because of springs G and the minimum light passes through the cross. This is the condition in which the apparatus is set up under construction before blocks B are soldered down.

The gaps forming the arms of the cross are set by heads K as required. The angle of two opposite arms are set up between limits operating heads I independently of the cross gaps.

In operation the adjustments are very smooth. A film of oil be-

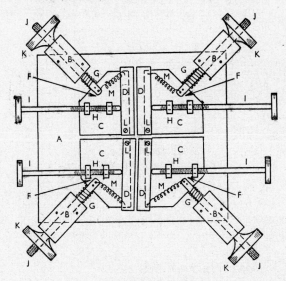

Fig. 2. Adjustable cross aperture.
Scale about 1½ size.

tween the segment pieces and base plate A holds the segment pieces
firmly under control and they do not rock when screws I are operated.

With the apparatus in place before the light source, the cross gaps
and angles can be adjusted to suit exactly a cross-shaped phase plate
in an objective. The plate A rotates in the axis of the instrument but
this movement is not shown in Fig. 2. (See 'Phase Microscopy',
page 403.)

Stage Attachment: Mr F. C. Wise's Design
In routine work like that encountered in Museums much searching of
slides may have to be done. When the slides are all of varying thickness a lot of time is wasted adjusting objective focus and condenser
position to accommodate thickness differences. In Wise's stage the
slide is pressed upwards against bars B so that the slide is always
located on the microscope by its top side. So long as the condenser
has sufficient focal length to reach through it, any thickness of slide
may be set up and the object mounted on its top surface is always
within range of the fine adjustment. The cover glass is still to be
accommodated but this causes no inconvenience.

A further advantage of this attachment is that the slide will give
way downwards should the objective touch it, hence it is an excellent
safety stage. (Fig. 3.)

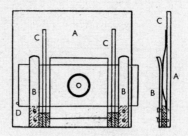

Fig. 3. Safety stage.
A. Base plate 2 mm. thick.
B. Bars 1 mm. thick.
C. Strips of watch spring.

The edges of all parts must be rounded with emery paper to prevent the slides digging into the metal as they slide in and out.

Sartory's Method of Making Cells

It is often necessary to support a cover glass a definite distance from a slide or to contain a liquid between the two. Usually cells for this purpose are built up from cement but these have an indeterminate depth.

Obtain an old sparking plug and dismantle it. The insulating part of several makes of plug will be found to be a wad of mica washers. These washers may be separate with a needle and chosen to be of any thickness when measured with a micrometer down to about a hundred microns or so. They can be stuck to the slide with any of the usual cements.

Mullinger's Illumination

This system employs Polaroid sheet to enable a quick change to be made between light ground and dark ground illumination. Such a change-over is often useful in analysis work where the effects of reagents as a means of differentiating objects are being tried.

An annulus is made up to fit the substage stop carrier as follows:

(1) Obtain a clear glass disk to fit the stop carrier.
(2) Stick upon this with gum pieces of Polaroid as shown in Fig. 4. The centre piece being cut to size in the same way as a centre stop is cut for a spot lens. ('Dark ground Illumination' p. 348.)

It is easiest to use a piece of paper and make a central stop then use this as a pattern. The rest of the disk of glass is covered with Polaroid with its polarization axis at a right angle to the centre piece.

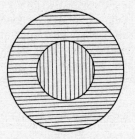

Fig. 4. Mullinger's disk.

An analyser is now made from a single piece of Polaroid and fitted in a revolving cap of cardboard over the eyepiece.

When the analyser is crossed with the centre area of the substage disk, dark ground illumination results. When the analyser is rotated through 90°, light passes through the centre area, but that from the outer area beyond the aperture of the objective is stopped, hence illumination by transmitted light results.

When using the apparatus it must be remembered that polarized light is being employed whatever the position of the analyser. This sometimes produces unexpected results.

The Polaroid sheet is best mounted between disks of clear glass. If balsam is used to secure it, it must not be heated beyond say 60° C. or the Polaroid is damaged. The material is best not cemented, but only clamped in place. The R.I. of balsam and Polaroid are different.

Simple Comparator Microscope

It is most difficult to compare microscopical objects when they are not side-by-side on the same slide. For industrial work twin microscopes are necessary so that an unknown object may be compared with a known one in a standard collection. The commercial comparator eyepieces which connect two separate microscopes together are excellent and expensive.

Messrs Sartory and Wise made a simple twin body microscope united by its eyepiece for work on museum specimens. The instrument was satisfactory in operation and is described below.

The focussing mechanism of an old stand was used to support twin tubes with centres about 57 mm. apart. An objective thread is made on the lower end of each, but refinements such as draw-tubes were omitted. The tops of the tubes were united with a block of hard-wood drilled to hold them tightly. On the wood was placed a disk of clear glass to which was cemented with balsam three prisms as shown in Fig. 5. The eyepiece was supported over the knife edge of the centre one. The field of view thus consisted of half from each body divided by the knife edge which appeared as a thin line. One body was made shorter than the other and a simple sliding focal adjustment fitted so that the magnification and focus of the bodies could be made the same.

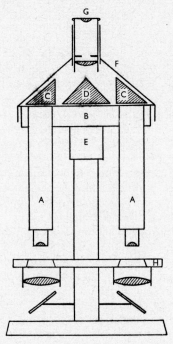

Fig. 5. Part-sectional view of simple comparator microscope.

The instrument is best used with powers of about 20 mm. but may be of any degree of elaboration desired.

Simple Apertometer for Dry Objectives

The apparatus consists essentially of a kind of gibbet (Fig. 6) standing on a base board, B, which should be not less than 40 cm. long—the exact size being not material so long as it is large enough. At the centre of one side is attached a wooden block, P. 12 cm. long and about 5 cm. square. To the upper end of the block is fixed a strip S, of the same material, 12 cm. long, 5 cm. wide, and 3 cm. in diameter. Blacken all the woodwork, and cut out two pieces of stoutish white

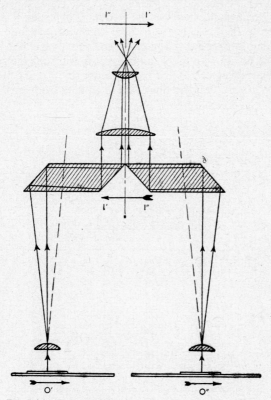

Fig. 5 (*a*). Commercial twin body comparator eyepiece.
O′ O″, Objects to be compared.
I′, I″   Composite image before and after eyepiece magnification.

card, C, about 20 cm. long and 7 cm. wide. Arrange these strips on
the base so that the exact centres of the abutting ends come immediately under the centre of the hole in the arm S, and provide guides
(small nails or something similar) so that the cards may slide to and
fro readily, whilst at the same time being kept in proper position.

The hole in the arm S, is for accommodating the objective to be
tested. To use the apparatus the objective is so arranged that its
front lens is 10 cm. from the cardboard slider. It may be supported
on washers or gripped in the hole by packing. Place the eye above the
objective, look through it from a distance above of about 25 cm. and
slide the cardboard pieces outwards until they are just visible at the
margins of the objective aperture. The inner edges of the cardboard
then mark the limit of angle at which light can enter the objective.

Measure the distance apart of these edges with a ruler in centimetres and refer to the table below for the numerical aperture associated with the air angle. (See 'Numerical Aperture', page 41.)

Anyone handy with tools could improve the appliance by making a sliding arrangement on the block with a screw to fix the objective at exactly 10 cm., but the simple affair as described works well.

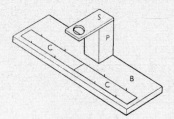

Fig. 6. Apertometer.

D.	A.	D.	A.	D.	A.
1	·1	11	·74	21	·90
2	·2	12	·77	22	·91
3	·29	13	·79	23	·92
4	·37	14	·81	24	·92
5	·45	15	·83	25	·93
6	·51	16	·85	26	·93
7	·57	17	·86	27	·94
8	·62	18	·87	28	·94
9	·67	19	·88	29	·94
10	·71	20	·89	30	·95

TABLE XXX

Stereoscopic Vision with the High-Power Binocular Microscope: J. J. Jackson's Method

Two segments of Polaroid are placed in the stop ring of the substage condenser, with their polarizing axes at right angles to each other. The join to be in a north and south direction. Fig. 7.

Fig. 7.

Two polaroid eyepiece caps arranged as follows:

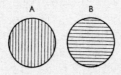

Fig. 8. A. Left-hand eyepiece filter 'crossed' with segment No. 2.
          B. Right-hand eyepiece filter 'crossed' with segment No. 1.

The apparatus is satisfactory for obtaining quickly an idea of the contours of a small object like a diatom or dust particles.

Henderson's Trough (*Fig.* 9)

This apparatus is cut from Perspex and may be made of any required size. The effect is that of a dark-ground illuminator without the disadvantage of a focussed light source which has to be changed as different levels of the trough are searched. Perspex is an excellent transmitter of light.

The trough (B) is sawed out of Perspex and all its surfaces polished with fine emery paper and Perspex polish. The whole is mounted on a blackened brass base plate (a). Over the central U-shaped opening are cemented glass covers (C) to form the water holding part. Murrayite is a suitable cement. The small U-shaped holes transmitting the light from the lens-fronted bulbs (F) are 6 mm. radius. The lamp holders (E) are standard flashlamp type.

This thickness of Perspex between the lamp and the water prevents overheating of the specimens.

The studs (D) may be made to project and so allow the trough to be used with either side facing the objective.

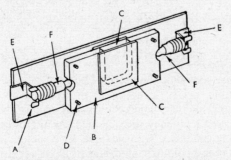

Fig. 9. Dark ground demonstration trough.

Young's Method of Plotting the Path of a Ray of Light Through a Spherical Surface (Fig. 10)

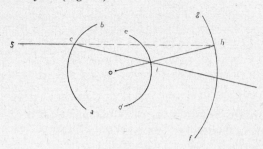

Fig. 10.

Draw the surface in question as arc *ab* with centre at *o* and radius of curvature *r*. Consider a specimen placed at *s* in a medium of RI (*n*) from which a ray of light proceeds, and draw ray *sc* produced across the paper. The RI of the lens material, that which appears on the right hand side of surface *ab*, is n^1 and is normally greater than *n*. Taking *o* as the centre, make a geometrical construction as follows:

$$\text{Calculate } \frac{n\,r}{n^1} \text{ and } \frac{n^1 r}{n}$$

Use these dimensions as radii and draw arcs *ed* and *fg*. Line *sc* will be found to cut *fg* at *h*.

Join *ho*, which will then cut *ed* at *i*.

Join *ci* and the path of a ray leaving *s* in a medium of RI (*n*) will be given by *sci*. The part of the path *ci* applies only to a medium of RI (n^1). When various media are in use, for instance immersion oil, on the left of *ab* and various solid materials on the right to form the

lens, the construction is the same but equations $\dfrac{n\,r}{n}$ must be re-worked with different values of *n* and n^1.

Comparator Projector

When comparisons must be made of photographs of any kind twin slide projectors can be arranged to project images on to a white screen. 50 mm. microscope objectives may replace the conventional lenses. Measurements may then be made on the screen directly and the apparatus calibrated from a stage micrometer. Images may be superimposed and coloured by means of filters for easy recognition.

CHAPTER 30

Electron Microscopy

Introduction to the Apparatus

THE STUDY which has become known as Electron Optics is really quite different from 'optics' in that the image-forming radiation is not light nor is it refracted and reflected like light. It will be seen later that certain parts of the electron microscope can be given names equivalent to the parts of the optical microscope, but these are only a convenient notation.

In the electron microscope, electrons are the kind of radiation used to form the magnified image of the object. An electron beam is usually considered to be a corpuscular emission very like a shower of beta particles from a radioactive substance. If, however, electrons are sent through a crystal, the layers of atoms of which act like an optical diffraction grating of great dispersion, they too are shown to give diffraction patterns. One must conclude from this that electrons behave like an electromagnetic radiation and a wavelength may be calculated for them. The wavelength varies with the generating voltage but is typically of the order of one thousandth that of light and is determined from the crystal scatter and diffraction.

In the section dealing with general microscopy it will be seen that microscopical resolution depends ultimately upon the wavelength of light used. When ultra-violet light is used with a special lens system transparent to it, an increase of resolution over that of visible light of five times is possible, and even this comparatively small improvement revealed many hitherto hidden structures. If the wavelength of the electron is a thousand times less, and if the radiation could be caused to form an image with a magnification suitable to the resolution theoretically possible with such a wavelength, atoms would be visible. As things are today it is not possible to build up the required amount of aberration-free magnification owing to refraction difficulties described later, but even at this early stage of development of the apparatus it is considered certain that very large organic molecules can be seen.

A certain group of substances through which the extremely narrow band of frequencies of visible light can pass are called 'transparent' but this property does not extend into the part of the spectrum occupied by electrons. Air and other gases are not transparent to electrons, nor are ordinary substances unless specially thin, therefore the whole microscope and specimen must be in a very good vacuum in order to work at all, and the specimens must be of the thinnest preparation, preferably suspended in the vacuum, or at the most placed on a specially thin film of collodion or other membrane described later. Because the specimens must be in a good vacuum they cannot contain any water because this would boil off at the low pressure and spoil the vacuum, therefore all organisms must be dead and dessicated. Ordinary biological stains appear when in the specimen as though it were packed with gravel, and have no differentiating power in such radiations, in fact they cease to act as stains at all when examined at this degree of resolution. At the present time it is impossible to use any technique like that of vertical illumination of an ordinary object, so replicas must be used, see 'Mirror Electron Microscopy', page 535, and page 547.

Properties of the Electron

The electron is a particle of electricity which has been given a negative sign and all atoms are made up of a nucleus which has a positive charge surrounded by varying numbers of electrons kept from falling into the positive nucleus by centrifugal force of revolution in orbits. In an electrical conducting material electrons, may be urged from one atom's orbit to that of another and so pass through the substance in the form of a current. In an electrically insulating material the electrons cannot be shifted about, therefore a current cannot flow. Because a stream of electrons, constitutes a current each individual electron can be affected by magnetic and electrostatic influences just as can a current-carrying wire. When sufficient voltage is applied to an electrical circuit the pressure on the electrons is sufficient to cause them to jump a gap as occurs in a motor car ignition system, and while the electrons are pouring across the gap they can be deflected by a magnet. Being a stream of charged particles they can also be deflected by means of a high voltage electrode placed nearby. The high voltage method is known as 'electrostatic' deflection; it is sometimes used in television tubes and often in cathode ray oscillographs. In the electron microscope it is usual to employ the electromagnetic system of deflection because great amounts of deflection are required, very much in the same way that high curvature is required in the optical objective and this condition is best arrived at by magnetic means. The electron instrument then becomes a series of iron-cored coils rigidly mounted, the purpose of which is to deflect the electron beam in such a way that an image is formed.

When a stream of electrons strikes some materials such as zinc sulphide, sodium glass, calcium salts and many others, they deliver their energy to the substance and light is emitted. In this way a screen of one of these substances painted on to the inside of a tube and exposed to electrons will glow and can be made to present a picture if the electrons are varied in intensity. The familiar television tube is more complicated than the microscope in its image presentation because it has to produce a moving picture from synchronized incoming signals. The microscope has its object within it, therefore its image is really only a modified shadowgraph. If a stream of electrons is proceeding along an evacuated tube, each electron travels normally in a straight line path, so if a pattern of some kind made out of an electron absorbing material be placed anywhere in the beam a shadow of it will be formed on the screen. Early demonstrations were made with a device called a Crookes tube (after Sir W. Crookes the inventor) which consisted of a glass tube about 250 mm. long with a metal electrode sealed into one end, and in about the middle of the tube a small Maltese cross hinged so as to become erected in the path of the beam when the tube was tilted. The whole was evacuated to a high degree and an induction coil connected between the end electrode and another wire sealed through the glass anywhere near the cross. The result was that cathode rays were generated and left the negative end electrode in straight lines rendering the soda glass at the distant end of the tube fluorescent under their bombardment. When the tube was tilted the cross was erected and a really sharp shadow of it was cast on the fluorescent end. This experiment showed that the rays travelled in straight lines and that they were ejected from the electrode to pursue their own straight courses independently of the position of the opposite electrode.

The first microscopes attempted were something like the present day X-ray one in which the specimen was placed very close to or in contact with the source of cathode rays. A shadowgraph was thus formed but as there was no image-forming device the magnification and resolution were low. The old tubes had cold cathodes and a trace of gas in the envelope to allow some electrical conduction. Power was obtained from an induction coil with a voltage approaching a quarter of a million, i.e. a spark capable of jumping 100 mm. or more in air at atmospheric pressure.

It soon became clear from experimental evidence that there is a haze of electrons surrounding any red hot metal and that these free electrons arise due to the heat of the metal, which so speeds up atomic movement that they are thrown out. They do not rise very far when the metal is exposed to gas at atmospheric pressure, but if the vessel is evacuated they have a path length of several centimetres.

If a charged electrode is placed within this range the electrons can be collected readily, forming an electrical circuit between the hot metal and the collector. If the collector has a hole in it many electrons pass through it, and so on down the tube for many metres in the form of a fairly well defined beam. This is the way it is done in the present-day electron microscope. The deflecting magnet coils are placed round the beam in order to focus it on the screen at the end of the tube.

The Magnetic Lens

When a coil of wire has an electric current passed through it a magnetic field of force surrounds it, and when the shape of the coil is according to Fig. 1 the direction of the lines of force are as shown dotted.

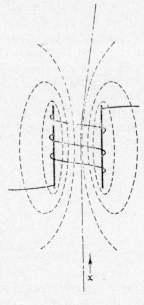

Fig. 1.

If an electron beam is passed into the system in the direction x, it experiences deviation as shown by the chain-dotted line. However, this amount of field curvature, whatever the size of the coil, is quite insufficient to provide the high degree of deviation required in a microscope. A magnetic objective lens is differently constructed. It has soft iron in its coil assembly, the function of which is to concentrate the magnetism, also the iron ring, Fig. 2, can be very small and modified by drilling and shaping to give a better field shape.

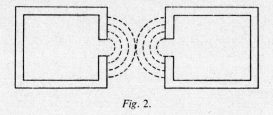

Fig. 2.

The deviation produced in an electron beam is now great, Fig. 3.

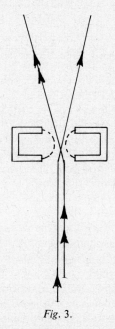

Fig. 3.

Unfortunately the magnetic lens of high power is uncorrected and shows similar aberrations to the optical lens.

Considerable improvement can be made in the image-forming qualities of the lens by inserting various shaped pole pieces into the aperture and by splitting up the pole pieces into disks separated by non-magnetic material. Fig. 4 shows a typical pole-piece section similar to that used in a commercial instrument. The cross-shaded

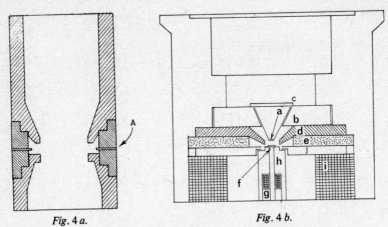

Fig. 4 *a.*   *Fig.* 4 *b.*

Diagrammatic Section of Electron Microscope Objective.
a. Object position. b. Stage. c. Object carrier. d. Pole piece. (i). e. Cooling apparatus.
f. Objective aperture. g. Astigmatic corrector. h. Pole piece (ii). i. Main windings.
j. Mount and cooling device.

part is soft iron of accurately known quality and uniformity machined and polished to optical accuracy, the diagonally shaded part is brass or bronze employed as a spacer between the iron parts, and part A is one of six or eight set-screws adjustable radially, which can be employed to correct astigmatism. In some microscopes a disk of platinum, a few tens of microns thick with a hole in its centre 60μ diameter is placed between the iron pole pieces to act as a diaphragm. The value of this fitting is somewhat doubtful and it is difficult to line up mechanically, but under some conditions, e.g. with thick specimens, an improvement in contrast is obtained, probably due to the operation of similar factors in the optical instruments.

Aberrations in the Electron Objective

As mentioned in an earlier paragraph, electrons may be deviated by electrostatic or electromagnetic fields, but as the ordinary instrument of today employs electromagnetic deviation for focussing almost exclusively, electrostatic focussing will not be discussed here, but will be mentioned under 'Historical Developments', page 536.

The field in the gap of an electromagnet varies in intensity, clearly being stronger in the regions near the iron than in the centre of the gap. Electrons experience deviation accordingly, also changes of velocity. Choice of pole-piece shapes can be made so that the electrons experience a nearly correct amount of deviation, but velocity errors remain and appear in the image like spherical aberration. Associated with this effect though not appearing as an aberration is the spiral deviation due to the fact that the force upon an electron due to a transverse magnetic field is at right angles to its direction of travel. In practice the effect is to impart a twist to the stream of electrons as they pass through the field of the lens. As the current is changed in the magnetic lens the image twists in and out of focus, but is not distorted.

Another effect which has no parallel in the optical microscope but which should not be confused with aberrations proper, is the scatter experienced by electrons passing through denser parts of the object which cause loss of velocity and so greater deviation, often preventing them from reaching the image or causing them to come to a focus below the proper focal plane. This effect will be mentioned in greater detail under the heading 'Interpreting the Image', page 533.

Spherical Aberration

All that can be done to offset this error is to work with small apertures so that the electron stream in use passes only through the central part of the pole-piece hole, very much as in the uncorrected optical instrument. The limitation of aperture here does not matter because no electron instrument so far designed can resolve better than one hundredth part of that theoretically possible with the equivalent wavelength of even the slowest electrons. Small apertures are a function of pole-piece shape in profile section, but great mechanical accuracy is demanded in the concentricity of all parts, metallurgical composition of the parts and mechanical location on the axis of the instrument.

The amount of current passing through the objective lens-coil can be varied thus altering the magnification (which is typically X 100), but this also affects the spherical aberration. It follows that variation of pole-pieces design, together with choice of current, can be used to make a lens with minimum aberration. Normally the lens coils are fed from a storage battery to ensure steadiness, and frequently are water-cooled against mechanical changes due to temperature. Aberrations which properly come under the heading of 'poor workmanship' rather than inherent design troubles are frequently due to dirt in and on pole-piece parts. These have to be handled with surgical cleanliness.

Chromatic Aberration

Fortunately this is not an important aberration in the magnetic lens because an electron beam has inherently a mono-energetic characteristic. Chromatic aberration appears electronically in the form of errors in focal plane position due to different amounts of deviation by the uniform magnetic field of electrons travelling at different speeds upon entering the lens field. This can be prevented by a very high degree of stabilization of the high tension voltage responsible for drawing the electrons through the instrument. A typical HT is 100 Kv and this would be stabilized to one part in 50,000 which is sufficient to eliminate chromatic aberration. (See page 533, 'Interpreting the Image.') A point filament in the gun is an advantage.

Astigmatism

Electromagnetic lenses are astigmatic for the same geometrical reasons as are optical ones and reference should be made to the appropriate sections dealing with optical microscope lenses. In the electron microscope it is not very serious mainly because only tiny fields of view are employed and the tiny apertures employed for other reasons help greatly in reducing astigmatism. However, some correction is desirable, especially when it can be got easily, and the simplest method is to insert iron grub screws radially in the pole-piece gap as shown in Fig. 4. These provide the necessary modifications to the shape of the magnetic field as they are screwed in and out. About six screws are required and they are adjusted on test with a particular objective and instrument combination. A good test specimen for this is one similar in some ways to the test plate of the optical microscope. If a tiny round hole in an ordinary collodion film is selected, astigmatism is apparent when the round hole periphery is not all in sharp focus at one time. The differences are most apparent across diameters, e.g. it is possible to bring into sharp focus the two ends, of say, the horizontal diameter or those of the vertical one, but not all together. Trial-and-error adjustment of the radial screws, and considerable patience will ultimately correct the lens. See also Fig. 4b.

Coma and Distortion

These aberrations are both only of importance in off-axis parts of the field. As no one is interested in taking large field of view electron micrographs (there are plenty of matters to try the intelligence beside this), these two aberrations may be disregarded. Again, the small aperture of the system reduces these errors to practically tolerable proportions without extra trouble.

The Source of Electrons

In the practical microscope of the present day a thin hot wire of tungsten is used as the source of electrons. This is bent in the form of an inverted 'V' with a sharp apex, mounted on stiffer wires, in turn fixed in a mounting plate which can be centred to the rest of the instrument. It is common for the operator to make up his own filaments out of smooth drawn tungsten wire supplied by the manufacturer of the microscope. The filament mounting varies with the manufacturer, but it is always more or less as described, because the point of the 'V' forms the true source from which the image is derived and this must be accurately centrable.

The electrons may be drawn away from this filament by the anode voltage of the instrument only, or there can be a grid placed near the wire which can be charged to a separate positive potential for the purpose of drawing a greater number of electrons from the filament for the main HT to form into a beam as in a cathode modulated oscillograph. This kind of operation has been called the 'biased source' kind and is the preferred method for two reasons, firstly, a denser electron beam can be obtained thus shortening exposure times and easing the focussing difficulties, and secondly, slight variations in electron emission are rounded off by the grid potential. The whole electron emitting assembly has become called the 'gun' after current electronics practice, and the tendency is to make the assembly as compact, mechanically stable and long lived as possible, thus doing away with some tricky adjustments. In older instruments though otherwise very good, there are frequent difficulties in obtaining sufficient mechanical stability with lab.-made filaments, and this became most apparent when dense electron beams and focussed point source illumination were employed.

Closely associated with electron emission is the high tension voltage which accelerates them through the microscope. As mentioned above, this must be very stable. In old instruments standard rectifier and smoothing circuits were used, but these require large inductance and capacitance values to obtain the required degree of smoothing and such circuits were never more stable than about one part in twenty thousand. The modern method is to derive the HT from a high frequency oscillator which can be stabilized much more easily by means of electronic circuits. The capacitors and inductors are also very much smaller because of the high frequency being handled. Stability of the order of one part in fifty or even one hundred thousand is possible. With such a degree of accuracy, variation of electron speed through the microscope and consequent chromatic aberration is eliminated.

It must always be remembered when setting up the instrument that a very much larger amount of energy can be put into an electron beam than can be put into a beam of light; this means in practice that a specimen can be burned up easily if too 'bright' a source be used. All matter is nearly opaque to electrons unless specially thin specimens are used, and when the electrons are absorbed in the thickness their energy is liberated as heat. With tubes that have variable intensity emission care must be taken not to turn up the wick to obtain easier-to-see spots for lining-up and focussing purposes as this will do damage to the specimen, and the spot will spread thus defeating the object of the increased strength of electron beam. On the whole the best definition in images of thin specimens is to be obtained from the low intensity parallel electron beam not increased in intensity by biasing grids. However, opinion is divided as to whether the lengthened exposures cause greater or lesser loss of definition due to mechanical vibration and electrical instability. Electrical instability is likely to have the greater effect when mechanical shocks are short and sharp, because displacement due to these will be of short duration compared with the total exposure time and so may not be recorded photographically. If, however, the mechanical movements are of the continuous kind such as those due to traffic, blurring of the image will certainly occur and short exposures with an intense beam are demanded. A weak condenser is employed with an intense beam, its function being more a source image shifter than a focusssng device. A condenser is not often used with an ordinary source because the electron beam is then parallel.

To Set up an Electron Microscope

As described in earlier sections, an electron microscope consists of a source of electrons, an arrangement of bias voltage for increasing the electron output beyond that normally obtained from the source, a condenser to position the image of the source (or the 'spot') in the plane of the object, an objective lens, an intermediate fluorescent screen with a centre perforation to indicate the image in the rear focal plane of the objective, a projector lens and a fluorescent screen to receive the final image. All these parts must be exceedingly carefully lined up, and an understanding of optical microscopes will help the operator considerably.

It is not practicable to line up an electron microscope in a factory so as to require no further adjustment, and designs vary as to how the adjustments are made. Some employ flat-faced flanges, vacuum grease sealed, while the modern tendency is to employ flexible plastic tubes to connect components so that all can be moved a small amount without destroying the connections. The method employed by each manufacturer is clear when seen, and all are experimental. As with optical microscopes, the electrical and mechanical axes are not coincident. This is due to slight machining irregularities in pole pieces, winding irregularities, and inhomogenity in materials, particularly pole pieces. It is fortunate that the twist in the electron beam due to current changes through the focussing coils can be used conveniently to ascertain the electrical axis of a magnetic lens.

The lining up of an electron instrument is not so easy to describe in steps as is the optical device because, besides being more complicated no two instruments are the same. However, the important points are as follows:

1. In all examples the first step is visual alignment of the filament source, cathode grid, condenser coil, objective coil, screen, and Wehnelt cylinder. This entails cleaning the parts carefully.
2. The filament image must be caused to pass right through the microscope to fall upon the centre of the fluorescent viewing screen for different values of objective current.

When these conditions are satisfied, the microscope will be properly lined up.

In order to arrive at this adjustment, proceed as follows:

1. Remove the object holder and set up by eye the remaining parts by spying through the instrument in the usual way.
2. Set up the filament point to be exactly under the hole in the cathode grid. This may be performed by eye, the method being to use the filament azimuth adjusting screws if fitted, or bend the filament wire if it has not been previously used (it will break if it has).
3. Energize the condenser and objective but adjust the objective so that its current is low, then as a low power objective it can cast an image of the filament on to the final fluorescent screen.
4. Centre this image by adjusting the azimuth screws of the condenser.
5. Set the objective current to its normal value, reset the gun azimuth screws so that the bright filament image falls through the intermediate screen hole.
6. Introduce a specimen, observe the spiral focussing effect on the fluorescent screen, note the centre of rotation and make slight adjustments to the objective azimuth screws to make this centre coincide with that of the screen.

When the microscope has a bias applied to its grid to produce a beam of greater intensity the condenser has a much greater effect upon the system and so must be centred early. The procedure in this case is as follows:

1. Line up the instrument by eye and set the filament under the centre of the grid hole (as above).

2. Obtain a circular image of the electron source on the intermediate focussing screen by adjusting the gun centring screws and the tilt screws, over a range of condenser currents.

3. Introduce an object into the microscope and adjust the condenser focus so that a large circular image of the source appears on the viewing screen. Varying the condenser current shows by image twist its centre. The gun should be centred to conform to this condenser axis and the condenser centration re-checked. Tilt adjustment is important in this case.

4. Further adjustment of these components will place the centre of rotation of the source at the centre of the viewing screen.

5. Adjust the objective current to the proper value and check that the centre of twist of the object image is coincident with that of the viewing screen. Touch up the objective centring screws to make it so.

It is quite common to find a diaphragm plate in electron objectives, put there to act as an anti-scatter device rather like the diaphragm often found in the thickness of the stage of old optical microscopes. These diaphragms are usually platinum disks a few tens of microns thick, perforated with a hole about 50μ diameter varying with the objective. The function of the device is not to limit the electron beam, therefore its centration is not so critical as may be thought. It should be sufficient to locate this carefully cleaned disk in its cleaned carrier collar concentrically with the pole piece. It must be chemically clean, but if it is off centre and is stopping too many of the image forming electrons it can become coated with a material which may affect the electrical characteristics of the whole objective. In general, if the image is seen to be more astigmatic with the objective diaphragm than without, its centration must be suspect. Little can be done about it except to try a new position in rotation. However, some objectives have been designed to have the concentration of this component adjustable within the objective. Some help in the direction of depth of focus can be got in thick specimens by attention to these details.

Magnification in Electron Microscopes

As with the optical instrument, the magnification must be determined by direct methods, the easiest in this case being a replica of a fine diffraction grating. A spacing of at least 600 lines per millimetre is required and probably the replica will have to be shadowed both to make it visible and to stabilize it for size. Several trials should be made, as replicas used in this way cannot be trusted to retain their spacing, so low intensity electron beams should be used to limit the heat produced in the object. For high powers crystals of copper phthalocyanine lattice spacing 9·9Å is used. This is better than using 'standard' particles.

The magnification employed on a job should, as usual, be as low as possible so that the object is bombarded and so changed the least amount. Of course, the detail sought must be made larger than the grain of the photographic emulsion, but ordinary fine grain plates are adequate and will record an image that can be enlarged ten times.

Positive Replicas

These are usually of the form of replicas of replicas and so are always of a second accuracy nature. Whenever possible a negative or ordinary replica should be taken, as thin as possible, but where the heavier film is demanded in the interests of stripping it should be of collodion because this is most easily dissolved away. Should a plastic negative be used the process is not so easy but the method is simple. A thick replica is taken from the job in the usual way, its surface flooded with another replica medium of different chemical characteristics and the collodion dissolved away by floatation in a solvent. The negative may first be metal shadowed as a method of reinforcement. Any replica may be stiffened mechanically by depositing in a normal direction upon it a layer of silicon oxide by evaporation under the shadowing bell.

Impression Replicas

Some industrial work upon strong objects may be done by pressing blocks of heated polystyrene upon a surface with a pressure of the order of tons per square centimetre. After cooling, the replica block is detached and its surface is covered in the evaporation chamber with silicon oxide or monoxide to such a thickness that the film can be handled when floated free of the polystyrene block. A carbon replica may also be deposited on the polystyrene. The waste polystyrene is removed with a saw and the remainder, near the wanted surface, dissolved in methyl bromide leaving the replica behind. Such a replica will be found difficult to wash because it sinks in methyl bromide, but it can be handled on a spatula in a petri dish over an oblique light.

Ordinary collodion replicas may be taken from a plastic press impression and will then be positives.

High Resolution Method

When very fine detail is to be examined it is necessary to cover the objects with a layer of shadowing before the coarser collodion supporting film is applied. The method is to spread the particles on a clean polished glass as mentioned above, shadow them, make a thin collodion replica-like film on top, and then strip the metal and collodion together from the glass. Evaporated metals do not show a macromolecular structure. This technique demands the highest skill and cleanliness in all its operations.

In some work a replica is required to be taken at a low temperature and it may be undesirable to raise the object to room temperature before stripping the film owing to possible changes of various kind including migration of molecules, also it may be desirable to take a replica *in vacuo*. If the object is placed in the evaporator, conditions may be set up there as required, and a layer of silicon monoxide about 3000A thick deposited normally upon it. The vacuum system may then be allowed to return to normal pressure, the silicon film having the 'fixed' structure upon it. Number 6010 epoxy resin mixed with amine in proportions 9:1 is then poured on to the silicon monoxide, allowed to set, and then stripped off the object complete with the oxide film. The film side of the resin block is now the 'replica' and may have replicas taken off it in ordinary ways to give positives.

Mounting Particles upon Supporting Films
The typical way of examining an object in an optical microscope is to place it or a suspension containing samples upon a glass slide, and then view it as a transparent specimen. In the electron instrument a fine film of collodion itself, supported upon a fine metal grid or stretched over a tiny hole in a plate, is used instead of the glass slide.

The grids employed are a few millimeters in diameter and about 200 meshes to the centimetre made by depositing metal electrolytically to the shape required and then dissolving away the supporting medium. The small-hole technique is the better for high power work because less distortion occurs due to electron bombardment than occurs in the meshes of a grid, therefore there is less chance of the film rupturing. On the other hand the wire mesh allows a much better chance of finding a suitable area to examine.

All supporting films experience shrinkage under electron bombardment to amounts varying between 20 and 60 per cent.

Collodion consists of 12 per cent of nitrocellulose dissolved in ether alcohol, and is used when diluted with four parts of amyl acetate. A few drops of this are allowed to fall upon the surface of water which must be distilled; they then spread, and in about two minutes harden, and can be used to act as a raft upon which the grids are rested. The film is cut roughly with scissors while still afloat, into convenient areas, which are gathered from the water by touching the upper surface of the grid and collodion with an object like a smooth piece of glass upon which contact, the whole will adhere to the glass sandwiching the grid between it and the collodion film. The slip of glass can be rolled out of the water in such a way that the loose flaps of the film edge wrap round the slip edge and secure the film in place against the effects of surface tension as the lot is lifted clear of the water. When dry the collodion film will be found closely adherent to the grids ready for use.

Several materials have been tried for making these supporting films and doubtless many more will be added to the list as development goes on. There follows a list of materials in common use: modifications of collodion employing different nitrocelluloses; Formvar; other plastics when a suitable solvent can be found; polystyrene; and vacuum deposited films of silicon or metal, particularly aluminium and beryllium. A recent technique is to deposit metal on to a liquid surface of sufficiently low vapour pressure, e.g. glycerine. Successful use has been made of silicon monoxide which is very stable in the electron beam; extruded films from slip planes which make very good foils, and accentuated oxide films on metals like aluminium after dissolving away the metal. The specimens to be studied are placed upon the supported membrane from a suspension in a pipette in the usual way, or they may be incorporated in the film liquid before this is poured to make the film. It is worth noting that a film prepared by pouring the liquid on to a solid surface like glass instead of a liquid will usually detach more easily by floatation (as described under 'Replicas', page 527) when the solutions are newly made up. As with staining techniques in optical microscopy the physical effects of ageing are not understood and may improve a process or ruin it.

Sections for Direct Study

In general the method of sectioning is unsuitable in electron microscopy because (i) the sections can rarely be made sufficiently thin for electron penetration even in high voltage experimental microscopes and (ii) there is too much probability of disturbance of macromolecular internal structure by the knife. Very little can be done about internal disturbance as it is nearly impossible to support the tiny structures to study which electron microscopy is demanded, with any impregnation substance which can be cut. Paraffin wax mixed with collodion is generally the best when this work must be attempted, but very promising results have been obtained with Araldite cut with a glass knife (see page 314), but the cut must be made when the Araldite or epoxy resin is at the right condition of polymerization. The question of adequate thinness centres around the mechanical rigidity of the microtome and the sharpness and even quality of the knife blade. Ordinary steel blades can be made to do the work so long as the microtome is of specially rigid design and the sliding or rocking gear perfectly adjusted. The specimen-advance mechanism normally fitted, however, will not do and a special holder of the kind that advances the specimen as a rigidly mounted block of metal warms from a very low to a room temperature, must be fitted. There is no difficulty in fitting such a block holder to any otherwise suitable instrument, and sections of thickness down to 200A can be obtained. See page 315.

Little more can be said about the sharpening of microtome knives than has been set down in a previous section in optical microscopy under this heading; the only advances in this direction towards electronic sections being some ultra high speed rotary cutting instruments, and even these are not very effective. They consist of an air turbine driven rotary head revolving at about 40,000 revolutions per minute with the knife blade projecting from one side. As the specimen in its block is advanced to the blade the sections fly off in flakes and are directed often by air streams to receiving grids. Paraffin wax is the usual supporting material for these experiments and the holder usually has to be cooled with solid carbon dixode during cutting to preserve solidarity.

Sectioning work cannot be recommended in electron microscopy because even when a section of suitable thickness is obtained the danger of recording artifacts is great. These may arise due to the remains of the embedding material in the structure, the scattering effects of the fixing chemical when organic material is examined and the mechanical disturbance of the parts by the knife. Some useful information has been obtained concerning the distribution of materials in sections by the method of taking a replica from a micro incineration.

Sections for limited metallurgical work may be made by polishing the specimen electrolytically until a hole appears. It is possible that the area surrounding the hole will present a region sufficiently thin to be transparent to the electron beam, but if not the holes may be painted out under an optical micromanipulator and the polishing continued until more holes appear in the generally thinner specimen. It is then more probable that one with a thin edge will be found. When polishing in this manner it is best to use a specimen prepared with thick edges like a tray so that a more even polishing action occurs on the thin centre area.

Staining for Electron Microscopy

It was observed at the beginning of this chapter that ordinary biological stains are of little use in electron microscopy but successful attempts have been made to increase the density of specimens to electrons in biological material by soaking in solutions of heavy metals such as osmium and 1 per cent lanthanum. Osmium enters the specimen and acts upon the fats thus making vacuoles and membranes stand clear in their surroundings. Lanthanum is opaque to electrons and may be dissolved in biological culture media; it tends to lodge in or on surfaces of organisms and when used at about 1 per cent dilution is not lethal.

Ordinary fixatives give trouble by leaving unwanted substances in the section producing very similar effects in the image to those due to ordinary stains, they also have a greater effect than is generally thought in the shrinkage they cause to cell walls. Measurements have shown that osmium (used as a fixative) causes up to 50 per cent shrinkage, and other usual ones mentioned in the optical microscopy chapters cause more or less this amount depending upon many variable factors such as type of specimen, strength of fixative and time of action. In order to avoid this trouble, water soluble fixatives have been tried and a soluble resin called Agvon used in two strengths, taking about four days to complete the fixing, has given promising results.

All sections shrink in thickness in the electron beam to an amount which may approach 50 per cent which may well be additional to that due to fixing.

The Metal Shadowing Technique

This subject has been dealt with in the sections in which optical replica techniques were treated, so here it is necessary only to add some important details affecting electron microscopy.

There are advantages in employing an evaporation set of large chamber size, say about 500 mm. in diameter, mainly for the following reasons:

1. The heated element is a long distance from the sides of the bell glass, therefore there is less likelihood of occluded gas being driven out of the walls while evaporation is taking place.
2. As the atoms leave the finite sized source, a long free path allows a better approximation to a point source to be seen from the position of the object being shadowed. This in turn leads to sharper shadows of fine detail.

3. A number of specimens may be treated at one time; this represents an important saving of time and material owing to the fact that a critical pumping may take two hours to arrive at and maintain a vacuum of the required 10^{-4} mm of mercury.

For effective shadowing it is essential that the mean free path of the atoms of metal leaving the filament be greater than the distance between the filament and specimen. This usually means that a vacuum of 10^{-4} mm mercury must be maintained.

Metals Suitable as Shadowing Media

It is essential for electron microscopy that the metals employed have the following properties:

1. They must not show a structure of their own, e.g. zinc evaporates into spherical globules, silver is very grainy and aluminium forms discrete particles.
2. They must deposit cleanly, i.e. not show a tendency to pile up into whiskers and give what in optical microscopy would be described as a rotten image (due actually to minute shadows of these whiskers overspreading the main shadows).
3. The metal must be dense enough to be opaque to electrons at such a thickness that fine detail is not plastered over. Typically 40A is sufficient.

Several metals satisfy these conditions, particularly gold, chromium and palladium. Of these metals some have advantages over others. Gold is good for high power work in fine detail, but if a thick layer is built up to shadow effectively a coarse replica too much energy is absorbed from the electron beam, and film distortion and granulation of the gold result. On the whole chromium is best for general use but does not provide such a clean shadow as gold. Palladium has similar qualities to chromium, evaporates fairly easily without excessive temperatures and gives a clean shadow comparable to that of gold.

Two methods may be employed to measure the amount of metal deposited upon the specimen. The amount required depends upon the thickness of covering and the angle of the shadow, and both must be determined by experiment for the job in hand. Typically about 40 milligrams of metal are actually evaporated during one shadowing operation, but of course, most of this is lost around the bell glass cover. A given amount by weight of shadowing metal may be deposited upon the tungsten filament of the evaporator, in which case the time of evaporation is not critical, or, the metal may be contained in some sort of trough arrangement built into the filament ribbon, in which case exposure is by time. Another method is to employ a filament made of the wire to be deposited, in which case the wire is flash fused or is left on steadily for a given length of time. The method of evaporating a given quantity of metal previously deposited on the filament is most accurate for critical electron microscopy.

There is need for research in all matters concerning shadowing techniques and as electron microscopy improves 'single atom layer' shadowing is required. Attempts have been made to use uranium in the form of sulphide. This very dense compound was prepared by heating uranyl nitrate with sulphur; it vaporizes easily, and very good shadowing has now been obtained. Theory indicates that a layer of only a few atoms thickness would provide adequate shadowing.

Evaporation techniques are used for other purposes than for shadowing. A thin film of certain metals such as beryllium, applied normally, act as a strengthening film to replicas, and may be used as a separator between films when subsequent dissolution of one of them is required as for example when producing positive replicas. Non-metals may be similarly used, typically, silicon oxide produced from needles of quartz attached to the tungsten filament, evaporated by flash heating of the filament to limit the heat absorbed by the specimen.

Of evaporation experiments there is no limit, but the important points to be observed are as follows:

1. The heat of the evaporation process must not affect the object being shadowed and this means that a flash-evaporation of an easily evaporated material in a large bell glass will give the most accurate results.
2. The angle with the horizontal at which shadowing is performed should be the maximum before which the shadows of the various parts of the wanted structure mix up with each other. For bacteria an angle of about 10^0 is typical. The angle at which shadowing is done must be accurately recorded for each specimen during trials, for it is generally not possible to identify specimens when different unknown shadowing angles have been used on them.

Interpreting the Image

In the previous sections dealing with aberrations some matters largely comparable with optical ones concerning image interpretation were discussed. There are, however, a number of details associated with electron microscopy only, which are important.

A simpler matter is that of replicas for although they look very realistic, in fact as good as stereo-photographs, it must be remembered that they are usually negatives. Truly a 'shadow' artificially applied allows electrons to pass and so produces a dark exposure area on the photographic plate which is realistic, but it is the replica itself which is the negative. This rarely matters any more than does the reversal of direction of movement in a compound optical microscope to a practised dissector, but it must be noted.

Of image-forming errors the greatest are due to the same mechanism as that which makes the image possible, namely electron scattering. In electron microscopy 'refractive index' has no meaning nor in the simple sense has electron opacity. The electron image is made possible because electrons from the illuminating beam are scattered out of it and are lost to the lens system, therefore the part of the specimen which scatters the most electrons appears the most dense in the final image though it may not be so in the physical sense.

Electrons are lost to the final image also because although the scattering process may not be sufficient to cause loss of electrons it may slow them down in speed and so cause abnormal deflection within the lenses, which in turn causes them to focus above or more usually below the image plane and so be lost in that way. Without electron scattering no image would be possible, but with it artifacts occur, so it will be seen that the only way to understand the image is to make many preparations of objects of known structure and of varying thickness and examine them at different microscope voltages. Clearly the greatest errors due to excessive scattering will occur with the thickest specimens and lowest voltages.

The penetration of the specimen by electrons is proportional to the HT voltage of the microscope. Voltages lower than 30 Kv are of no practical[1] use because penetration of typical objects like ordinary biological cells is not possible, and even the supporting films absorb many electrons. Ordinary cell work demands between 100 and 200 Kv to get penetration while still allowing sufficient scatter to form an image. Very high voltage tubes are difficult to construct both because insulation becomes difficult and deflection magnet currents become very high if high magnification is required.

If crystaline material is in the specimen, diffraction-type of scattering takes place, and a considerable amount is of the inelastic kind where loss of electron velocity occurs and more marked errors in interpretation of density are likely to follow for the reasons outlined above. Such materials give different results for different orientations

---

[1] See 'Mirror Electron Microscopy'

and it is quite easy for a thin extended crystal to diffract out nearly the whole of the electrons passing through it. Diaphragms should be inserted in the microscope with care so that they do not clip the beam (it will be remembered that they do not limit 'numerical aperture', which expression has no meaning in electron optics).

As with optical microscopy, common sense and gradually accumulating experience with a variety of objects in a variety of microscopes are the only guides to correct interpretation. No opportunity should be lost of taking photographs of known structures, and of the same structure in different microscopes (usually only possible with 'spread' grids). It took a length of time approaching a hundred years to settle some controversial matters of structure visible in the optical microscope so workers must expect some errors to be shown up as development proceeds with the electron type.

Reflection Electron Microscopy

It is possible to mount a shadowed replica or other specimen in the microscope on edge so that the electron beam glances from its surface and goes on to form an image on the screen. Naturally great aberrations occur with the method, as detail must be blurred in an elongated manner, but good results can be obtained on such structions as fibres dried down on to glass where the shape of the object is known, and mainly its detailed contours are wanted. There is no comparable optical method as here the image obtained depends upon electron absorption by the surface irregularities and so is really a shadowgraph. (See also Addendum, page 550.)

Mirror Electron Microscopy

The experimental method of 'Mirror Electron Microscopy' in which electrical potential distributions on a surface may be made visible by their effect upon impinging electrons of 7 to 35 Kv energy has given encouraging results in studies of grain boundaries. The specimen is given a fraction of a volt electrical charge which has the effect of repelling electrons near the specimen surface back to a screen surrounding the electron gun. The resulting image on the screen is difficult to interpret but consists of an 'outline' caused by a combination of the charge distribution on the surface if this charge is not uniformly distributed, and the topography of the surface. The resolution of the method is $0 \cdot 3\mu$ and so compares with that of optical microscopy. Insufficient work has been done to determine the possibilities of the method but encouraging photographs have been published by L. Mayer in the *Journal of Applied Physics*, Volume 26, Number 10, October 1955. The ordinary electron microscope will

not work at such low voltages therefore much experimental construction is required in the instrument.

Historical Developments

Sufficient mention has already been made of the experiments with
cathode rays which lead up to the construction of a microscope. If
the fundamental matters concerning electrical discharge in high
vacuo are to be studied, reference must be made to books on physics
as the subject is very involved.

An important matter from the microscopist's point of view is
whether a microscope should have electrostatic or electromagnetic
focussing methods. The electrostatic focussing arrangement consists of
three disks mounted symmetrically upon an axis a definite distance
apart charged to different potentials. When correctly spaced the field
near the axis is of spherical curvature, Fig. 5.

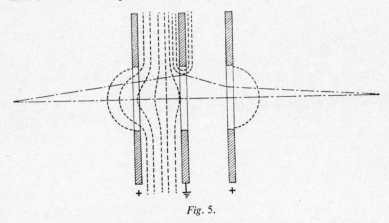

Fig. 5.

This design has the following advantages:

1. It is comparatively easy to construct and adjust.
2. The main high tension voltage fluctuation has much less effect
   upon the image than a similar fluctuation in a magnetic instrument because the electrostatic deflecting voltage is derived from
   the same source as the main HT, and when the main HT fluctuates
   downward, causing a reduction of electron speed, the power of the
   objective lens also drops, thus deflecting the slower moving electrons less, so keeping the final image in much the same place.
   (In the magnetic instrument there is of course no connection
   between the main HT of the microscope and the current in the
   coils.)

3. When the focussing voltage is changed the image expands and contracts radially about the optic axis.

There are, however, considerable disadvantages as follows:

4. The aberrations, particularly spherical, are much greater, and astigmatism cannot be corrected with apparatus like radial set-screws and quadrupoles.

5. The magnification is fixed.

6. The magnification is limited by the voltage it is possible to apply to plates so closely spaced (fractions of a millimetre) as demanded by the geometry of the objective.

7. Electrical breakdown discharges through dust and other particles cause image distortion and mechanical damage just as in electrostatic cathode ray tubes.

Although these disadvantages exist the Trüb-Tauber microscope was put on to the market with an electrostatic objective, the other lenses being electromagnetic, in order to obtain the advantage associated with a high degree of stability of high tension voltage (2 above). It appears unnecessary to go in for this complication now that the electronics of high voltage supply is better developed.

It was usual before so late a date as 1950 for laboratories to consider constructing their own electron microscopes, but this is now not seriously considered as much standardization of construction and production has taken place in all parts of the world. The components of the electron optical system are not yet quite standard in arrangement as they have been in optical microscopes for 150 years. Experiments are still made with fourth and fifth lenses in the train, but it is likely that the simple arrangement of condenser, objective and projector will become standard as methods of correction of the aberrations are developed.

Optical and Electron Methods Compared
It has become increasingly necessary in recent years to tie up the results of optical and electron microscopy in order to establish the truth of observations, particularly as elaborate methods often must be used to get the specimen into a state in which it can be introduced into an electron microscope. One typical method is to obtain a thin section of a specimen, say 1μ thick, place it upon an ordinary grid and examine and record the features in the electron microscope. The grid may then be dissolved away, then the embedding medium is dissolved out, say it is methacrylate in carbon tetrachloride, the specimen is then stained in ordinary stains and examined optically

on a slide using the grid impression marks as a guide to the field of view required.

It must be remembered that all sections tend to draw away from the grid wires, lose thickness due to all causes up to 60 per cent (20 to 50 per cent in Araldite) and receive a contamination deposit from the apparatus of about $0 \cdot 03$ to $0 \cdot 3A$ per second of exposure in a current typically $0 \cdot 01$ to $0 \cdot 2$ amperes per square cm.

Standard Methods of Preparing Objects for Electron Microscopy: The Replica Technique

In an earlier section, replica techniques for optical microscopy were described, therefore the basic details concerning the method are not repeated here.

Electron replica work is characterized by greater delicacy. Optical replica films may in general be thick enough to be stripped off a surface which is as rough as an ordinary machined part, but the electron film cannot be made substantial enough to be removed directly from such a surface and still be transparent to electrons. Likewise the ordinary optical etch for metallurgy specimens is too deep, giving too great adherence to the specimen for a sufficiently thin electron replica to be stripped off. When such very thin replicas are used there is considerable danger of them distorting when under electron bombardment in the microscope, and while they are being stripped. It is the usual practice to reinforce the films with a deposit which is transparent to electrons, and to choose a replica medium which is on the stiff brittle side rather than the tough but plastic one.

Electron replicas must not be such that their molecular structure is comparable in size with the detail one wishes to record and this requirement rules out some good media for some critical jobs. All replicas are shadowed both to obtain contrast and to strengthen the film. Depending upon the nature of the surface from which a replica is to be taken, the replica will be either a plain one simply following the contours of the surface, or it may incorporate particles which were loosely connected with the surface. These particles can be deliberately placed upon the surface to be taken up by the replica film, or they may be contaminants. Such a replica is usually known as a 'complex' one or in America a 'pseudo' replica. When a surface is too rough to allow a replica of sufficient thinness for electron microscopy to be stripped from it, a thicker, more optical-like replica is taken and a positive taken off that, of sufficient thinness. It sometimes is impossible to strip off a replica and still retain in a subsequent positive, materials adhering to its under surface. In this case it is usual to dissolve the subject from under the replica. When this is done a thin film can be used and there is a greatly reduced risk of mechanical distortion.

A standard 'background' must often be employed in order to produce flat smooth films on which to support specimens or upon which to scatter particles to be picked up by a replica film. Fortunately, polished plate glass is excellent as a background as it gives a replica structure which cannot be resolved in any microscope. The glass plate chosen must not be cleaned in harsh chemicals, or etching results, and common domestic detergents are the best cleaners.

A useful needle for handling electron microscope replicas and other parts is made after the style of a diatomist's needle but employing instead of the bristle a polar bear's whisker.

Replica Materials

For ordinary work, not of the highest resolution, collodion is the best material at present. It shows molecular structure at the highest resolutions, but it is soluble in several convenient chemicals such as ether alcohol and amyl acetate. Its film is reasonably tough and not very liable to stretch and deform, while it may be started and pulled off by employing relatively rough methods. The commercial material Formvar is much used as the basis of strippable lacquers, and it makes excellent replicas, tough, though liable to stretch, which stand up well to the electron beam. It is best dissolved in ethylene dichloride at a dilution of about one part Formvar to six of solvent. Formvar films polymerize in the beam, are not so easy to strip as collodion ones so should be used on smooth surfaces.

A long list of replica materials and methods is not included here because they are legion, and every worker arranges matters to suit the job. Much use is being made at the present time of the carbon replica which is deposited on the specimen, probably itself supported on a substance, from an open carbon arc in air a few centimetres distant. When the specimen is dissolved from beneath the carbon, excellent rendering of surface detail is obtained. The carbon films are fairly tough in the microscopical sense and may be shadowed in the ordinary way for direct viewing with platinum.

The materials mentioned above, plus perhaps any soluble plastics, will, when handled as the job demands, perform any replica task. This applies similarly to the methods of handling replicas, described below.

Techniques for Obtaining Replicas

There is little difference between optical and electron methods. A suitable dilution of the material is poured on to the surface under study, allowed to stand for a few seconds, then poured off leaving an even liquid film behind which dries. Replicas may only be taken in warm dry rooms or else artifacts due to moisture condensation at the moment of contact between the volatile solvent and the job will be produced, just as when lacquering bright metal parts in a cold workshop. No replica film should be artificially dried, room air drying is adequate. The trickiest job is separating the replica from the job without breakage and distortion, and after much experience it has been found that some kind of floatation method is best. The film may be started from the job at a point remote from the area of interest by means of a scalpel, needle, or piece of sticky tape, but the upper surface of the prepared replica must not be handled. In favourable cases the replica may be pulled straight off without more ado, but often it is necessary to allow a hydraulic wedge to aid the separation. To obtain this assistance the unwanted edge of the job with the film freed from the surface should be immersed in distilled water in such a manner that the detaching replica floats on the surface skin of the water while it is being pulled off. With care the upper surface of the replica can be kept dry and the whole film left floating on the water. With a collodion replica, detachment frequently occurs without more mechanical effort than that needed to free the waste edge. When a material has to be dissolved away from beneath a replica the method is similar in that the replica is left floating on the solvent.

The method of lifting replicas follows that of optical section lifting generally and any smooth flat spatula or glass slide passed under the floating film will serve to move it from bath to bath for washing. A replica must not be taken from the bath and placed upon the supporting grid or distortion will follow. The correct method is to drop the mounted cleaned grid upon the floating replica then remove both by means of a strip typing paper, when the paper and the replica film trap the grid between them. The whole can then be lifted from the water (remembering that the wanted side of the replica is downwards), the paper cut away and the grid left with the replica adhering.

Large areas of replica should always be cut up into sizes about 15 mm by 4 mm by means of a sharp knife before separation from the job. However, there is always a tendency for these to wrinkle upon water, particularly when collodion is used. Less tendency to wrinkling is the main advantage of plastic materials.

Satisfactory replicas may with experimental care be taken from some unlikely surfaces, e.g. a growth of bacteria on a smooth gelatine or agar surface, the surface residents from the skin of fruits, and fungoid hyphae extending over glass may all be taken as complex replicas.

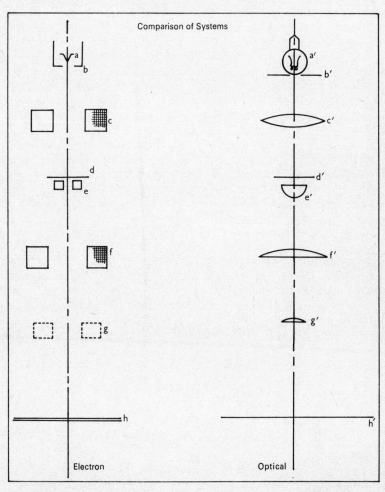

Comparison of Systems

Electron

Optical

a. Electron source 'V' filament.
b. Wehnelt cylinder.
c. Condenser.
d. Object carrier.
e. Objective assembly.
f. Projector lens.
g. 2nd projector for large fields.
h. Fluorescent screen.

a′. Light source.
b′. Diaphragm.
c′. Condenser.
d′. Object stage.
e′. Objective.
f′. Field lens.
g′. Eyelens.
h′. Projected image.

The Scanning Electron Microscope

In this apparatus the incident beam of electrons or 'probe' is caused to scan the specimen after the fashion of a raster in a television tube. A collector electrode is played near the surface being scanned and is connected through an amplifier to the modulator electrode of an ordinary cathode ray tube in a television system. All is synchronized.

The probe must be very small in diameter actually about the same as the penetration of the specimen by the electron beam, typically a few hundred A°, driven by a voltage of a few Kv. All must be adjusted on test. Other kinds of radiation may be collected, i.e. fluorescence, slow secondary electrons and fast primary electrons depending upon the nature of the specimen and the probe voltage, some may be used usefully but mostly they represent noise. The advantage of this system is not particularly high magnification but a great improvement in resolution over the optical instrument. Specimens may be prepared simply as dry mounts and can be easily shadowed with aluminium, gold or the other standard materials. The electronic amplification and incident beam variability are excellent tools for sorting out structure. The specimens may also be viewed in a light microscope without change of preparation.

CHAPTER 31

The Proton Microscope

I N THE previous chapter dealing with electron microscopy the main properties of the electron were described sufficiently for the working of the microscope to be understood. It was there mentioned that the electron is a particle which revolves in an orbit around a positive nucleus in an atom. The proton is the important positively charged particle, one or several of which are in the nucleus round which the electrons rotate. In the atomic nucleus there are other particles beside protons but it is the charged proton which can be beamed and deflected in a similar way to the electron to make a source of radiation for a microscope. The simplest way of obtaining protons for microscopical purposes is to employ the hydrogen gas atom, strip its one electron off it and then draw away the remaining proton by means of a high negative voltage into the gun of the microscope.

The work on the proton microscope is experimental construction of the apparatus, and is mostly French. The advantage of protons over electrons as the image-forming radiation is that the proton being nearly two thousand times as heavy as the electron and travelling at a greatly reduced speed, experiences less aberration as it passes through the focussing apparatus. The proton has a shorter wavelength than the electron and as with the electron the full resolution possible cannot be attained, but as the electron instrument works at the equivalent in a light system of a mere pin-hole aperture, loss of resolution due to frequency of radiation is far from arising as ultimate resolution cannot be reached in either case. Theory indicates that the aberrations resulting from the use of proton radiation will be so much less that the increase of overall resolution due to higher lens aperture and reduced aberrations of other kinds will be about ten times that of the electron system.

The same difficulties with regard to the object being in a vacuum, the dissipation of heat within the object and damage to the object by the impact of the corpuscles of radiation, exist as with the electron system, and some microscopists are of the opinion that owing to the fact that protons are heavy and create about a thousand times more ionization than electrons in materials they try to penetrate, a practicable microscope, i.e. one with adequate penetration would be too destructive of the specimens. However, the French workers have progressed sufficiently to construct a large rack mounted instrument to commercial specification. No results of actual studies of specimens have yet been published.

The Ion Source

The generation of protons with suitable characteristics for use in a microscope is not so easy as the generation of electrons. Much research is still going on upon the subject and fortunately some aspects of the problem are also encountered in the ion source design of high-energy accelerators for physics purposes. It is necessary that the protons be mono-energetic, that is, great differences in velocity cannot be allowed or the equivalent of chromatic aberration would occur, also molecular ions are not wanted.

Several kinds of source have been used:

1. Ionization of hydrogen by means of an oscillating column of electrons developed in a gas tube between two hot electrodes about 2 cms apart. After some hundreds of oscillations the electrons are absorbed unless they have previously caused an ionization event.

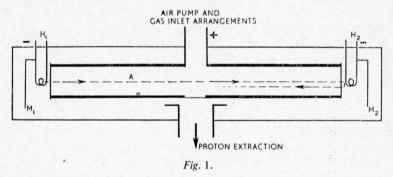

Fig. 1.

The apparatus is shown in Fig. 1. A tube A contains the supply of gas to be ionized, at a pressure of about 0·001 mm of mercury, two heaters H_1 and H_2 have a potential difference of 300 V to the gas tube, mirrors M_1 and M_2 act as electron reflectors. Electrons starting from either heater are attracted into the tube A, pass through it until they are repelled by heater H_2 then return to near H_1 and so continue oscillating until absorbed. When an ionization event takes place the proton energy is always 0·03 electron volts, therefore a good mono-energetic source is provided. It does, however, contain about 30 per cent of molecular ions. An extractor electrode is placed beside tube

A, charged to such a voltage that the oscillating column is not affected by its field. Electrons are prevented from striking parts of the apparatus by surrounding the tube with magnets and grids.

2. Magnetic lenses have been used in association with 1. above, arranged to separate out particles of unwanted mass by imaging the wanted ones.

3. The gas in a tube is ionized by high frequency current of the order of 200 Mc/s. In this apparatus no heaters are required, the gas pressure is about 8μ mercury, 1 ma of proton beam is obtainable and there are practically no molecular ions produced. The wanted protons are extracted through a very small aperture of the order of fractions of a millimeter.

All these forms of proton source have been used on microscopes

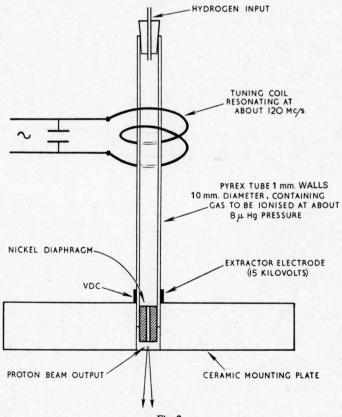

HYDROGEN INPUT

TUNING COIL RESONATING AT ABOUT 120 Mc/s

PYREX TUBE 1 mm. WALLS 10 mm. DIAMETER, CONTAINING GAS TO BE IONISED AT ABOUT 8μ Hg PRESSURE

NICKEL DIAPHRAGM

EXTRACTOR ELECTRODE (15 KILOVOLTS)

VDC

PROTON BEAM OUTPUT

CERAMIC MOUNTING PLATE

Fig. 2.

and accelerators but type 3 is the one likely to be developed though it requires more apparatus.

Proton Lenses

The mathematics associated with the flight of charged particles in electrostatic and electromagnetic fields is beyond the scope of a book dealing mainly with optical matters, but it is found that an electrostatic lens designed for electrons will behave in a similar manner with respect to aberrations, upon any other charged particles (see page 536), but the diameter of the central electrode diaphragm has major effect upon the focal length of the lens, and its accuracy of shape seriously affects certain aberrations; different focal lengths occur with different particles. The independence of the electrostatic system of charge and mass of the image-forming particles makes it superior to a magnetic system. The lenses are usually similar in structure to that shown on page 536 but the outer electrodes are dished with convex sides toward the centre one, insulated to withstand 100,000 volts, of focal length about 4 mm and with a very accurately circular hole in the central diaphragm about $\frac{1}{2}$ mm diameter. A potted construction is employed to ensure stability and all that has been said regarding the surgical cleanliness and accuracy of mechanical location of electron lenses applies with equal force to proton ones.

The most far-reaching effect of designing electrostatic lenses for microscopes is that any charged particle may be handled by the system. It has been calculated that the theoretical resolving power of the proton microscope is $1 \cdot 5 \times 10^{-8}$ cms, but theory indicates that if the mass of the illuminating particles can be increased by employing say argon or zenon resolution may also be increased by a factor of 12 or so over that possible with protons, such resolution brings the structure of the atom within range of the instrument. Thermal agitation of the objects and recoils of particles struck by illuminating particles would almost certainly spoil the results but some increase in resolution over that possible with protons may well bring internal structure of large molecules within range.

Aberrations in the Proton Microscope

Spherical aberration is the same in principle as in the electron microscope and for the same reason cannot be corrected, because there is no means of balancing it out by material with certain opposite characteristics as in a light system. It is of course possible to choose a system which has the minimum spherical aberration but one has to use very small apertures of the order of $0 \cdot 002$ in order to reduce axial spherical aberration to a few millimicrons for unless the system can be made as good as this the benefits of the proton system are not realized. Chromatic Aberration is due to protons entering the microscope with different initial speeds thus experiencing different degrees of refraction in the lens. Protons are also altered in speed by passage through the specimen but that is necessary in order to detect the presence of the specimen. With present-day ion sources it is possible to obtain protons sufficiently monochromatic to bring chromatic errors to within the limits of spherical errors, typically 1 m. micron of chromatic spread can be expected.

Elipticity errors have no optical counterpart except perhaps in general bad workmanship, but in the electrostatic lens the circularity of the hole in the central electrode diaphragm is vital and should be no greater in error than 1 micron for instruments constructed with all other aberrations minimized. One micron of elipticity is likely to produce aberration of the order of 1 m. micron.

Relativity aberration applies particularly to high speed electrons in an electron microscope where the electrons have a speed which is not negligible compared with that of light. In the proton instrument the partical speeds are very much less even though the voltage may be some hundreds of KeV, so relativity considerations do not apply.

Practical Operation of the Proton Microscope

Objects for study by electron microscopy can be mounted upon a membrane, and if a suitable material is chosen (see 'Electron Microscopy') the effects of electron absorption in the membrane can be separated from that in the object, but in proton microscopy the specimen must be mounted in space as no suitable membrane seems likely to be discovered. Specimens may be spread upon a very fine grid, and the natural grid formed by the secondary structure of a diatom has been suggested. In electron microscopy a membrane thickness of 10 m. microns is workable, not causing too much chromatic aberration and absorption; the loss of energy in terms of electron volts can be measured and calculated and is found to be about $6 \cdot 3$ eV with a 10 m. micron membrane in a beam of 100,000 eV electrons. Protons behave similarly but the results are of a different order of magnitude; for example a 100,000 eV proton is travelling with the speed of a 50 eV electron and penetration is very small. Experiments conducted with higher energy protons in accelerators indicate energy losses of the order of 900 eV per 10 m. microns for 300,000 eV beams. Bearing in mind that the main advantage of the proton microscope lies in its better corrections it is clear that any membrane (or specimen) must be about $0 \cdot 01$ m. micron thick in

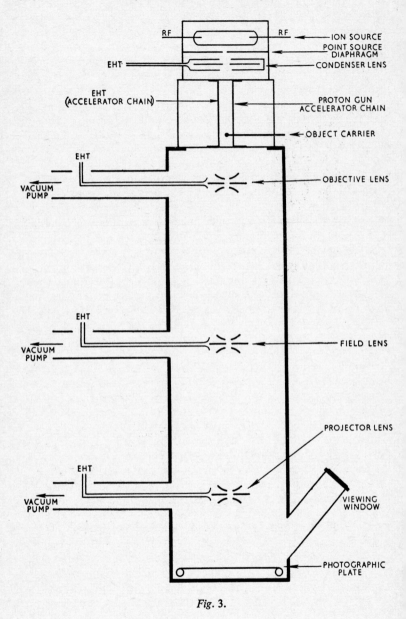

Fig. 3.

Diagram of the Proton Microscope

order that acceptable chromatic aberration figures be achieved. With charged particle illuminated microscopes it is easiest to calculate chromatic and spherical aberration in terms of lost speed and consequent greater deviation in the lens, i.e. in electron volts of lost energy in the beam, this quantity can usually be measured by the normal methods of physics.

In outward appearance the proton microscope is similar to the electron type except for the more complicated ion source. They are usually inverted, massive, continuously pumped devices with several pumping points placed near the lenses. The high electrostatic voltages are fed within the vacuum enclosure for convenience and safety. The pumping arrangements and mechanical assembly are complex but follow the lines of those described under 'Electron Microscopy'.

Addendum

The Reflection Electron Microscope
The principle of this method of study of flat surfaces is that electrons scattered by the surface from an electron beam which passes at a grazing incidence say about 3° to the surface under study, are collected and brought to a focus by the microscope in the usual way. Great foreshortening in the image is unavoidable but can be reduced by using a less oblique angle of incidence at the cost of resolution. The method is limited by chromatic aberration and velocity dispersion of the incident beam, and in the best conditions the image is difficult to interpret. The incident beam (or probe) becomes broad and scattered due to space charge if the voltage is too low.

Special apparatus is necessary for this work and all is experimental, though interesting maps of surfaces mainly of metals and crystals have been published.

Much work is continually being done on the subject of electron probes and reference must be made to the recent research papers. Artifacts are very common and convincing when seen in a contrasty photograph particularly of a regular structure.

Appendix

Microscope construction has changed greatly during the last forty years and this century has seen the passing of the large brass English microscope with which all the foundation stones of modern biology and natural history were laid. It is of little practical use to attempt to employ the productions of Powell and Lealand, and Ross and others, for modern laboratory work, because the methods of microscopy and the design of associated apparatus have changed so much. The old ten-inch tube English microscope was made that way because the distance of distinct vision in the normal human being is 10 inches (250 mm.) so it was thought reasonable to standardize on this. It has the advantages that the stage is in focus for the unused eye, less eyepiece power is required for a given amount of magnification allowing more light and a longer eye-point, and flatter objective fields were possible. Later, the invention of the Wenham and Powell prisms established the long tube for many years because only the 250 mm. breeches tubes have sufficient length to allow the eyes to converge comfortably on the objective.

The first of these brass instruments which microscopists consider to be proper working examples appeared about 1850 and had focussing adjustment by means of a stem driven up and down within a heavy sheath by means of a rack and pinion. The sheath had the stage and the inclination axles bolted to it while the stem carried at its top a horizontal bar which held the body about 100 mm. forward of the sheath, over the stage. All this structure was of massive brass and gun-metal, all possible parts were cast to save expensive machining, the surface was finished with a draw file and rotten-stone and water, giving a fine velvet-bright finish obtainable in no other way.

All microscopes were built by hand at the bench and drawings were not used, it was the practice to give the workmen the castings and leave them to clean them up and drill holes (with spade drills) where they estimated they should be. The parts of different examples, of say the Ross No. 1 microscope, cannot be interchanged for this reason. Some specialization was permitted, for example one man made all Powell and Lealand stages and another made all Ross fine adjustments. Brass was commonly used for bearing surfaces and because brass on brass will not stick or jump when used as a bearing, a peculiar smooth greasy movement was obtained on these old instruments.

The days of the great engineer scientists were also days of carefully constructed apparatus and beautiful finish. A worker would not possess an ugly microscope if he could possibly afford an elegant English one— and all English work in this trade was then elegant and carefully finished as if each craftsman were making the instrument for his own sideboard. All-bright metal work was the rule and the lacquer put on one hundred years ago is often found quite sound and the brasswork still beautiful today. Andrew Ross made his lacquer by extracting from shellac with acetone (known then as pyro-acetic ether), the hard resinous components, leaving behind the waxy, soft components which, if included, would soften the lacquer. The soaking out was done cold and took about seven days to complete. The acetone was then allowed to evaporate until the solution was about as thick as present-day thin varnish. It was then diluted as required with methylated spirit and applied to the job with a brush. This lacquering business was, and is, tricky. The brass must be raised to a little above room temperature so that when a spirit mixture is applied to it, moisture does not condense on the surface. The lacquer is then applied quickly with a brush or the job is dipped. The art of putting a second coat of lacquer over the first is dependent upon the complete baking dry of the first coat at about 60° C., but even then it must be applied sparingly and quickly after the job has cooled to room temperature. In this way—with smoothing with rotten-stone, polishing with whitening, and careful lacquering—were the beautiful results achieved and, except where the art is practised by a few amateurs, it will never be seen again. Side by side were the ugliest objects and the finest made by the Victorians, but in the sphere of optics the advances were great and the ugliness non-existent.

During the period 1860 onwards until the invention of the petrol engine, small and large microscopical societies flourished in all our main towns and it was never below the famous men of the time to belong to these. Great advances in technique were made simply by people playing with things and telling their friends about the results. In the era post-petrol-engine, the societies could not withstand the attack on people's interests and later the wireless set nearly finished off all individual effort in the field of natural history. It is true that wireless sets and petrol engines may replace natural history with natural curiosity, but it is doubtful if either has advanced the happiness of people as a whole so much as the discoveries of the early microscopists, most of whom were amateurs.

Today, there are professional societies of all kinds set up to 'advance' science and the bank balances of the far-seeing few who started them with a few friends and turned them into water-tight professional institutions after they had obtained sufficient members. But almost none of them pays any attention to practical laboratory arts and it is to the great credit of the Quekett Club and Royal Microscopical Society in London, and the American Microscopical Society in the USA, that they still hold together as vigorous groups and teach the practical arts of microscopy to all comers without demanding qualifications before they start.

Glossary

Abbe: Name of a famous German optician of the late nineteenth century. The name used to describe an effective non-corrected substage condenser due to Abbe.

Aberration: Uncorrected errors in a lens system due mainly to the use of lenses made of one kind of glass with spherical curvature; not used to describe poor workmanship. (See Chromatism, Spherical aberration, Coma, Astigmatism.)

Accommodation: The ability of the eye to adjust itself to different conditions of focus to form images of objects at varying distances and subtended angles. (See Depth of focus, Depth of field.)

Achromatic: Free from colour. Used to describe an optical system which has corrections applied such that two spectrum colours originally due to lens dispersion come to a focus in the same image plane giving an image nearly free from colour fringes.

Acuity: Ability of the eye to resolve detail with or without the aid of spectacles. A property of the retina. A measure of the quality of the eye in general.

Akehurst changer: A sliding condenser changer named after the inventor.

Analyser: Optically, a birefringent substance used in such a position in an optical system that it passes light polarized in the opposite direction to that admitted by the polarizer. If the specimen has optical activity, light passes through the analyser. When no optical activity is present the field is dark.

Angström Unit: A unit used in measuring the wavelength of light; equal to 10^{-10} metre.

Annulus: A ring-shaped space made up of diaphragms to pass a ring-shaped cross section beam of light. Used in phase contrast and differential illumination systems.

Apertometer: A device for measuring the optical aperture of an objective.

Aperture: The area of a lens through which light passes. A hole or diaphragm in an optical system.

Aplanatic: An optical system in which all zones of the lens bring light to one focal plane. Such a system may not be achromatic.

Apochromatic: Used to describe an optical system in which the corrections are such that three colours of the spectrum are brought to the same focus in the image plane. The sizes of the images in the various colours may not all be the same, therefore it is usual in a microscope to employ a specially designed eyepiece known as 'compensating' to equalize the image sizes. The resulting image is for practical purposes free from colour fringes.

Arm: The bar of metal extending from the racking stem of a Ross type microscope to hold the body.

Artifact: An appearance in an image or photograph due to causes within the optical system, not a true representation of the features in the object. Stray light, improper adjustment, too small apertures, too high magnification, all cause artifacts.

Airy Disk: Similar to Confusion Circle. The disk due to object glass diffraction, presented as the image of a point of light in an optical instrument. (After Sir George Airy.)

Aspherical: A surface not forming part of a sphere but being part of a solid of revolution. Usually applied in optics to a paraboloidal section lens or mirror. Irregular shapes are not referred to as aspherical.

Astigmatism: A distortion in an image where elements in one plane are in different focus from those in another.

Back lens: The combination of elements forming the last lens in an objective assembly. The component nearest the eyepiece in a microscope objective when it is fitted to the microscope.

Balsam: In optics always Canada balsam. A clear gum resin which hardens after heating; much used to cement lenses together and for mounting objects. Refracitve index nearly the same as crown glass, $1 \cdot 5$. (See tables.)

Barrel: Any tube housing optical components.

Barrel distortion: In photography, distortion of field giving expanded centre regions. (Not applicable to microscopy.)

Batten holder: The type of electric lamp bulb holder which screws to a wall or 'batten' of wood.

Bezel: A thin upstanding rim of metal left after turning a lens mount, later to be rolled over on to the lens to secure it in place by its edge. Any piece of metal so used.

Birefringent: The property of a substance which has two refractive indices. Same as 'double refracting' and 'polarizing'.

Blind: Descriptive of a lens assembly which has become unusable owing to deterioration of its components. A lens which fails to produce an image.

Bright field: (See Light-field.)

Bullseye: A large uncorrected deeply curved lens used to enlarge the image of a lamp flame. It is of no use as a lamp condenser proper.

Camera lucida: A fitting attached to an eyepiece to deflect an image of the object on to a sheet of paper for drawing purposes.

Cavity slide: Same as 'Well slide.'

Cell: (i) The socket or recess into which a lens fits. (ii) A lens mount which has turned upon it screw threads for fitting to an instrument. (iii) A cavity in a slide, or a built-up box to contain a specimen.

Cement: Any material, opaque or clear, used to unite lenses or objects.

Cement cell: A circular box built up on a slide to hold a specimen or liquid.

Centring nosepiece: A device of varying design placed between an objective and the microscope body by means of which the optical axis of the objective can be adjusted with respect to the mechanical axis of the body.

Chromatic: An optical system uncorrected for colour aberration. (See 'Achromatic'.)

Clear: To soak a specimen intended for mounting in a gum resin or in an oil, in another liquid, typically cedar wood oil, to ensure complete penetration and consequent transparency of the specimen.

Coddington lens: A solid, nearly spherical lens with a deep groove milled around its periphery and filled with black wax to form a built-in diaphragm. (Old fashioned.)

Coherent: Descriptive of light which originates from one source. Several beams may be produced from one elevation of a source by means of beam division by refraction and reflection: they are then coherent.

Coma: An aberration in an optical system in off-axis zones due to spherical aberration. From 'comet', i.e. the appearance of the image of a point instead of being round has a tail of light drawn from it radially with respect to the optical axis of the system.

Compensating eyepiece: An eyepiece for use mainly with objectives of the apochromatic correction, which magnifies more in the red than in the blue light of the spectrum.

Compensator: A quartz component used in a petrological microscope for phase-changing purposes.

Complementary: Applied to colours. A colour which when added to another produces white is its complementary colour. Complementary colours are usually mixed but in some cases may be monochromatic. A term also employed when illuminating a stained specimen with coloured light to produce maximum contrast with the background.

Compound: Applied to a microscope; meaning an instrument consisting of an objective and an eyepiece combination separated by a distance greater than the focal length of either (typically six inches). A normal microscope as opposed to a magnifying glass.

Compressorium: A mechanical device for adjusting the distance apart of two glass plates for the purpose of securing a specimen, usually a living creature.

Condenser: A lens or assembly of lenses arranged to parallelize or to focus a beam of light for illumination purposes. Condensers are known by their functions, i.e. substage cond., lamp cond., etc.

Condenser changers: Metal slides of standard size upon which are mounted microscope substage condensers for easy interchange on the microscope.

Cone: The geometric shape of a bundle of rays leaving a lens and proceeding to a focus. The 'width' of a cone is the diameter of its base compared with its height, a wide cone having a broad base. (A loose term.)

Confusion circle: The area occupied by the image of a point source of light, rendered in the form of a disk instead of a point, due to diffraction and aberration in the optical apparatus.

Conjugate: In microscopy, 'conjugate foci'. A lens by its nature bends light by nearly a fixed amount, therefore light which is divergent when it enters the lens arrives at a focus farther from the lens than it would have had it been parallel on entry. Vice versa for convergent light. The two focal points for a beam of light not parallel on either side of the lens are 'conjugate foci' and are infinite in number. (A loose term not used in technical optics.)

Cornea: The transparent soft horny membrane covering the front of the eye.

Correction Collar: A ferule on an objective by turning which the distance between the front lens and the rest of the combination can be varied.

Coverglass: A thin piece of high quality clear glass used to cover a specimen on a slide, mainly for the purpose of converting the surface of the mount into an optically smooth surface through which the objective can work.

Crazed: Decomposition of a lens surface or of a mount in which the material breaks up into irregular small areas.

Crimping: A method of securing two or more rods together by slipping a sleeve over them and then squeezing corrugations into the whole joint by means of pliers with corrugated jaws.

Critical: Used to describe a carefully set up optical condition in which all adjustments are as perfect as possible. (An expression due to E. M. Nelson.)

Dark-field: Same as Dark-ground.

Dark-ground: In a microscope; a method of rendering an object self-luminous from below by means of oblique light which does not directly enter the objective.

Dark-well: A concave, nearly hemispherical cup of blackened metal designed to be brought up under a specimen by means of a substage attachment, for the purpose of providing a black background for a specimen being examined under top light or vertical illumination.

Davis diaphragm: An adjustable iris diaphragm placed immediately above an objective for the purpose of controlling its aperture.

Deep: A high-magnification eyepiece. A lens in which the radius of curvature is comparable with the diameter. A curve of small radius.

Depth of field: A term with the same meaning as 'Depth of focus', but applied to the photographic image when a microscope is not present in the system.

Depth of focus: The distance between upper and lower points in an object as seen in its image, through which a clear image can be seen at one focal adjustment. Usually used qualitatively.

Diaphragm: Any aperture or stop controlling the shape of a light beam.

Dipping tube: A finger-operated pipette used for picking up objects in liquid.

Direct light: The illumination of a transparent specimen by causing the light to pass on one optical axis from the lamp through substage condenser, specimen and microscope to the eye or photographic palte.

Direct vision: A type of spectroscope in which the operator looks through the instrument at the source of light to be analysed.

Disk: As in common usage but also applied to the focal plane of an aerial image, e.g. Ramsden 'disk'.

Dispersion: The division of white light into its component colours by the action of a lens or prism. A quality of the glass or other material used in optical components. The amount of spectrum-spread in a spectroscope.

Displacement: The lateral shift of an image off the axis of an optical instrument due to the presence of an apparatus. Not an aberration.

Distortion: An effect due to an optical instrument, where an image produced is not a faithful representation of the object. Different kinds of distortion are known by descriptive names such as 'barrel' and 'pincushion' distortion as rough descriptions of the presented shape of rectangular objects.

Double image: The twin images produced by a material having the property of birefringence where one image is displaced laterally.

Doublet: A type of short focus hand magnifier consisting of two deeply curved plano-convex lenses, one considerably larger than the other, placed one above the other with a diaphragm between, set for minimum spherical aberration. (Old fashioned.) A cemented acromatic pair.

Draw filing: A method of finishing metal work where a fine file is drawn along the job sideways, the hands holding the tip and tang of the file.

Draw-tube: A tube carrying the eyepiece, fitted within the main body of a microscope so that the distance eyepiece-to-objective (i.e. the 'tube length') can be varied.

Drawing eyepiece: Same as Camera lucida.

Drum: The cylindrical part of a screw type measuring instrument upon which the calibration is engraved.

Dry bearing: A bearing which requires no liquid or other lubricant.

Dry lens: Usually applied to a microscope objective or condenser. An objective which employs air as the medium between its front lens and the coverglass of the object. Similarly for a condenser.

Dry mount: An object mounted on a slide without having been impregnated with a mountant. An object mounted in air.

Electronic: Any apparatus which employs transistors, tubes or valves in which the properties of the flight of electrons in space are utilized.

Embed: To fix solidly an object, usually in wax, for the purpose of supporting it to resist the force of the knife when cutting sections.

Empty magnification: A condition where the magnification applied by an optical instrument is greater than that required to reveal all the detail to a particular eye which the instrument's aperture can resolve.

English microscope: A microscope constructed with a 10 inch long tube. The expression is now out of date. (10 inches = 250 mm.)

Equivalent focus: The focal length of a simple thin lens which is equal in magnitude to that of a combination of lenses. The term only has meaning when used for computation purposes.

Erector: A device consisting of prisms or lenses placed between the objective and eyepiece of a microscope or telescope for the purpose of orienting the image so that it appears in its natural position both with regard to its situation and direction of movement, with respect to the object.

Exit pencil: The diameter of the beam of light leaving an optical instrument measured by looking at the eyepiece from a distance and observing the diameter of the disk of light seen. (See also 'Exit pupil'.)

Exit pupil: The diameter of the emerging pencil of light from an optical instrument as determined by looking at the illuminated area of last lens of the system from a distance of about two feet.

Extinction: The condition of darkness caused by analyser and polarizer being in opposite planes, i.e. 'crossed', or adjusted similarly to accommodate rotation of the plane of polarization caused by a specimen.

Eye glass: A simple magnifying lens intended to be supported in front of the eye by a framework secured to the wearer's head.

Eye lens: The component of an eyepiece combination nearest to the eye when in normal use.

Eye shade: An opaque screen extending from the microscope body across the path of vision of the unused eye for the purpose of helping the observer to keep the unused eye open.

Eyepiece: The final combination of lenses in a microscope or telescope which presents the image to the eye or camera.

Eyepiece camera: A small camera which is supported mechanically by the eyepiece fittings of an instrument.

Eyepiece caps: Shaped brass covers with a central perforation which fits outside the eyelens or eyepiece in order to keep the eye in the correct place with respect to the eyepiece axis and eyepoint. Not all eyepieces are so fitted; those which are, are called 'capped' eyepieces.

Eyepiece micrometer: A scale which fits into an eyepiece in the focus of the eye lens, which, when calibrated can be used to measure objects directly. Screw adjustable wires may be fitted to cross the scale as markers, or they may have themselves a screw micrometer calibration.

Eyepiece pointer: A needle mounted adjustably in an eyepiece for the purpose of indicating to an observer a point of interest in the image.

Eyepoint: Used to describe the distance between an eyepiece eye lens and the plane of the image, i.e. an eyepiece has a long or short eyepoint.

False image: Any image rendered untrue by artifacts. Not usually applied to a distorted image, but to an image improperly formed or inaccurately focussed.

Figure: The shape of the surface curve of a lens or mirror. Usually used to describe a surface which is not spherical. 'Figuring' is the last polishing operation on a lens or mirror when its curves are made exact.

Finger: A needle or bristle adjustably mounted on an objective or micromanipulator for manipulating microscopical objects while in the field of view.

First Surface mirror: A reflector which has its exposed surface silvered, i.e. impinging light is reflected from the first surface of contact only thus avoiding multiple reflections from other surfaces.

Fixation: The killing of a structure to prevent any visible changes in it at the time of killing or later.

Focus: The plane in which rays of light meet to form an image. The action of adjusting an instrument to obtain this condition.

Foggy: Used to describe an unclear image due to scattered light within the optical system; also a photographic plate affected by extraneous light.

Foot: The part of a microscope which stands on the table, and upon which the inclinable part is mounted.

Fringe: The coloured surround to an image in white light due to uncorrected chromatic aberration. Diffraction lines surrounding an image or appearing in an interference apparatus.

Gang: To connect controls or adjustments together by any means so that one movement operates many parts, keeping them all in correct adjustment relative to each other.

Gap-stage: A microscope stage constructed with a cut-out extending from the centre perforation to the front edge so that the slide can be held readily in the fingers to feel the clearance between slide and objective. The gap also enables a condenser to be withdrawn on its changer slide without racking down. Known also as a 'horse-shoe' stage.

Gash: Any oddment of apparatus (slang), e.g. gash lens, gash slides, etc.

Gate: A slit as in a spectroscope or monochromator which can be adjusted to pass one colour band in a spectrum.

Getter: A material fused electrically in an evacuated space for the purpose of absorbing or precipitating the last traces of air.

Glass: Old fashioned reference to any lens, e.g. 'glasses' for spectacles, 'object glass' for objective. Properly, an artificial hard transparent substance made from silica, soda and various additives.

Goniometer: Any device which rotates, and measures the angle turned.

Grass: The grass-like appearance on a cathode ray tube trace due to thermionic noise in the amplifying circuits. It is equivalent to the hissing noise in a loudspeaker. (Slang.)

Grating: Rulings producing alternate transmission and absorption strips used in optical experiments.

Greenough binocular: A type of twin-lens binocular microscope named after the inventor.

Hardening: The process of making a specimen tough to resist distortion by the knife of a microtome.

High power: In microscopy, meaning any magnification employed on a microscope which demands the use of an objective of shorter focal length than 6 mm.

Histological microscope: A compound microscope of simple mechanical construction intended only for low power work such as recognizing tissue features in animal and plant specimens.

Horse-shoe stage: See 'Gap Stage.'

Homogenous: With reference to an optical system, one where lenses and other components are united optically by a liquid medium of substantially the same refractive index as the glass or other refractive material. In a glass system cedar oil makes a homogenous arrangement (RI $1 \cdot 5$). In a quartz system a mixture of oils must be used (RI $1 \cdot 56$). Water immersion systems (RI $1 \cdot 33$) are not homogenous with glass and quartz.

Hot: Expression used to indicate a high level of radioactive contamination. Any dangerous area. A specially adjusted device designed to give high performance.

Iceland spar: Perfect crystals of carbonate, or calcite, found in Iceland and used to make polarizers and analysers. Strongly birefringent.

Image: The optical appearance of an object produced by means of lenses or mirrors.

Immersion: Any optical system in which a part is connected optically to another by a drop or layer of liquid of definite refractive index.

Impregnation: The process of filling a material completely with another material.

Infra-red: Electromagnetic radiation of wavelengths 800 to 1,000 milli-microns, invisible; in the spectrum region between red visible light and heat radiation proper.

Initial magnification: The magnification provided by a microscope objective without the aid of any eyepiece.

Injection: The technique of making vessels visible in tissue by expanding them with a coloured liquid introduced under moderate pressure.

Interference: The effect produced when two or more trains of coherent light waves mix together causing cancellation and reinforcement at some phases and frequencies, hence light and dark and sometimes coloured fringes.

Invert: To reverse the relations of the parts of an object as seen in an image of it, i.e. it is reversed both up to down and left to right, not only in one plane.

Iris: A kind of circular aperture diaphragm adjustable after the manner of the eye, employed to control the diameter of a beam of light.

Jam: A sticky compound manufactured for the purpose of applying to machine belts to prevent them slipping. Sometimes used to smooth out jumpy adjustments on optical instruments. (Slang.)

Junk: Any laboratory odds and ends not waste. (Slang.)

Knurl: The roughening of an instrument control knob to give a grip for the fingers. The tool which applies the roughness.

Köhler illumination: A method of microscope illumination where the lamp condenser is so placed as to become the apparent souce of light, i.e. the lamp image is positioned upon the lower lens of the substage condenser.

Lacquer: In optics, the clear or pale gold transparent coating composed of shellac and gums which is applied to stationary metal surfaces to preserve their brightness.

Lag: To insulate against heat loss caused by radiation and convection.

Lamp: Any canister enclosing a source of light.

Lamp condenser: The parallelizing lens fitted to a lamp.

Lamp diaphragm: The aperture determining the diameter of the emerging beam of light from a lamp.

Lampblack: Finely divided carbon as found inside paraffin lamp chimneys.

Lap: To grind a hard substance on a plate charged with an abrasive. The grinding plate itself.

Lash-up: Any temporary erection of apparatus. (Slang.)

Lens: A piece of glass (or rarely another substance) shaped to produce an image, or to affect the corrections of an image.

Lens hood: A projection surrounding a lens for the purpose of keeping out extraneous light.

Lens paper: A specially made tissue paper for cleaning lenses.

Lever fine adjustment: A form of microscope fine adjustment in which the body tube or nosepiece only is raised against a spring by means of a micrometer-operated lever of the first order.

Lieberkühn: A metal reflector surrounding an objective for the purpose of reflecting up-coming light back on to the surface of a specimen.

Light: An electromagnetic radiation existing in the waveband 7700 Angström Units to 3,600 AU (400 to 800 millimicrons) capable of rendering objects visible to the human eye and capable of being bent or refracted by clear substances.

Light-field: The condition in a microscope where the illuminating light passes into the objective thus giving a bright ground in the instrument. (See Dark-ground.)

Limb: The part of a microscope upon which are mounted the course adjustment, substage fittings and the stage.

Line: An old measure of length, i.e. the width of a line drawn with a quill pen, equal to 1/12 inch (say 1·5 mm.).

Long focus: A low power objective or any combination having a long working distance.

Long tube: The name given to an old microscope with 10 in. (250 mm.) tubes. An objective corrected for use with such tube length.

Lycopodium powder: The even sized spores of the club moss *Lycopodium elevatum.*

Macerate: The process of soaking a specimen in water, etc., in order to bring about a change such as softening.

Mains: Used in the electrical sense meaning the public domestic distribution system.

Male: Used in the engineering sense meaning a part of a joint which enters another part (the female). Optically, the convex grinding tool of a pair.

Medium power: Applied usually to objectives meaning those of focal length in the range 12 mm. to 6 mm. (A loose term.)

Mesh: In engineering, the condition of gear wheels or racks and pinions when fitted together, i.e. they are accurately or not accurately 'in mesh'.

Melt: Optically, glass already made but re-melted for casting purposes.

Metal: Optically, the liquid glass as made in the earthenware furnace pot.

Micro Furnace: An apparatus which mounts on a microscope, in which metals, etc., are fused by electric current, within range of the objective (see text).

Micron: The normal microscopical unit of length equal to 1/1,000 mm.

Microphotography: The process of making a reduced size photograph of a large object.

Mil: One thousandth part of an inch (engineering term).

Mill: In instrument work, the grooving applied axially to a knob to give a grip for the fingers. Originally the straight grooves were formed by cutting with a milling cutter but they are now usually formed with a straight knurling tool.

Milli Curie: A thousandth part of a Curie. A unit of quantity of radio-activity. One Curie $3 \cdot 7x10^{10}$ disintegrations per second.

Moderator: A device for controlling the intensity of light without affecting its colour.

Moist stage: A device like a large compressorium in which a specimen can be studied while being kept damp. (See text 'Growing Stages'.)

Monochromatic: Light which consists of one frequency of vibration only. In practice, a narrow band of frequencies. A lens system constructed to work with a definite frequency of light.

Monochromatic: Light which consists of one frequency of vibration only. In practice, a narrow band of frequencies. A lens system constructed to work with a definite frequency of light.

Monochromator: An instrument for producing light of one frequency only. Usually a device which selects a required frequency from a continuous spectrum.

Monocular: An optical instrument intended for use with one eye.

Myopia: The condition of short-sightedness where the curvature of the eye lens is too great thus focussing the image in front of the retina at normal distances of vision. By holding an object very close to such an eye the rear conjugate focus can be lengthened thus allowing an image to be seen clearly.

Nelsonian microscopy: A classical method of setting up a microscope, due to E. M. Nelson.

Nicol prism: A composite prism of Iceland spar arranged to reflect out the extraordinary ray of light leaving the ordinary ray plane polarized. Used as a polarizer or as an analyser in polariscopic apparatus. (After the inventor.)

Nose-piece: The part of a microscope which carries the objective.

Object: The article being studied.

Object glass: Same as objective.

Objective: The lens of a microscope or telescope which is nearest the object when the instrument is set up.

Oblique light: In a microscope, light applied to an object from below obliquely, but entering the objective, thus producing light field conditions.

Ocular: Same as Eyepiece.

Oil: Optically, immersion oil; almost always cedar oil is meant, for connecting an objective to a slide (or synthetic products).

Opal: An electric light bulb with a frosted surface or thin translucent coating.

Optical activity: The property of a material to affect the plane of polarization of light.

Ordinary ray: The ray of light which passes through a birefringent substance with least deviation; usually the greater in intensity.

Orientation: The position of a crystal axis with respect to the optical axis of the instrument.

Pack: Used to describe built-in apparatus in the form of one unit of assembly, e.g. 'Ultrapack' describing a commercial darkground unit complete with lamp.

Pair: Wrongly used to describe the components of a simple achromatic combination.

Paraffin: In modern parlance always the mineral burning oil is meant. The wax is referred to as 'paraffin wax' and refined oil as 'medicinal paraffin'.

Parfocal: Description of objectives or eyepieces so set up on their respective mounts that they all focus in the same plane when interchanged.

Pencil: Optically, a beam of light of small cross sectional area.

Penumbra: A halo or surrounding light effect in an image, e.g. the ring of light (or darkness) surrounding the image in a phase contrast system.

Persistence: The effect which remains on the retina after the image has been removed.

Perspex: The trade name of a light, transparent plastic available in the form of sheet or rod.

Phase: In physics, the relationship of amplitude with time, usually restricted to matters which vary regularly and repetitively.

Photomicrography: The art of recording the magnified appearance of a microscopical object by means of photography.

Pillar: Name given to an old-fashioned microscope fine adjustment where the body was carried on guides moving on a solid pillar attached vertically to the rear of the stage. The device was spring-loaded against a micrometer control screw at the top of the pillar, the spring acting so as to oppose the microscope body weight.

Pinch: The name given to that part of an electric light bulb or radio valve where the wires supporting the electrodes pass out of the glass envelope.

Pincushion distortion: Distortion in an image where the centre regions of a rectangular object are constricted. See the opposite condition 'Barrel distortion'.

Pitch: Optician's pitch. Used to secure temporarily glass to supporting plates during grinding and polishing operations. (For composition, see text.)

Pith: Elder pith is used as an embedding medium for rough microtomy, i.e. pith embedding.

Play: Backlash in a mechanical system; poor mechanical fit, i.e. 'a lot of play' in (say) a bearing.

Pot colour: The colour of glass as it comes from the melting crucible.

Potted: An assembly of parts immersed in a hardened synthetic resin.

Post Office Register: An electrical relay fitted with a counting attachment which records each impulse up to a maximum speed of about five operations per second.

Power: Same as 'objective' (old fashioned term). The amount of linear magnification of a microscope or telescope.

Profile microscope: A projection instrument designed to produce a shadow section of an object usually a screw thread.

Radial microscope: An experimental instrument designed to rotate upon its base and to be inclinable in all azimuths. It was used in oblique light experiments. (Obsolete.)

Rads: A unit of atomic radiation dose, nearly equivalent in most circumstances to a Roentgen which is that amount of X or gamma radiation which produces 1 ESU of electricity in 1 gramme of air.

Raft: Used to describe the tray containing eyepieces and objectives in a microscope cabinet (slang).

Ramsden disk (or circle(: The aerial disk of light formed above the eyelens of a microscope or telescope, in reality an image of the objective back lens.

Regulation: The amount of voltage change with current load in an electrical apparatus, e.g. a transformer.

Residual colour: The colours seen in an image formed by an achromatic system due to outstanding uncorrected colours of light focussing in different planes. (See Secondary spectrum.)

Resolution: The ability of a microscope to separate details in the object studied. It is a function of aperture not magnification only, e.g. the resolving power of one objective may be higher than that of another giving the same magnification.

Revolving Nose-piece: An objective changer which carries several objectives which can be revolved into place on the microscope.

Revolving Stage: A stage which rotates in the optical axis of the microscope. It need not rotate the full 360 degrees to be so called.

Ribbon: A series of microtome sections adhering to each other by their edges in the order in which they were cut.

Ribbon filament lamp: A low voltage high current electric bulb in which the filament takes the form of a ribbon of tungsten about 8 mm long and 2 mm wide, thus providing a structureless source.

Ring: An old name for a microscope substage or the part taking its place beneath the stage proper.

Ring illuminator: A ring of flashlamp bulbs, 100 mm. in diameter used for low power top lighting.

Ross fine adjustment: An old form of microscope fine adjustment in which the nosepiece only of the microscope is moved axially by means of a second order lever operated by a micrometer screw, with the fulcrum in front of the body tube.

Ross microscope: A classic design of stand of about 1860 in which the body tube was held forward over the stage on an arm. Focussing was by racking the stem upon which the arm was mounted.

Rotten image: A microscope image destroyed in quality by employing more magnification than the aperture of the system permits. (The term is not employed to describe a bad image due to aberrations.)

Rotten-stone: A soft calcareous freestone found in Europe, used to polish brass plates. Employed much by clockmakers.

Safety stage: A stage fitted with a top plate which gives way downwards against a weak spring when the objective touches the specimen.

Secondary spectrum: The outstanding traces of colour fringing left in the image formed by an apochromatic system due to the fact that although three colours of light come to the same focal plane, slight errors still exist for the other spectrum colours. They may focus above or below the plane of the corrected ones depending upon lens design.

Shadbolt's turntable: After Shadbolt. (See 'Turntable.)'

Shallow: A low power eyepiece.

Shear: A force tending to divide an object by transverse application against one part, the other part being held. A member sustaining such a force is said to be 'in shear'. A type of reference image displacement in an interference microscope system (see text).

Shim: A thin piece of material used to raise a part to a given height; or to space components permanently during assembly.

Shutter: Improperly used to describe a diaphragm behind a microscope objective. (See 'Davis diaphragm'.) A movable screen used to control the time during which light is admitted to an optical system.

Shy: Used to describe a fitting made too small for its purpose; a part which does not fit because it is too small (slang term).

Slide: (i) A piece of clear glass 3×1 in. (76×25 mm.) and about 1 mm. thick used to support microscopical objects during examination. (ii) Mechanical guides used to secure accurate linear movement.

Slip: Same as slide (i).

Slit: A narrow rectangular aperture employed to admit a light beam of that shape.

Smear: (i) A cylological preparation made by pressing a soft substance on to a slide with a smooth knife blade thus forming a film of the substance. (ii) A contaminated swab drawn across an agar plate to start a bacteriological culture.

Soft X-rays: X-rays generated at low tube voltage below approximately 30 KV. Radiation generated at approximately 200 KV and above is known as 'hard'.

Speeded Cine Film: Cine photography where the exposures are taken with a normal camera but at a slow rate of operation. Presentation of results is made at greater than normal speeds of projection. A form of 'time lapse' photography suitable for recording fast-changing phenomena.

Spherical aberration: The error in focus of a lens with spherical curvature due to rays from the outer zones meeting the axis of the lens nearer to the lens than do the rays from the inner zones. The amount of aberration is measured linearly along the optical axis.

Spot lens: A simple lens with its centre zones blacked out, used for simple dark ground illumination in a microscope.

Spread slide: A slide containing specimens spread at random upon it, e.g. an emulsion of diatoms dropped upon a slide, dried and mounted without arrangement.

Stabilize: Used electrically to mean levelling ripples and undulations in the output of a power supply.

Stage: The table of a microscope upon which objects to be examined are placed.

Stage fine adjustment: An old but very efficient microscope fine adjustment in which a hinged heavy metal stage plate carrying the specimen was tilted by a micrometer screw located in a solid part of the stage support, against spring loading. An arc of movement resulted but no backlash or shake occurred. A modern focussing stage.

Stage forceps: Gimbal-mounted tweezers used on a stage to grip and rotate irregular objects.

Stage micrometer: A glass slide accurately ruled in graduations usually $0 \cdot 01$ mm. and $0 \cdot 1$ mm. apart.

Stage plate: Any plate of material placed upon a microscope stage to support a special object or to project the stage, or to provide rotation.

Stage vice: A miniature vice used to hold small regular metal specimens.

Stand: The mechanical supporting structure of a microscope.

Star test: A test for spherical aberration and coma in which a tiny aperture in an opaque substance is used as an object. Similarly a bright light reflected from a small globule of mercury may be used.

Stem: The upright metal piece of a Ross type microscope upon which is fixed the coarse adjustment rack for focussing purposes.

Strain: Mechanically, the amount of deformation in a structure due to load.

Structureless source: A source of light which when examined optically shows no pattern, e.g. a lamp flame. (A coiled filament lamp shows the structure of the filament.)

Substage: A rack or screw focusing carrier for apparatus under a micro-scope stage.

Substage lamp: An electric lamp for mounting in or below a substage.

Surface plate: A specially flattened metal or glass sheet upon which other objects are tried for flatness.

Sweat: In engineering, to cause solder to flow round parts to be joined by holding the whole assembly in a flame.

Test: The hard shell or outer envelope of a rhizopod composed of either chitin or small particles of grit.

Test plate: A device for testing microscope objectives consisting of a silvered slide with scratches ruled across thus forming an object with broad dark and light bands ideal for showing up aberrations. (Due to Abbé.)

Time lapse: A system of cine photography in which single exposures are taken at time intervals which are great compared with the exposure time of each frame. When projected at normal speed a condensed history of the subject is presented.

Top light: A method of microscope illumination in which light is directed on to the objective side of the specimen by means of a lens or mirror thus making the specimen self luminous.

Top plate: The upper surface of a mechanical stage upon which the specimen is placed.

Traverse: The side-to-side motion of a mechanical stage of a microscope usually measured in millimetres with a vernier.

Triplet: An achromatic combination employing three lenses, often uncemented.

Tripod foot: A form of microscope foot where the body is slung between two legs of a heavy tripod the third leg forming the rear support.

Tube length: The distance from an objective at which the eyepiece should be placed to obtain least spherical aberration. It is measured mechanically between the objective screw shoulder and the eyepiece diaphragm.

Turntable: A freely revolving heavy round table about 100 mm. diameter with clips to secure a slide, employed to paint rings on mounts to secure coverglasses and to build up thicknesses of varnish to protect a specimen from pressure.

Ultra Microscopy: A method of making very small objects visible, though not resolvable into shapes by rendering them self-luminous by light which does not itself enter the microscope. Related to 'Dark ground'.

Ultra-violet: Electro magnetic radiation between the wavelengths 100 and 400 millimicrons, not visible to the eye, but still usually referred to as UV 'light'.

Understage: Same as Substage, but also used to describe a condenser carrier fixed to the underside of the stage.

Valve: (i) Electrically, a device for controlling the flow of electrons. Same as American name 'Tube'. (ii) One half of a diatom shell.

Visible light: That part of the electromagnetic spectrum to which the human eye is sensitive, roughly of the wavelengths 400 to 800 millimicrons. (See Light.)

Vital staining: A method of colouring a living organism.

Water of Ayr: A commercial name for a fine grained sharpening stone intended for use with water as the cutting compound.

Wedge: (i) Optically, a tapered strip of glass usually dark coloured in neutral for the purpose of moderating a light. Usually used in pairs after the fashion of engineering folding wedges. (ii) A similar wedge-shaped strip of crystalline quartz used in polariscopic apparatus for light phase shifting purposes.

Wedge fine adjustment: A microscope or projector fine focussing control where a tapered cylinder of hard steel is advanced by a screw thus slowly pushing aside a rider engaged on the taper. Satisfactory for heavy loads when used with modern lubricants (molybdenum disulphide) otherwise liable to excessive backlash.

Well slide: A microscope slide with a depression ground in it to hold liquids.

Wenham binocular: A classical stereoscopic binocular microscope in which the rays from the objective are divided by a Wenham prism and diverted up breeches tubes to two eyepieces.

Whip: In engineering, flexture of a mechanical part under vibration or load.

Wheeled animalcules: An early name for rotifera.

Working distance: The distance in linear measurement between the objective frontmost part and the specimen, measured in the medium in which the objective is supposed to work.

Zone: The annular-shaped areas into which a lens is divided for the purpose of computation of corrections, e.g. 'centre', 'middle' and 'outer' zones (not usually precisely defined).

Abbreviations

A	Angström unit (1/1,000 micron, measure of length)
AC	alternating current
AIEE	Associate of the Institution of Electrical Engineers
ASA	axial spherical aberration
AVO	Trade name of electric instrument maker of Avocet House, Vauxhall Bridge Road, London
BA	British Association (type of screw thread)
C	Centigrade
CGS	centimetre-gramme-second (system of measurement)
CINo	colour index number (Society of Dyers and Colourists)
CLA	centre-line average height method (measure of metal surface irregularity). See British Standards Specification 1134
DG	dark-ground illumination
ESU	electrostatic unit (electricity)
EHT	extra high tension (the very high tension applied to cathode ray tubes, over about 4500 volts)
F	Fahrenheit
FRMS	Fellow of the Royal Microscopical Society
FRS	Fellow of the Royal Society
GEC	General Electric Company
H	Henry (unit of inductance)
HT(HT)	high tension (high voltage electricity up to 500 volts)
ins	inches (English measure of length); 1 in. = 25·4 mm.
Kv	kilovolts = 1,000 volts (electrical measure)
mm	millimetres (metric measure of length)
MRI	Member of the Royal Institution of Great Britain
mil	same as British 'thou' = 1/1000 in. = 25·4 microns (μ)
mph	miles per hour (statute)
NA	numerical aperture (optical measurement)
PTFE	polytetrafluorethylene (electrical insulator)
PVA	polyvinyl chloride acetate copolymer
QMC	Quekett Microscopical Club
R	Reaumur
RI	Refractive index (an optical quantity)
RMS	(i) Royal Microscopical Society
	(ii) root mean square (method of measuring surface irregularities, see British Standards Specification 1134
SEI	Salford Electrical Instruments Limited (Salford, Lancs.)

swg	standard wire gauge (measure of diameter of wires and thickness of sheets) (Imperial measure)
Vdc (ac)	volts direct current (alternating current)
w	watt (unit of electrical power)
X	the transverse or horizontal direction in mathematics
Y	the vertical direction or axis in mathematics
Z	the direction at right angles to the X and Y directions in mathematics
μ	symbol representing micro $\left(\dfrac{1}{1,000,000}\right)$ or micron $\left(\dfrac{1}{1,000,000}\right)$ of a meter

Acknowledgments

I am greatly indebted to the Company of Charles Baker of Holborn for permission to use directly a very large amount of their literature on the interference microscope; the descriptions were so clear and accurate that improvement was not possible. Messrs Charles Baker acknowledge the help given them in their work on interference microscopy by Dr E. J. Ambrose, Dr S. Iversen, Dr John R. Baker, Mr J. Smiles, Dr S. L. Cowan, Dr A. J. Hales, and Dr R. N. Singh.

For the photomicrographs on Plates 12 and 13 we are indebted to Dr Ambrose, and for Plate 14 (D) to Dr J. M. Mitchison and Professor M. M. Swann and the *Quarterly Journal of Microscopical Science*. We are grateful for facilities kindly afforded us for photomicrography by Dr Ambrose at the Chester Beatty Research Institute and by Professor Causey and Mr Edwards, Department of Anatomy, Royal College of Surgeons of England.

I would like to record my thanks to the Engineer-in-Chief General Post Office for permission to publish work done by me at the Research Station, Dollis Hill, and for the use of the plates illustrating Contour Microscopy and Optical Bench Micrographic Apparatus.

I am also indebted to the President and Officers of the Quekett Microscopical Club for permission to use material already published in the Journal, and to the many members of the Club who have given invaluable assistance in microscopical studies.

Index